Recent Advances in Green Solvents

Recent Advances in Green Solvents

Editors

Reza Haghbakhsh
Sona Raeissi
Rita Craveiro

Basel • Beijing • Wuhan • Barcelona • Belgrade • Novi Sad • Cluj • Manchester

Editors

Reza Haghbakhsh
Chemical Engineering
University of Isfahan
Isfahan
Iran

Sona Raeissi
School of Chemical and
Petroleum Engineering
Shiraz University
Shiraz
Iran

Rita Craveiro
Department of Chemistry
Universidade Nova de Lisboa
Caparica
Portugal

Editorial Office
MDPI
St. Alban-Anlage 66
4052 Basel, Switzerland

This is a reprint of articles from the Special Issue published online in the open access journal *Molecules* (ISSN 1420-3049) (available at: www.mdpi.com/journal/molecules/special_issues/Solvent_Applied).

For citation purposes, cite each article independently as indicated on the article page online and as indicated below:

Lastname, A.A.; Lastname, B.B. Article Title. *Journal Name* **Year**, *Volume Number*, Page Range.

ISBN 978-3-0365-8615-1 (Hbk)
ISBN 978-3-0365-8614-4 (PDF)
doi.org/10.3390/books978-3-0365-8614-4

© 2023 by the authors. Articles in this book are Open Access and distributed under the Creative Commons Attribution (CC BY) license. The book as a whole is distributed by MDPI under the terms and conditions of the Creative Commons Attribution-NonCommercial-NoDerivs (CC BY-NC-ND) license.

Contents

Preface vii

Reza Haghbakhsh, Sona Raeissi and Rita Craveiro
Editorial for Special Issue: "Recent Advances in Green Solvents"
Reprinted from: *Molecules* **2023**, *28*, 5983, doi:10.3390/molecules28165983 1

Wan Nur Aisyah Wan Osman, Nur Athirah Izzati Badrol and Shafirah Samsuri
Biodiesel Purification by Solvent-Aided Crystallization Using 2-Methyltetrahydrofuran
Reprinted from: *Molecules* **2023**, *28*, 1512, doi:10.3390/molecules28031512 4

Manuela Panić, Mia Radović, Marina Cvjetko Bubalo, Kristina Radošević, Marko Rogošić and João A. P. Coutinho et al.
Prediction of pH Value of Aqueous Acidic and Basic Deep Eutectic Solvent Using COSMO-RS Profiles' Molecular Descriptors
Reprinted from: *Molecules* **2022**, *27*, 4489, doi:10.3390/molecules27144489 18

Changli Ma, Taisen Zuo, Zehua Han, Yuqing Li, Sabrina Gärtner and Huaican Chen et al.
Neutron Total Scattering Investigation of the Dissolution Mechanism of Trehalose in Alkali/Urea Aqueous Solution
Reprinted from: *Molecules* **2022**, *27*, 3395, doi:10.3390/molecules27113395 32

Jong Woo Lee, Hye Yoon Park and Junseong Park
Enhanced Extraction Efficiency of Flavonoids from *Pyrus ussuriensis* Leaves with Deep Eutectic Solvents
Reprinted from: *Molecules* **2022**, *27*, 2798, doi:10.3390/molecules27092798 44

Giovanni Ghigo, Matteo Bonomo, Achille Antenucci, Chiara Reviglio and Stefano Dughera
Copper-Free Halodediazoniation of Arenediazonium Tetrafluoroborates in Deep Eutectic Solvents-like Mixtures
Reprinted from: *Molecules* **2022**, *27*, 1909, doi:10.3390/molecules27061909 55

Maira I. Chinchilla, Fidel A. Mato, Ángel Martín and María D. Bermejo
Hydrothermal CO_2 Reduction by Glucose as Reducing Agent and Metals and Metal Oxides as Catalysts
Reprinted from: *Molecules* **2022**, *27*, 1652, doi:10.3390/molecules27051652 71

Hamed Peyrovedin, Reza Haghbakhsh, Ana Rita C. Duarte and Alireza Shariati
Deep Eutectic Solvents as Phase Change Materials in Solar Thermal Power Plants: Energy and Exergy Analyses
Reprinted from: *Molecules* **2022**, *27*, 1427, doi:10.3390/molecules27041427 83

Graciane Fabiela da Silva, Edgar Teixeira de Souza Júnior, Rafael Nolibos Almeida, Ana Luisa Butelli Fianco, Alexandre Timm do Espirito Santo and Aline Machado Lucas et al.
The Response Surface Optimization of Supercritical CO_2 Modified with Ethanol Extraction of *p*-Anisic Acid from *Acacia mearnsii* Flowers and Mathematical Modeling of the Mass Transfer
Reprinted from: *Molecules* **2022**, *27*, 970, doi:10.3390/molecules27030970 100

Nur Hidayah Zulaikha Othman Zailani, Normawati M. Yunus, Asyraf Hanim Ab Rahim and Mohamad Azmi Bustam
Experimental Investigation on Thermophysical Properties of Ammonium-Based Protic Ionic Liquids and Their Potential Ability towards CO_2 Capture
Reprinted from: *Molecules* **2022**, *27*, 851, doi:10.3390/molecules27030851 119

Bartosz Nowosielski, Marzena Jamrógiewicz, Justyna Łuczak and Dorota Warmińska
Novel Binary Mixtures of Alkanolamine Based Deep Eutectic Solvents with Water
—Thermodynamic Calculation and Correlation of Crucial Physicochemical Properties
Reprinted from: *Molecules* **2022**, *27*, 788, doi:10.3390/molecules27030788 . . . . **136**

Mikhail Kostenko and Olga Parenago
Adsorption of *N,N,N′,N′*-Tetraoctyl Diglycolamide on Hypercrosslinked Polysterene from a Supercritical Carbon Dioxide Medium
Reprinted from: *Molecules* **2021**, *27*, 31, doi:10.3390/molecules27010031 . . . . **158**

Hsin-Liang Huang, P.C. Lin, H.T. Wang, Hsin-Hung Huang and Chao-Ho Wu
Ionic Liquid Extraction Behavior of Cr(VI) Absorbed on Humic Acid–Vermiculite
Reprinted from: *Molecules* **2021**, *26*, 7478, doi:10.3390/molecules26247478 . . . . **175**

Preface

In previous decades, industries have mostly developed at the cost of environmental neglect, resulting in numerous environmental issues and, ultimately, a distressing global crisis. Therefore, in the 21st century, governments and world organizations are attempting to legislate sustainable development laws for industrial advances. "Green chemistry", as a rather novel field of chemistry and chemical engineering, is one of the key routes to assist researchers in sustainable development. Among the most valuable assets of green chemistry, "green solvents" are the primary candidates to replace the industry workhorses, i.e., conventional harmful solvents. In recent decades, green solvents have been studied intensively by scientists and researchers, and the number of published articles has been increasing exponentially. The most important green solvents, such as supercritical fluids, ionic liquids, and deep eutectic solvents, have been investigated in various fields and for numerous applications; yet, much remains unknown, and there is great room for further investigations into this class of solvents. This Special Issue aims to cover the most recent advances in the interdisciplinary area of green solvents.

Reza Haghbakhsh, Sona Raeissi, and Rita Craveiro
Editors

Editorial

Editorial for Special Issue: "Recent Advances in Green Solvents"

Reza Haghbakhsh [1,2,*], Sona Raeissi [3] and Rita Craveiro [2]

1. Department of Chemical Engineering, Faculty of Engineering, University of Isfahan, Isfahan 81746-73441, Iran
2. LAQV, REQUIMTE, Departamento de Química da Faculdade de Ciências e Tecnologia, Universidade Nova de Lisboa, 2829-516 Caparica, Portugal; rita.craveiro@campus.fct.unl.pt
3. School of Chemical and Petroleum Engineering, Shiraz University, Shiraz 71345-51154, Iran; raeissi@shirazu.ac.ir
* Correspondence: r.haghbakhsh@eng.ui.ac.ir or r.haghbakhsh@fct.unl.pt

1. Introduction

Today, environmental conservation is one of the most urgent targets. Fortunately, this goal is taken into consideration in the general policies of most countries. Accordingly, worldwide development needs to be aligned with environmental considerations [1]. In all countries, the industrial sector inevitably has the most impact on sustainable development. Green chemistry, by considering environmental issues affecting the planet and its living creatures, provides useful guidelines for any kind of sustainable development, especially in the industries. This can be achieved by screening and recommending novel ideas, methods, processes, etc. In this way, green chemistry can be considered as a general scientific field, with 12 principles. The fifth principle of green chemistry is "Safer Solvents and Auxiliaries", which directly emphasizes the importance of green and environmentally friendly solvents [2]. Following this principle, in the past few decades, the general idea of replacing conventional hazardous and harmful solvents with green solvents has been highlighted in scientific communities and research centers in industries. Various types of green solvents, such as supercritical fluids, ionic liquids (ILs), deep eutectic solvents (DESs), etc., have shown high levels of potential in many applications. The number of published studies on the topic of green solvents has significantly increased year on year [3]. Therefore, "green solvents" can be considered a hot topic of green chemistry, deserving more specific investigations in scientific publications. Because of this, *Molecules* has devoted a Special Issue to recent advancements in the interdisciplinary area of green solvents. In this Special Issue, fundamental as well as application-based studies and innovative techniques regarding green solvents were covered. We were delighted to welcome interesting, high-quality, and valuable studies in this field.

2. Contributions

This Special Issue includes eleven original research articles covering various aspects of green solvents, including fundamental knowledge and applications.

Huang et al. [4] studied the performance of ionic liquids for the extraction of heavy metals. They experimentally investigated the efficiency of 1-butyl-3-methylimidazolium chloride in extracting Cr(VI) from a Cr-contaminated simulated sorbent in soil remediation.

Kostenko and Parenago [5] considered the application of supercritical CO_2 for impregnation in place of organic solvents, in order to prepare sorbents based on hyper crosslinked polystyrene and the chelating agent N,N,N′,N′-tetraoctyl diglycolamide. They proposed an environmentally friendly and green solvent for this process.

Nowosielski et al. [6] carried out fundamental research by investigating important physical properties (density, speed of sound, refractive index, and viscosity) of both pure and aqueous solutions of a number of deep eutectic solvents. They studied DESs made of tetrabutylammonium chloride + 3-amino-1-propanol and tetrabutylammonium bromide + 3-amino-1-propanol or 2-(methylamino)ethanol or 2-(butylamino)ethanol.

As a comprehensive study, Zailani et al. [7] reported the results of their research on the synthesis of a series of ammonium cations coupled with carboxylate anions producing ammonium-based protic ionic liquids, and their densities, viscosities, refractive indices, thermal decomposition temperatures, glass temperatures, and CO_2 absorption was also reported.

In an application-based study, da Silva et al. [8] proposed the use of supercritical CO_2 to extract non-volatile compounds from *A. mearnsii* flowers. They concluded that the extracted essence showed antimicrobial activity. They also showed the presence of p-anisic acid, a substance with industrial and pharmaceutical applications.

Also, in the field of energy, Peyrovedin et al. [9] showed that green solvents can be useful in solar energy plants, and in addition to showing how they make the process eco-friendly, they also elaborated on its good performance. They studied the feasibility of using a number of deep eutectic solvents as phase change material (PCM) for solar thermal power plants with organic Rankine cycles.

In another article, Chinchilla et al. [10] presented applications of green solvents in the field of catalysts. They proposed high-temperature water reactions to reduce CO_2 by using an organic reductant and a series of metals and metal oxides as catalysts.

Ghigo et al. [11] studied deep-eutectic-solvent-like mixtures, based on glycerol and different halide organic and inorganic salts, as new high-potential media in the copper-free halodediazoniation of arenediazonium salts. They reported the experimental results of the reaction and also presented a computational investigation to understand the reaction mechanism.

Lee et al. [12] investigated the applications of deep eutectic solvents in extractions from natural leaves. They compared the efficiencies of various ratios of choline chloride and dicarboxylic acids for the extraction of flavonoid components from *Pyrus ussuriensis* leaves with respect to conventional solvents.

In a fundamental research article, Ma et al. [13] studied the atomic structure of cellulose that was dissolved in alkali/urea aqueous solutions. They used trehalose as the model molecule with total scattering as the main tool to study three kinds of alkali solution, consisting of LiOH, NaOH, and KOH. They reported interesting findings on the most probable all-atom structures of the solution, the hydration shell of trehalose, the penetration of ions into glucose rings, and the molecule interactions of urea with hydroxide groups.

Panić et al. [14], by proposing a new modeling strategy, presented their experimental and modeling investigations into the development of a simple and straightforward model to estimate the pH values of deep eutectic solvents over a wide range of values. They developed the model according to a large number of deep eutectic solvents, using artificial intelligence techniques for modeling.

In the last article in this Special Issue, Osman et al. [15] studied the application of a new green solvent for biodiesel purification via Solvent-Aided Crystallization (SAC). Biodiesel purification is an important experimental step in biodiesel synthesis, and it would be more beneficial to consider green solvents for this task. In their work, they also investigated the technological improvements in the purification of biodiesel via SAC and compared the performance of a new green solvent with conventional solvents in the production of high-purity biodiesel.

3. Conclusions

The concept of green solvents is not new, and supercritical fluids and ionic liquids have led the way in this field. Green solvents are dynamic and continuously thriving, with new solvents and new categories of solvents being added over time. Deep eutectic solvents, as a recently introduced category of green solvents, are highly significant and may encompass huge numbers of members. Currently, new ILs and DESs are being introducing to the scientific community at a rapid pace. Consequently, wide ranges of investigations, from fundamental research for new members to application-based research for well-known members, are all vital and important from their own perspectives. One of the aims of this Special Issue was to highlight the wide scope of research being carried out on green solvents and to involve a variety of researcher and reader orientations. We have included a range of studies, from fundamental research on physical properties and reaction kinetics to application-based studies such as essence extraction, impregnation via supercritical fluids, and biodiesel purification. We have also covered the even more specific field of energy for solar power plants using green solvents.

Author Contributions: R.H., S.R. and R.C. conceived, designed, and wrote this editorial. All authors have read and agreed to the published version of the manuscript.

Funding: This research received no external funding.

Data Availability Statement: Not applicable.

Acknowledgments: The authors are grateful to the University of Isfahan, Universidade Nova de Lisboa, and Shiraz University for providing facilities.

Conflicts of Interest: The authors declare no conflict of interest.

References

1. Haghbakhsh, R.; Raeissi, S.; Duarte, A.R.C. Group contribution and atomic contribution models for the prediction of various physical properties of deep eutectic solvents. *Sci. Rep.* **2021**, *11*, 6684. [CrossRef] [PubMed]
2. Anastas, P.; Eghbali, N. Green Chemistry: Principles and Practice. *Chem. Soc. Rev.* **2010**, *39*, 301–312. [CrossRef] [PubMed]
3. Domínguez de María, P.; Guajardo, N.; González-Sabín, J. Recent granted patents related to Deep Eutectic Solvents. *Curr. Opin. Green Sustain. Chem.* **2022**, *38*, 100712. [CrossRef]
4. Huang, H.L.; Lin, P.C.; Wang, H.T.; Huang, H.H.; Wu, C.H. Ionic Liquid Extraction Behavior of Cr(VI) Absorbed on Humic Acid–Vermiculite. *Molecules* **2021**, *26*, 7478. [CrossRef] [PubMed]
5. Kostenko, M.; Parenago, O. Adsorption of N,N,N′,N′-Tetraoctyl Diglycolamide on Hypercrosslinked Polysterene from a Supercritical Carbon Dioxide Medium. *Molecules* **2022**, *27*, 31. [CrossRef] [PubMed]
6. Nowosielski, B.; Jamrógiewicz, M.; Łuczak, J.; Warmińska, D. Novel Binary Mixtures of Alkanolamine Based Deep Eutectic Solvents with Water—Thermodynamic Calculation and Correlation of Crucial Physicochemical Properties. *Molecules* **2022**, *27*, 788. [CrossRef] [PubMed]
7. Zailani, N.H.Z.O.Z.; Yunus, N.M.; Rahim, A.H.A.; Bustam, M.A. Experimental Investigation on Thermophysical Properties of Ammonium-Based Protic Ionic Liquids and Their Potential Ability towards CO_2 Capture. *Molecules* **2022**, *27*, 851. [CrossRef] [PubMed]
8. da Silva, G.F.; de Souza Júnior, E.T.; Almeida, R.N.; Fianco, A.L.B.; Santo, A.T.E.; Lucas, A.M.; Vargas, R.M.F.; Cassel, E. The Response Surface Optimization of Supercritical CO2 Modified with Ethanol Extraction of p-Anisic Acid from Acacia mearnsii Flowers and Mathematical Modeling of the Mass Transfer. *Molecules* **2022**, *27*, 970. [CrossRef] [PubMed]
9. Peyrovedin, H.; Haghbakhsh, R.; Duarte, A.R.C.; Shariati, A. Deep Eutectic Solvents as Phase Change Materials in Solar Thermal Power Plants: Energy and Exergy Analyses. *Molecules* **2022**, *27*, 1427. [CrossRef] [PubMed]
10. Chinchilla, M.I.; Mato, F.A.; Martín, A.; Bermejo, M.D. Hydrothermal CO_2 Reduction by Glucose as Reducing Agent and Metals and Metal Oxides as Catalysts. *Molecules* **2022**, *27*, 1652. [CrossRef] [PubMed]
11. Ghigo, G.; Bonomo, M.; Antenucci, A.; Reviglio, C.; Dughera, S. Copper-Free Halodediazoniation of Arenediazonium Tetrafluoroborates in Deep Eutectic Solvents-like Mixtures. *Molecules* **2022**, *27*, 1909. [CrossRef] [PubMed]
12. Lee, J.W.; Park, H.Y.; Park, J. Enhanced Extraction Efficiency of Flavonoids from *Pyrus ussuriensis* Leaves with Deep Eutectic Solvents. *Molecules* **2022**, *27*, 2798. [CrossRef] [PubMed]
13. Ma, C.; Zuo, T.; Han, Z.; Li, Y.; Gärtner, S.; Chen, H.; Yin, W.; Hanm, C.C.; Cheng, H. Neutron Total Scattering Investigation of the Dissolution Mechanism of Trehalose in Alkali/Urea Aqueous Solution. *Molecules* **2022**, *27*, 3395. [CrossRef] [PubMed]
14. Panić, M.; Radović, M.; Bubalo, M.C.; Radošević, K.; Rogošić, M.; Coutinho, J.A.P.; Redovniković, I.R.; Tušek, A.J. Prediction of pH Value of Aqueous Acidic and Basic Deep Eutectic Solvent Using COSMO-RS σ Profiles' Molecular Descriptors. *Molecules* **2022**, *27*, 4489. [CrossRef] [PubMed]
15. Osman, W.N.A.W.; Badrol, N.A.I.; Samsuri, S. Biodiesel Purification by Solvent-Aided Crystallization Using 2-Methyltetrahydrofuran. *Molecules* **2023**, *28*, 1512. [CrossRef] [PubMed]

Disclaimer/Publisher's Note: The statements, opinions and data contained in all publications are solely those of the individual author(s) and contributor(s) and not of MDPI and/or the editor(s). MDPI and/or the editor(s) disclaim responsibility for any injury to people or property resulting from any ideas, methods, instructions or products referred to in the content.

Article

Biodiesel Purification by Solvent-Aided Crystallization Using 2-Methyltetrahydrofuran

Wan Nur Aisyah Wan Osman [1,2], Nur Athirah Izzati Badrol [1] and Shafirah Samsuri [1,2,*]

1. Chemical Engineering Department, Universiti Teknologi PETRONAS, Seri Iskandar 32610, Malaysia
2. HICoE-Centre for Biofuel and Biochemical Research (CBBR), Institute of Sustainable Buiding, Universiti Teknologi PETRONAS, Seri Iskandar 32610, Malaysia
* Correspondence: shafirah.samsuri@utp.edu.my

Abstract: The previous biodiesel purification by Solvent-Aided Crystallization (SAC) using 1-butanol as assisting agent and parameters for SAC were optimized such as coolant temperature, cooling time and stirring speed. Meanwhile, 2-Methyltetrahydrofuran (2-MeTHF) was selected as an alternative to previous organic solvents for this study. In this context, it is used to replace solvent 1-butanol from a conducted previous study. This study also focuses on the technological improvements in the purification of biodiesel via SAC as well as to produce an even higher purity of biodiesel. Experimental works on the transesterification process to produce crude biodiesel were performed and SAC was carried out to purify the crude biodiesel. The crude biodiesel content was analyzed by using Gas Chromatography–Mass Spectrometry (GC-MS) and Differential Scanning Calorimetry (DSC) to measure the composition of Fatty Acid Methyl Esters (FAME) present. The optimum value to yield the highest purity of FAME for parameters coolant temperature, cooling time, and stirring speed is −4 °C, 10 min and 210 rpm, respectively. It can be concluded that the assisting solvent 2-MeTHF has a significant effect on the process parameters to produce purified biodiesel according to the standard requirement.

Keywords: 1-butanol; 2-methyltetrahydrofuran; biodiesel; coolant temperature; cooling time; green solvent; stirring speed; solvent-aided crystallization

Citation: Wan Osman, W.N.A.; Badrol, N.A.I.; Samsuri, S. Biodiesel Purification by Solvent-Aided Crystallization Using 2-Methyltetrahydrofuran. *Molecules* 2023, 28, 1512. https://doi.org/10.3390/molecules28031512

Academic Editors: Reza Haghbakhsh, Rita Craveiro and Sona Raeissi

Received: 10 November 2022
Revised: 31 January 2023
Accepted: 31 January 2023
Published: 3 February 2023

Copyright: © 2023 by the authors. Licensee MDPI, Basel, Switzerland. This article is an open access article distributed under the terms and conditions of the Creative Commons Attribution (CC BY) license (https://creativecommons.org/licenses/by/4.0/).

1. Introduction

Nowadays, the depletion of fossil resources is not something new as the global population keeps on rising. This has led to the discovery of renewable fuels, such as biodiesel. Biodiesel has attracted a lot of interest as a future fuel because of its copious resources and environmental considerations [1]. The bio-based fuel business has seen an accelerated surge in sales and has become a driving force to create novel green technologies. These were influenced by government laws and concerns about ecological sustainability and the depletion of natural raw materials. Biodiesel's initial design was careful and methodical, emphasizing the industry in terms of long-term viability. Nowadays, this biofuel is easy to integrate into existing facilities and cars, and the industry sector has devoted a lot of effort to researching and promoting the fuel's capabilities.

In Malaysia, fossil fuels accounted for 95% of the overall primary energy output in the year 2006 [2]. This includes natural gas, petroleum, coal, peat renewables, and hydroelectricity. Primary energy is generally raw energy that has not been engineered or converted in any way. Malaysia is presently a fast-expanding country; thus, this prevalent tendency is likely to continue speculating for the next 20 years. On top of that, the study also claimed that Malaysia is presently the world's largest exporter of palm oil, despite being the oil's second-biggest producer after Indonesia [2]. On that account, Malaysia endeavoured to gain leverage in the expanding biofuel sector by encouraging palm oil-based biodiesel development upon recognizing its profitability. Due to this, Malaysia has been recognized as one of the countries that proactively encourages commercial operations for the use of biodiesel as a fossil fuel substitute [2].

The authors also stated that the biodiesel sector in Malaysia shows no activity until the Eighth Malaysia Plan, in the year 2001, established the Fifth Fuel Policy [2]. Renewable energy has been designated as the fifth source of electricity generation in Malaysia under the proposed legislation. The Malaysian biodiesel sector is also largely supported by the National Biofuel Policy. The legislation concentrates on biodiesel commercialization, utilization, study, development, and exportation, yet it excludes upstream parts of the industry growth. Biodiesel production and deployment are expected to keep on increasing, particularly in rapidly developing countries where economic development is accelerating. Malaysia expects to supply one million tonnes of biodiesel by the end of 2020, increasing 80% production compared to the previous year (2019) [1].

The process of separating contaminants from biodiesel is crucial to ensure that the developed fuel fulfils all required standards, delivering improved performance as well as preserving the engine from degradation [3]. Glycerol, soap, water, a catalyst used, and triglycerides are mostly residues that must be separated from crude biodiesel obtained. Purification is known to be one of the most essential stages in biodiesel production. Water washing, ion exchange adsorbents, and membrane-based adsorbents are the foremost often utilized technologies for the purification of biodiesel [4]. This purification method is critical in maintaining efficacy in engine performance. According to Arenas et al. [4], free fatty acids at high concentrations can develop deposit accounts in storage tanks and even injectors, hence reducing the lifespan of engines. In addition, the high water content can corrode the engine of automobiles. Therefore, the purification of crude biodiesel can be challenging as it contributes to the rise in biodiesel operating expenses. This opens up a discussion on the possible alternatives to the conventional method of biodiesel purification.

Purification of biodiesel is undoubtedly one of the important steps in biodiesel production. The main goal of the production process is to achieve high-quality fuel with hardly any contaminants that could sabotage its excellence. The impurities that could be present in biodiesel are glycerol, alcohol (namely, methanol), soap, free fatty acids, residual salts, metals, and production catalysts [5]. It is clear that the densities of biodiesel and glycerol are disparate enough to have them separated by gravitational settling and centrifugation [6]. Having different polarities is another determinant on the account that the separation between the ester and glycerol is rapid. Glycerol must be purified as it contains a large part of biodiesel impurities, and it would deposit at the bottom of the fuel tank causing the fouling of the injector [7]. The complete elimination of glycerol represents the exceptional quality of biodiesel. Another polar substance, methanol, is necessary to be removed as it has a low flash point which can be an inconvenience in terms of transportation, storage, and utilization [7]. In addition, they also mentioned that methanol is also a result of corrosion to pieces of aluminium and zinc [7].

Various techniques have been applied for the application of biodiesel purification in order to overcome the limitation of high water usage on the earlier method explained. Recently, a new method had been introduced known as SAC. This method is carried out under low temperatures compared to other biodiesel purification techniques. Hence, it could prevent the biodiesel from becoming volatile during or after the purification process. This is supported by studies mentioning that the biodiesel would be volatile at higher temperatures, in the range of 340–375 °C, which were obtained from thermal analyses of thermogravimetric analysis (TGA) and differential scanning calorimetry (DSC) [8,9].

The basic principle of SAC is to selectively reduce the viscosity of melts to alter the crystallization kinetics by the insertion of assisting agent with adequate quantities into the solvents [10]. Once the assistant solvents are injected, rapid crystallization occurs in a low-viscosity sample solution. This method is able to overcome the biggest difficulty in separating biodiesel–glycerol, where both are hard to separate [11]. This is due to these solvents creating high-viscosity crude melts that are difficult to distinguish by conventional methods, which had appealed to a notion that permits layer crystallization to extract these compounds.

Samsuri et al. [12] concluded that SAC could effectively remove undesired glycerol, methanol, and soap components, leaving a sample obtained known as purified biodiesel. Thus, it is an operative practice for a waterless approach to refine biodiesel in a more ecologically friendly way than other common purifying procedures, while being able to reduce the cost required for wastewater treatment afterwards. As a result, it is indeed critical to evaluate whether it is feasible for a certain solvent to be appropriate for each system besides not knowing the effects of crystallization during operation. Recent findings showed that SAC is highly influenced by the following parameters: concentration of solvent, cooling temperature and time, and stirring rate [12]. The optimum parameter is obtained by using the analysis technique of response surface plot analysis. Surface plots can be used to evaluate targeted response values and the connection of the operational parameters. It is found that biodiesel with a purity of 99.375% is obtained as the optimum condition by using the following parameters: concentration of solvent of 1.5 wt%, cooling temperature of 12.7 °C, cooling time of 35 min and stirring rate of 175 rpm. However, this study used 1-butanol as the assisting solvent.

1-butanol has a poor separation performance as an assisting solvent for SAC. This statement had been proven by Ahmad and Samsuri [11]. They analyzed the effect of different concentrations of 1-butanol in order to evaluate the optimum quantity of 1-butanol required for the biodiesel purification process via SAC. They used ultrasonic irradiation to aid this process and findings showed that the purity of biodiesel reduced as the concentration of 1-butanol increased. Conversely, inadequate 1-butanol could cause impure crystals forming resulting in nucleation, where the crystals might form alongside the whole chemical freezes. Therefore, they claimed that high-purity biodiesel may be achieved at lower cooling temperatures and intermediary 1-butanol concentrations, or with a longer response time if excess 1-butanol is employed.

Therefore, sustainable solution by using green alternatives in the purification process has been studied and researched to gain biodiesel satisfactory with its standard to lessen the ecological implications of using solvents in chemical processing. The use of environmentally sustainable solvents or green alternatives to traditional goods has recently gained a lot of interest, citing environmental advantages and worker safety as reasons. Green chemistry had been introduced as a way in managing effluent produced from chemical processes, specifically from the processing industry [13]. The sole purpose is to focus on the environmental effect of chemistry and eradicate environmental pollution through concerted, long-term preventative efforts. This concept led to the proposal of a low-toxicity alternative solvent with broad synthetic applications for the processing sector.

In this experiment, solvent 1-butanol is substituted with 2-Methyltetrahydrofuran (2-MeTHF) as an alternative assisting agent for crystallization and a better replacement in terms of environmental aspects for the said organic solvent. 2-MeTHF is derived from corn cobs and oat hulls [14]. According to Choi et al. [15], the global production of corn-grain has increased by 40% over the past decade and reached over 1 billion tons of production recently. This would enhance the production of corn residue which is stated about 47 to 50% of their residues are wasted [15]. On the other hand, it was reported that about 23 million tons of oat was globally produced in 2018 with oat hull waste representing 25 to 35% of the entire production [16]. Both of the residues need to be treated; hence, both of them have been recognized as safe and environmentally friendly solvents since they can be obtained from biomass feedstocks to which an exposure limit on humans up to 6.2 mg/day is permitted [17].

2. Results
2.1. Characterization of Crude Biodiesel
2.1.1. Differential Scanning Calorimetry

The DSC curve as in Figure 1 represents the temperature relationship on the heat flow as the outcome of calorimetric measurements for the biodiesel sample. The DSC graph demonstrated one exothermic peak, indicating the crystallization peak. The onset temperature is the temperature at which crystallization begins, the peak temperature indicates the temperature at which the maximum reaction rate occurs, and the end set temperature represents the temperature at which the process ends [12].

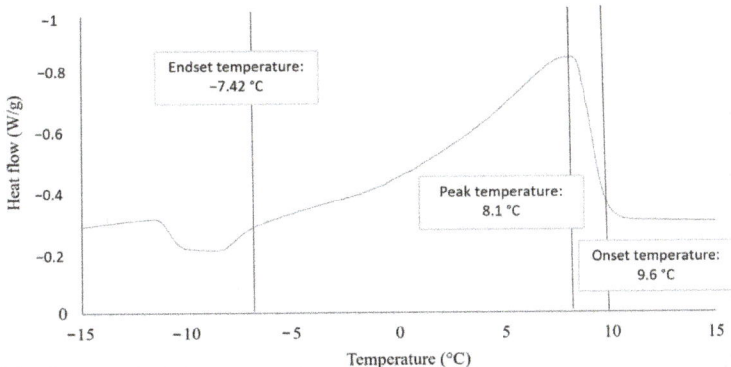

Figure 1. Graph of temperature vs. heat flow for biodiesel sample.

2.1.2. Gas Chromatography–Mass Spectroscopy

The crude biodiesel obtained after 24 h of gravity settling is analyzed using GCMS analysis to examine its quality in terms of FAME purity and the properties of biodiesel. Besides the sample of crude biodiesel, 16 biodiesel samples based on the different parameters for SAC had also been studied for GCMS characterization. The properties that can be obtained from the results are systematic name, retention time, correction area of individual components and the sum of the correction area. Figure 2 shows the abundance versus retention time graph for the chromatogram of GC-MS analysis for the crude biodiesel.

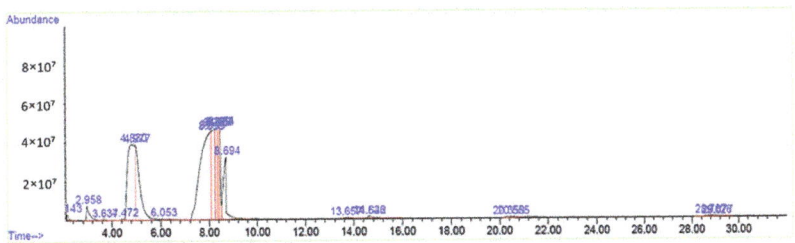

Figure 2. GC-MS chromatograph of biodiesel.

2.2. Effect of Coolant Temperature in SAC

The cooling time and stirring rate were kept constant at 15 min and 140 rpm, respectively. The temperature of the coolant in the chiller is adjusted within the parameter range of the experiment. The parameter range for coolant temperature is −4 °C, −6 °C, −8 °C, −10 °C and −12 °C. The coolant used is a 50% (v/v) ethylene glycol solution with water [12]. Figure 3 shows the plotted graph using GC-MS data for FAME purity against coolant temperature while Table 1 showed the observation of the effect of coolant temperature in SAC. For the coolant temperature parameter, at a constant 140 rpm and cooling time of 15 min, a coolant temperature of −4 °C indicates the optimum value to yield the highest purity of FAME content which is 100% purity.

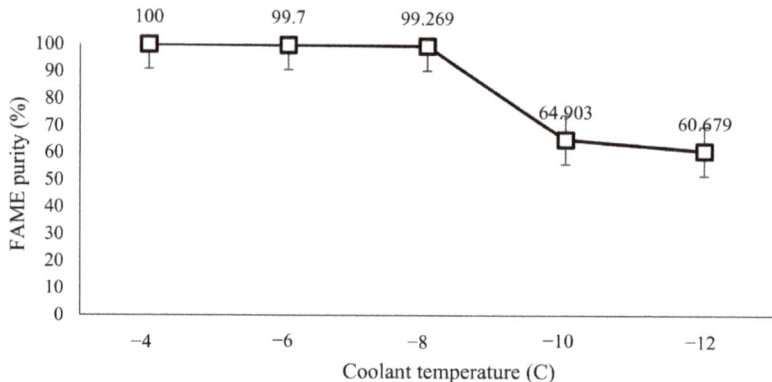

Figure 3. FAME purity against coolant temperature.

Table 1. Observation for the effect of coolant temperature in SAC.

Temperature (°C)	Stirring Speed (rpm)	Cooling Time (min)	Diagram	Observation
−4	140	15		The glycerol layer is not completely crystallized.
−6				The colour of the biodiesel layer appears to be cloudy.
−8				
−10				The biodiesel layer appears to be viscous. Only a little biodiesel is obtained. The glycerol layer appears to be thick.
−12				

2.3. Effect of Cooling Time in SAC

For this part of the experiment, the coolant temperature and stirring rate were kept constant, at −8 °C and 140 rpm, respectively. The parameter range for cooling time is 5 min, 10 min, 15 min, 20 min, and 25 min. Figure 4 shows the plotted graph using GC-MS data for FAME purity against cooling time, while Table 2 showed the observation of the effect of cooling time in SAC. For the cooling time parameter, at constant −8 °C and 140 rpm, a cooling time of 10 min indicates the optimum value to yield the highest purity of FAME content, which is 99.993% purity.

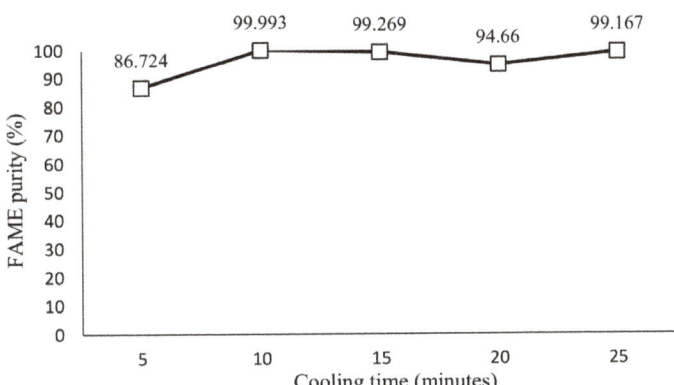

Figure 4. FAME purity against cooling times.

Table 2. Observation for the effect of cooling times in SAC.

	Parameter		Diagram	Observation
Cooling Time (min)	Stirring Speed (rpm)	Temperature (°C)		
5	140	−8		Glycerol is crystallized. A thin white layer of glycerol is formed.
10				Glycerol is crystallized. A white layer of glycerol is formed.
15				The colour of the biodiesel layer appears to be cloudy.
20				The colour of the biodiesel layer appears to be cloudy. The glycerol layer appears to be very thick.
25				

2.4. Effect of Stirring Speed in SAC

For this part of the experiment, the coolant temperature and cooling time were kept constant, at −8 °C and 15 min, respectively. The parameter range for stirring speed is 120 rpm, 130 rpm, 140 rpm, 175 rpm and 210 rpm. Figure 5 shows the plotted graph using GC-MS data for FAME purity against stirring speed while Table 3 showed the observation of the effect of stirring speed in SAC. For the stirring speed parameter, at a constant −8 °C and cooling time of 15 min, a stirring speed of 210 rpm indicates the optimum value to yield the highest purity of FAME content, which is 99.606% purity.

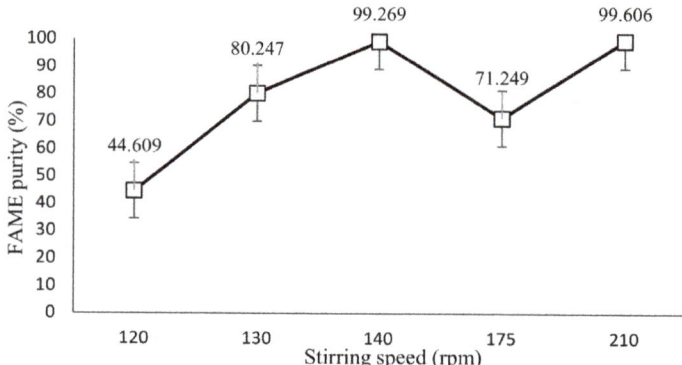

Figure 5. FAME purity against stirring speed.

Table 3. Observation for the effect of stirring speed in SAC.

Stirring Speed (rpm)	Parameter		Diagram	Observation
	Cooling Time (min)	Temperature (°C)		
120	15	−8		The colour of the biodiesel layer appears to be cloudy. The glycerol layer appears to be very thick.
130				The pale colour of the biodiesel layer is formed. A thin white layer of glycerol is formed.
140				The colour of the biodiesel layer appears to be cloudy.
170				The biodiesel layer appears to be viscous. The glycerol layer appears to be thick. The colour of the biodiesel layer appears to be cloudy.
210				The biodiesel layer appears to be very viscous. Only a little biodiesel is obtained. The glycerol layer appears to be thick. The colour of the biodiesel layer appears to be cloudy

3. Discussion

3.1. Characterization of Crude Biodiesel

3.1.1. Differential Scanning Calorimetry

From Figure 1, the value obtained from the graph for onset temperature is 9.6 °C, the peak temperature is 8.1 °C and the end set temperature of biodiesel obtained is −7.42 °C. The temperature range for the following section of the experiment was determined using the SAC approach employing the crystallization point of biodiesel from the analysis. It is in line with the finding from a previous study conducted by Samsuri et al. [12], where the starting point of crude biodiesel crystallization was 9.45 °C, which was maximal at 8.4 °C, thus showing the highest rate of reaction. Towards the end, the temperature dropped to −5.18 °C, indicating the end of the experiment.

3.1.2. Gas Chromatography–Mass Spectroscopy

According to Figure 2, the total composition of FAME percentage obtained from the crude biodiesel is 99.12% with the total amount of unsaturated fatty acid and saturated fatty acid form at 58.54% and 40.76%, respectively. The fatty acid available in the crude biodiesel is Dodecanoic acid, Methyl tetradecanoate, 9-Hexadecanoic acid, Hexadecanoic acid, 9-Octadecenoic acid, Methyl strearate and Eicosanoic acid. The highest correction area obtained from an individual component is from 9-Octadecenoic acid. Table 4 shows the tabulated results for the systematic name (Library/ID), trivial name, types of fatty acids, retention time and the percentage of FAME composition for the crude biodiesel.

Table 4. Data of GC-MS results for crude biodiesel.

Systematic Name	Trivial Name	Types of Fatty Acid	Retention Time (min)	Composition of FAME (%)
Dodecanoic acid	Lauric	Saturated	2.143	0.41
Methyl tetradecanoate	Myristic	Saturated	2.959	1.6
9-Hexadecanoic acid	Palmitoleic	Unsaturated	4.473	0.14
Hexadecanoic acid	Palmitic	Saturated	4.828	18.31
Hexadecanoic acid	Palmitic	Saturated	4.977	12.19
Hexadecanoic acid	Palmitic	Saturated	6.053	0.27
9-Octadecenoic acid	Oleic	Unsaturated	8.321	48.64
9-Octadecenoic acid	Oleic	Unsaturated	8.376	3.19
9-Octadecenoic acid	Oleic	Unsaturated	8.456	6.48
Methyl strearate	Stearic	Saturated	8.692	6.86
Eicosanoic acid	Arachidic	Saturated	14.641	0.61
Hexadecanoic acid	Palmitic	Saturated	20.358	0.42
	Total Unsaturated Fatty Acid			58.45
	Total Saturated Fatty Acid			40.67
	Total Fatty Acid			99.12

3.2. Effect of Coolant Temperature in SAC

The trend line in Figure 3 shows a slight decrease in trend, from $-4\ °C$ to $-8\ °C$, until it drops downs steeply, at $-10\ °C$, until $-12\ °C$. The highest percentage of 100% purity is at the highest temperature, which is at $-4\ °C$; meanwhile, at $-10\ °C$, the FAME yield purity obtained is the lowest, which is at 60.68%. The crystallization temperature indicated by the onset temperature of this experiment is found to be at $9.6\ °C$ and expected to end (as estimated) by the end set temperature of $-7.42\ °C$. Hence, biodiesel is predicted to crystallize during conducting this experiment as all the parameters are lower than the crystallization temperature. As the coolant temperatures of $-10\ °C$ and $-12\ °C$ are much lower than the end set temperature, it is expected that this operating condition would yield a low purity of FAME. In comparison to biodiesel produced from solvent 1-butanol, the study mentioned that the highest biodiesel purity was achieved, 99.375%, when the coolant temperature was set at $12.7\ °C$ [12]. This is because their onset and endset temperatures obtained from their DSC analysis were $9.45\ °C$ and $-5.18\ °C$, respectively, with a peak temperature of $8.4\ °C$. Nevertheless, this study was able to achieve even higher biodiesel purity which is 100% at a coolant temperature of $-4\ °C$ [12]. Hence, it is concluded that the use of 2-MeTHF as a solvent for SAC is able to produce higher biodiesel purity than 1-butanol despite the coolant temperature used.

In reference to the FAME purity versus coolant temperature graph, a higher yield is obtained at a temperature farther than the end set temperature and closer to the crystallization temperature. This can be explained by Ahmad et al. [18], who explained that FAME is more likely to be trapped within the solid layer developed by glycerol and other contaminants when the such temperature is approaching the crystallization point. When the heat transfer rate is slower at higher coolant temperatures, the solid can form in a more orderly pattern, leaving the pure methyl ester to concentrate in the solution [19]. The solid development rate is larger at lower temperatures of coolant, resulting in more methyl ester retention into contaminating solids. This can be further proven by research from Yahya et al. [20], who stated that the rate of ice crystals or solid development is governed by the temperature of the coolant.

3.3. Effect of Cooling Time in SAC

The trend line from Figure 4 shows the FAME yield to be increasing from 5 min to 10 min, which is from 86.72% to 99.99% purity. From 10 min until 20 min, the FAME yield is found to decrease slightly before it increases at 25 min with the FAME yield of 99.17%. The highest purity obtained is at 10 min with 99.99% of FAME purity which can be considered to be pure biodiesel. Considering the result obtained from DSC, the crystallization temperature obtained is at 9.6 °C, and the end set temperature is at −7.42 °C. The experiment is carried out at a temperature close to the end set temperature, which is −8 °C. As the experiment is conducted at a temperature much lower than the crystallization temperature, the solid layer from the contaminants is expected to be formed in the inner vessel. In comparison to biodiesel produced from solvent 1-butanol, the study mentioned that the highest biodiesel purity was achieved, 99.375%, when the cooling time was set to 35 min [12], which was longer than the optimum cooling time found in this study. The highest biodiesel purity found in this study is 99.99%, at a cooling time of 10 min, with which even higher biodiesel purity was obtained at a shorter cooling time compared to the study with 1-butanol. Hence, it is concluded that the use of 2-MeTHF as a solvent for SAC is able to produce higher biodiesel purity than 1-butanol despite the cooling time used.

In addition, it can be examined from Table 2 that during 5 and 10 min of cooling time, a white layer of glycerol is formed. Clear yellowish liquid biodiesel can also be seen formed in the vessel. Subsequently, during 15, 20 and 25 min of crystallization time, the liquid layer of biodiesel appears to be cloudy. The glycerol layer also appears to be thick over time. During 25 min of crystallization time, the layer of glycerol can be observed to be the thickest, resulting in a small volume of biodiesel formed. A larger yield of pure methyl ester was attained by using a prolonged cooling period [19]. In addition, as stated by Ahmad and Samsuri [11], for crystallization to occur, a longer crystallization time is preferable. However, as the FAME purity drops after an increasing amount of time, it can also be deduced that an increase in cooling time would also cause the growth of solid from the methyl ester to be reduced. The authors also stated that this may have been caused due to the saturation of the solute in the liquid phase inducing contamination of the solid. Consequently, the best range for cooling time is from 10 to 15 min, as proven by the purity of FAME at a constant temperature of −8 °C. Thus, this cooling time is not too prolonged for the separating process to take place.

3.4. Effect of Stirring Speed in SAC

Figure 5 shows an increase in the trend line from 120 rpm to 140 rpm from a value of 44.61% to 99.27%. At 175 rpm, the value of FAME purity decreases to 71.25%, and it increases to its highest purity at 210 rpm at 99.61%. In order to improve the formation of a solid, an aid for the solution movement is essential [18]. The parameter of this experiment is affected by the rate of stirring speed that is set by the laboratory mixer. As stated by Mohammed and Bandari [21], in maintaining a continuous temperature distribution and system flow, a gradual motion is essential. Therefore, a steady increase in stirring speed is chosen (120 rpm, 130 rpm and 140 rpm). After that, there is a disparity in stirring speed as the increment between the value is high (140 rpm, 175 rpm and 210 rpm). Consequently, this describes the irregularity of the trend line after 140 rpm. In comparison to biodiesel produced from solvent 1-butanol, the study mentioned that the highest biodiesel purity was achieved, 99.375%, when the stirring speed was set at 175 rpm [12]. Under similar stirring speeds, it is found that this study produced lower biodiesel purity (71.25%) compared to the one with 1-butanol. The highest biodiesel purity found in this study is 99.61% at a stirring speed of 210 rpm, which was a higher stirring speed used compared to the study with 1-butanol. Nevertheless, it is concluded that the use of 2-MeTHF as a solvent for SAC is able to produce higher biodiesel purity than 1-butanol despite the stirring speed used.

Furthermore, the efficiency of the purification can be observed from the graph by the purity of FAME. The highest value FAME yield can be seen at the highest stirring speed, which is 210 rpm. The contaminant in the biodiesel is circulated at a high flowrate, causing high separation between the solute and the solution. As Jusoh et al. [22] showed in their research, the formation of a high shear force, which could separate the solute from the solution, is imposed by a high circulation flowrate. Low separation is produced at a low stirring rate resulting in low purity of FAME because the solution moves more slowly. For the stirring rate of 175 rpm, there is a sudden drop in FAME purity. Although high stirring can yield good separation of contaminants, the moderate flow would also be prone to scrape away the solid developed on the vessel wall. This causes the impurities to mix with the liquefied biodiesel, resulting in low purity of FAME. This is researched by Mohammed and Bandari [21], who stated that stirring vigorously could prolong the solidification process and lower the liquid phase's final concentration.

4. Materials and Methods

4.1. Materials Used

For this experiment, palm oil was purchased from a nearby supermarket. Meanwhile, methanol and KOH were obtained from the UTP laboratory. About 1000 mL of oil with 12.75 g of KOH as catalyst and 225 mL of methanol as solvent was used in the transesterification process. Meanwhile, for the SAC process, ethylene glycol and water were used as a coolant in the chiller. Crude biodiesel from the transesterification method that already completed the gravity settling process was used as feed for the SAC process. Assisting solvent, 2-MeTHF was added to the crude biodiesel for the purification process.

4.2. Transesterification Process

The experimental setup for the transesterification process is referred to the study conducted by Ahmad and Samsuri [11] as shown in Figure 6. To begin with, 1000 mL of palm oil was poured into a round-bottom flask. Next, the flask was heated at a reaction temperature of 60 °C, which is controlled by the heating mantle. At the same time, 12.75 g KOH was dissolved in 225 mL of methanol. After that, the solution of methanol and KOH was poured into the heated oil in the flask and stirred for 10 min. The product obtained from this transesterification process is known as crude biodiesel. Later, 1 mL of the product was extracted into a glass vial for DSC and GCMS analysis.

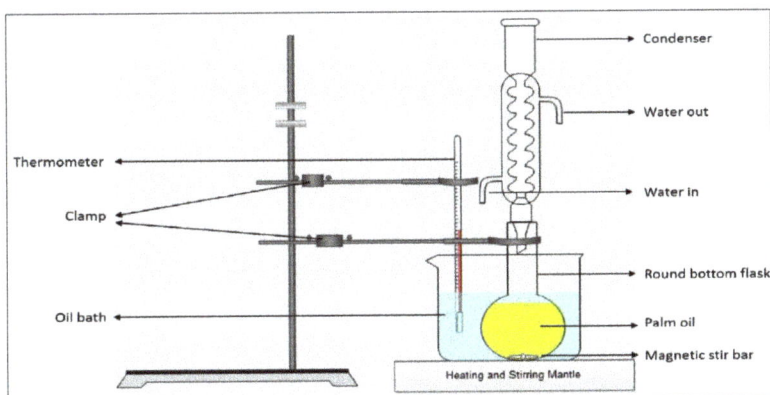

Figure 6. Transesterification method setup.

4.3. Solvent-Aided Crystallization

The experimental setup for the SAC process is referred to the study conducted by Ahmad and Samsuri [11] as shown in Figure 7. Firstly, the chiller was turned on to cool down the coolant temperature. The desired temperature was set before conducting the experiment. The range temperature for the whole experiment is between −4 °C and −12 °C. After that, about 500 mL of crude biodiesel with 1 wt.% of 2-MeTHF were fed into a cylindrical vessel (11 cm × 24 cm). The vessel was placed inside the chiller which was filled with coolant once the desired temperature was reached. Next, the stirrer was switched on and left until the expected cooling time.

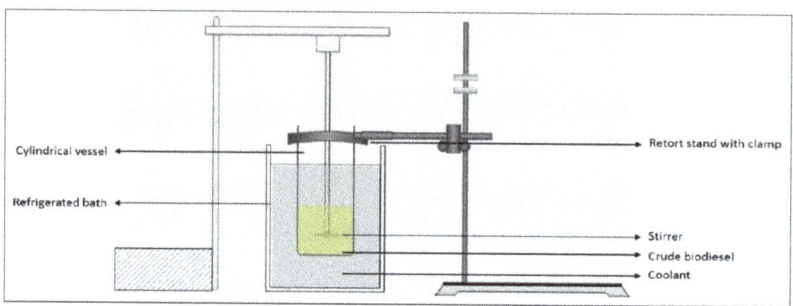

Figure 7. Solvent-aided crystallization method setup.

Solid contaminants are formed on the inner surface of a vessel, leaving pure biodiesel in liquid form. Subsequently, pure biodiesel was poured from the vessel to drain it out and detach the solid contaminants from the surface of the vessel. The solid contaminant was left to melt completely at room temperature. Thereupon, a sample of purified biodiesel was taken for GCMS analysis. The entire procedure was repeated under different operating conditions which are the temperature of the coolant, cooling time and stirring speed. The parameter range for coolant temperature is −4 °C, −6 °C, −8 °C, −10 °C and −12 °C, while cooling times are 5, 10, 15, 20 and 25 min, and stirring speeds are 123 rpm, 134 rpm, 140 rpm, 175 rpm and 210 rpm. All of the experiments were repeated twice, and average results were calculated for better data collection.

4.4. Characterization of Biodiesel

4.4.1. Differential Scanning Calorimetry

DSC is a device used to determine the amount of energy required to achieve a zero-temperature differential between a component and an inert reference substance by subjecting the two specimens to comparable temperature regimes in a contained manner [23]. The heat capacity or enthalpy of a sample of known mass is measured as changes in heat transfer. This analysis is suitable for glycerol and biodiesel, which are highly viscous melts [11].

Calibration of trials is used to record the temperature change and correlate it to the enthalpy change in the sample. The crude biodiesel sample is brought into equilibrium between −15 °C and 30 °C, at the rate of 5 °C min^{-1} for DSC measurement. In this research, it is vital to determine the crystallization point of the biodiesel sample to determine the lowest point of cooling temperature for the SAC process [11]. The crystallinity of materials is linked to the change in enthalpy by the energy required from the melting transition to proceed [23].

In this study, in DSC analysis, the sample was equilibrated, at 30 °C, and cooled immediately, at −15 °C, at a rate of 5 °C min^{-1}. Afterwards, the sample was maintained for 1 min and heated to 30 °C at a rate of 5 °C min^{-1}. Therefore, the procedure for transesterification was now complete. The remaining crude biodiesel was further used for gravity settling for 24 h before proceeding with the DSC analysis.

4.4.2. Gas Chromatography–Mass Spectroscopy

GC-MS is an analytic technology which combined gas–liquid chromatography separation features with mass spectrometry detection techniques to identify distinct compounds inside a test sample [24]. The mass of the analyte fragments is being used to identify these compounds. In academic research, this device facilitates the characterization and detection of newly synthesized or derivatized compounds by studying the new components [24]. Retention time (RT) is the time required for the compound to pass through the injection port to reach the detector [25].

In this study, the GC-MS device used in this experiment is PerkinElmer Clarus 600 Gas Chromatograph (GC). A flame ionization detector (FID) and an Elite 5-MS column with a dimension of 30 m × 250 µm × 0.25 µm of film thickness were installed in the GC. This device is used twice in this experiment, once after the reaction of transesterification (initial content of biodiesel) and lastly after conducting SAC (final content of biodiesel). During GCMS analysis, the oven temperature was set at 150 °C and held for 1 min. Afterwards, the temperature was raised to 240 °C, at 5 °C min^{-1} ramping speed, and was maintained for 5 min.

To determine the biodiesel purity, the percentage composition of individual FAME was computed using the following Equation (1). The biodiesel purity computation was then performed for all prominent peaks. For each SAC trial run as well, GC-MS would be used to determine the yield of FAME over all purified biodiesel samples. The purified biodiesel was left to melt after being treated to SAC and was collected for GC-MS analysis to determine the percentage of FAME composition to define its purity using the same mentioned formula.

$$\text{Percentage composition of FAME (\%)} = \frac{\text{Peak area of individual component}}{\text{Summation of correction area}} \quad (1)$$

5. Conclusions

Biodiesel is a non-toxic and biodegradable diesel alternative that is synthesized by the process of transesterification. 2-Methyltetrahydrofuran (2-MeTHF) can be used in chemical synthesis as an alternative to organic solvents for this project. In this context, it is used to replace solvent 1-butanol from a conducted previous study. The process parameters for SAC, which are different coolant temperatures, cooling time and stirring speed, are studied and analyzed for optimization. The chemical composition of biodiesel was taken into account when purified by the SAC process. The optimization process is considered successful once the optimum parameter value produces the highest purity of biodiesel, thus indicating that the biodiesel is free of contaminants. The optimum value to yield the highest purity of FAME for parameters coolant temperature, cooling time and stirring speed is −4 °C, 10 min and 210 rpm, respectively. Hence, by the proposed optimize parameter, it can be taken into account that SAC is effective in the purification of biodiesel. In conclusion, experimental research on the SAC method can assist in improving biodiesel purification.

For future study, a techno-economic feasibility study (TEFS) and cost benefit analysis will be carried out in order to estimate the cost as well as energy for this SAC system for implementation in industrial applications. The energy cost of the process can vary depending on factors such as the source of the feedstock, the type of equipment used (refrigerated, stirrer), and the efficiency of the process. Additionally, the benefits of the biodiesel purification strategy, such as reducing greenhouse gas emissions and decreasing dependence on fossil fuels, may outweigh the energy costs. The analysis would be needed to determine whether the biodiesel purification strategy is a worthwhile endeavor. In addition, process simulation software will be used for the determination of the scale-up process, including the equipment's size, design, operation, and process parameters optimization.

Author Contributions: Conceptualization, W.N.A.W.O. and S.S.; methodology, W.N.A.W.O. and S.S.; validation, N.A.I.B., W.N.A.W.O. and S.S.; formal analysis, N.A.I.B.; investigation, N.A.I.B.; resources, N.A.I.B. and S.S.; data curation, N.A.I.B., W.N.A.W.O. and S.S.; writing—original draft preparation, N.A.I.B. and W.N.A.W.O.; writing—review and editing, W.N.A.W.O. and S.S.; visualization, W.N.A.W.O. and S.S.; supervision, S.S.; project administration, S.S.; funding acquisition, S.S. All authors have read and agreed to the published version of the manuscript.

Funding: This research was funded by the Ministry of Education Malaysia via FRGS (Cost Centre: 015MA0-094, Reference Code: FRGS/1/2019/TK10/UTP/03/3) and Universiti Teknologi PETRONAS via YUTP-FRG (Cost Centre: 015LC0-378), and facilities support from HICoE–Centre for Biofuel and Biochemical Research (CBBR).

Institutional Review Board Statement: Not applicable.

Informed Consent Statement: Not applicable.

Data Availability Statement: Not applicable.

Acknowledgments: The authors would like to acknowledge the assistance from the Ministry of Education Malaysia via FRGS (Cost Centre: 015MA0-094, Reference Code: FRGS/1/2019/TK10/UTP/03/3) and facilities support from HICoE–Centre for Biofuel and Biochemical Research (CBBR) and Chemical Engineering Department. Support from the Ministry of Education Malaysia through the HICoE award to CBBR is duly acknowledged. The support from Universiti Teknologi PETRONAS through YUTP-FRG is also acknowledged.

Conflicts of Interest: The authors declare no conflict of interest.

Sample Availability: Samples of the compounds, if available, are available from the authors.

References

1. Zulqarnain; Yusoff, M.H.M.; Ayoub, M.; Jusoh, N.; Abdullah, A.Z. The Challenges of a Biodiesel Implementation Program in Malaysia. *Processes* **2020**, *8*, 1244. [CrossRef]
2. Lim, S.; Teong, L.K. Recent trends, opportunities and challenges of biodiesel in Malaysia: An overview. *Renew. Sustain. Energy Rev.* **2010**, *14*, 938–954. [CrossRef]
3. De Jesus, S.S.; Ferreira, G.F.; Maciel, M.R.W.; Filho, R.M. Biodiesel purification by column chromatography and liquid-liquid extraction using green solvents. *Fuel* **2019**, *235*, 1123–1130. [CrossRef]
4. Arenas, E.; Villafán-Cáceres, S.; Rodríguez-Mejía, Y.; García-Loyola, J.; Masera, O.; Sandoval, G. Biodiesel Dry Purification Using Unconventional Bioadsorbents. *Processes* **2021**, *9*, 194. [CrossRef]
5. Atadashi, I.; Aroua, M.; Aziz, A.A.; Sulaiman, N. Refining technologies for the purification of crude biodiesel. *Appl. Energy* **2011**, *88*, 4239–4251. [CrossRef]
6. Atadashi, I.; Aroua, M.; Aziz, A.A.; Sulaiman, N. High quality biodiesel obtained through membrane technology. *J. Membr. Sci.* **2012**, *421-422*, 154–164. [CrossRef]
7. Atadashi, I. Purification of crude biodiesel using dry washing and membrane technologies. *Alex. Eng. J.* **2015**, *54*, 1265–1272. [CrossRef]
8. Leonardo, R.; Valle, M. Evaluation of the Volatility Characteristics of Diesel/Biodiesel Blends Using Thermal Analysis Techniques. 2019. Available online: https://www.semanticscholar.org/paper/Evaluation-of-the-volatility-characteristics-of-%2F-Leonardo-Valle/2cad370eb33d87c714eeca3c5759352c053afc40 (accessed on 13 January 2023).
9. Goodrum, J. Volatility and boiling points of biodiesel from vegetable oils and tallow. *Biomass-Bioenergy* **2002**, *22*, 205–211. [CrossRef]
10. Eisenbart, F.; Ulrich, J. Solvent-aided layer crystallization—Case study glycerol–water. *Chem. Eng. Sci.* **2015**, *133*, 24–29. [CrossRef]
11. Ahmad, M.; Samsuri, S. Biodiesel Purification via Ultrasonic-Assisted Solvent-Aided Crystallization. *Crystals* **2021**, *11*, 212. [CrossRef]
12. Samsuri, S.; Jian, N.L.; Jusoh, F.W.; Yáñez, E.H.; Yahya, N.Y. Solvent-Aided Crystallization for Biodiesel Purification. *Chem. Eng. Technol.* **2019**, *43*, 447–456. [CrossRef]
13. Watanabe, K. The Toxicological Assessment of Cyclopentyl Methyl Ether (CPME) as a Green Solvent. *Molecules* **2013**, *18*, 3183–3194. [CrossRef] [PubMed]
14. Smoleń, M.; Kędziorek, M.; Grela, K. 2-Methyltetrahydrofuran: Sustainable solvent for ruthenium-catalyzed olefin metathesis. *Catal. Commun.* **2014**, *44*, 80–84. [CrossRef]
15. Choi, J.Y.; Nam, J.; Yun, B.Y.; Kim, Y.U.; Kim, S. Utilization of corn cob, an essential agricultural residue difficult to disposal: Composite board manufactured improved thermal performance using microencapsulated PCM. *Ind. Crop. Prod.* **2022**, *183*. [CrossRef]

16. Redaelli, R.; Berardo, N. Prediction of fibre components in oat hulls by near infrared reflectance spectroscopy. *J. Sci. Food Agric.* **2007**, *87*, 580–585. [CrossRef]
17. Bao, W.-H.; Wang, Z.; Tang, X.; Zhang, Y.-F.; Tan, J.-X.; Zhu, Q.; Cao, Z.; Lin, Y.-W.; He, W.-M. Clean preparation of S-thiocarbamates with in situ generated hydroxide in 2-methyltetrahydrofuran. *Chin. Chem. Lett.* **2019**, *30*, 2259–2262. [CrossRef]
18. Ahmad, M.A.; Letchumanan, A.; Samsuri, S.; Mazli, W.N.A.; Saad, J.M. Parametric study of glycerol and contaminants removal from biodiesel through solvent-aided crystallization. *Bioresour. Bioprocess.* **2021**, *8*, 54. [CrossRef]
19. Samsuri, S.; Amran, N.A.; Zheng, L.J.; Bakri, M.M.M. Effect of coolant temperature and cooling time on fractional crystallization of biodiesel and glycerol. *Malays. J. Fundam. Appl. Sci.* **2017**, *13*, 676–679. [CrossRef]
20. Yahya, N.; Zakaria, Z.Y.; Ali, N.; Jusoh, M. Effect of Coolant Temperature on Progressive Freeze Concentration of Refined, Bleached and Deodorised Palm Oil based on Process Efficiency and Heat Transfer. *J. Teknol.* **2015**, *74*. [CrossRef]
21. Mohammed, A.R.; Bandari, C. Lab-scale catalytic production of biodiesel from waste cooking oil—A review. *Biofuels* **2017**, *11*, 409–419. [CrossRef]
22. Jusoh, M.; Nor, N.N.M.; Zakaria, Z.Y. Progressive Freeze Concentration of Coconut Water. *J. Teknol.* **2014**, *67*. [CrossRef]
23. Wierzbicka-Miernik, A. Fundamentals of the Differential Scanning Calorimetry Application in Materials Science. Available online: http://www.imim.pl/PHD/www.imim-phd.edu.pl/contents/Relevant%20Articles/Fundamentals%20of%20the%20Differential%20Scanning%20Calorimetry%20application%20in%20materials%20science%20A%20Wierzbicka-Miernik.pdf (accessed on 6 November 2022).
24. Chauhan, A. GC-MS Technique and its Analytical Applications in Science and Technology. *J. Anal. Bioanal. Tech.* **2014**, *5*, 222. [CrossRef]
25. Sneddon, J.; Masuram, S.; Richert, J.C. Gas Chromatography-Mass Spectrometry-Basic Principles, Instrumentation and Selected Applications for Detection of Organic Compounds. *Anal. Lett.* **2007**, *40*, 1003–1012. [CrossRef]

Disclaimer/Publisher's Note: The statements, opinions and data contained in all publications are solely those of the individual author(s) and contributor(s) and not of MDPI and/or the editor(s). MDPI and/or the editor(s) disclaim responsibility for any injury to people or property resulting from any ideas, methods, instructions or products referred to in the content.

Article

Prediction of pH Value of Aqueous Acidic and Basic Deep Eutectic Solvent Using COSMO-RS σ Profiles' Molecular Descriptors

Manuela Panić [1], Mia Radović [1], Marina Cvjetko Bubalo [1], Kristina Radošević [1], Marko Rogošić [2], João A. P. Coutinho [3], Ivana Radojčić Redovniković [1,*] and Ana Jurinjak Tušek [1]

[1] Faculty of Food Technology and Biotechnology, University of Zagreb, Pierottijeva Ulica 6, 10000 Zagreb, Croatia; mpanic@pbf.hr (M.P.); mradovic@pbf.hr (M.R.); mcvjetko@pbf.hr (M.C.B.); krado@pbf.hr (K.R.); ana.tusek.jurinjak@pbf.unizg.hr (A.J.T.)
[2] Faculty of Chemical Engineering and Technology, University of Zagreb, Marulićev Trg 19, 10000 Zagreb, Croatia; mrogosic@fkit.hr
[3] CICECO—Aveiro Institute of Materials, Department of Chemistry, University of Aveiro, 3810-193 Aveiro, Portugal; jcoutinho@ua.pt
* Correspondence: irredovnikovic@pbf.hr

Citation: Panić, M.; Radović, M.; Cvjetko Bubalo, M.; Radošević, K.; Rogošić, M.; Coutinho, J.A.P.; Radojčić Redovniković, I.; Jurinjak Tušek, A. Prediction of pH Value of Aqueous Acidic and Basic Deep Eutectic Solvent Using COSMO-RS σ Profiles' Molecular Descriptors. *Molecules* 2022, 27, 4489. https://doi.org/10.3390/molecules27144489

Academic Editors: Reza Haghbakhsh, Sona Raeissi and Rita Craveiro

Received: 13 June 2022
Accepted: 11 July 2022
Published: 13 July 2022

Publisher's Note: MDPI stays neutral with regard to jurisdictional claims in published maps and institutional affiliations.

Copyright: © 2022 by the authors. Licensee MDPI, Basel, Switzerland. This article is an open access article distributed under the terms and conditions of the Creative Commons Attribution (CC BY) license (https://creativecommons.org/licenses/by/4.0/).

Abstract: The aim of this work was to develop a simple and easy-to-apply model to predict the pH values of deep eutectic solvents (DESs) over a wide range of pH values that can be used in daily work. For this purpose, the pH values of 38 different DESs were measured (ranging from 0.36 to 9.31) and mathematically interpreted. To develop mathematical models, DESs were first numerically described using σ profiles generated with the COSMOtherm software. After the DESs' description, the following models were used: (i) multiple linear regression (MLR), (ii) piecewise linear regression (PLR), and (iii) artificial neural networks (ANNs) to link the experimental values with the descriptors. Both PLR and ANN were found to be applicable to predict the pH values of DESs with a very high goodness of fit ($R^2_{independent\ validation}$ > 0.8600). Due to the good mathematical correlation of the experimental and predicted values, the σ profile generated with COSMOtherm could be used as a DES molecular descriptor for the prediction of their pH values.

Keywords: artificial neural networks; COSMO-RS; deep eutectic solvents; multiple linear regression; piecewise linear regression

1. Introduction

Green chemistry presents a way of creating and applying chemical products and processes that reduce or eliminate the use or production of substances that are hazardous to human health and the environment [1]. A growing area of research in green technology development is devoted to the design of new, more environmentally friendly solvents whose use would meet technological and economic requirements. Requirements for alternative solvents include a reasonable price, non-toxicity to humans and the environment, non-flammability, biodegradability, and possibility of regeneration or recovery [2,3]. Currently, known green solvents are water, carbon dioxide, bio-solvents, ionic liquids, and deep eutectic solvents. In the last decade, deep eutectic solvents (DESs) have received enormous attention in the academic community and the number of articles published has increased exponentially.

DESs were first described by Abbott et al. in 2003 as a mixture of a hydrogen bond donor (HBD) with a hydrogen bond acceptor (HBA), which exhibited much lower melting points than the pure compounds due to the formation of hydrogen bonds between constituent compounds [4–6]. Lately, DESs have shown great potential for industrial application thanks to their acceptable costs, the versatility of their physicochemical properties, and simple preparation. They also often present low cytotoxicity and good biodegradability. The properties that have gained them the environmentally friendly label are low volatility (reduced air pollution), nonflammability (process safety), and stability (potential for recycling and reuse). The number of structural combinations encompassed by DESs is tremendous; thus, it is possible to design DESs with unique physicochemical properties for a particular purpose. The physicochemical properties, such as the viscosity, density, and pH value, of DESs are crucial for industrial application of these solvents in terms of equipment materials, mass transfer, filtration, or pumping [7].

The pH values of aqueous solutions affect the enzyme activity, extraction efficiency, and stability of biologically active molecules. As such, the pH value is an important property of a solvent and, especially for DES design, one of the critical parameters. Though several papers have analyzed the pH behavior of DESs, there are still gaps in the understanding of how DES-forming compounds influence its pH value [8,9]. Despite this, some general conclusions can be outlined. For example, DESs containing organic acids (i.e., malic acid or oxalic acid) are, as expected, more acidic than those containing polyalcohols or sugars. The role of the water content in DESs regarding the pH behavior is still not entirely clear; however, it was observed that an increase in pH values with an increasing water content was reported for DESs with extremely low pH values while the pH values of DESs with pH in the higher range of values (lower acidity region) decreased with an increasing water content [7].

So far, the search for an ideal DES for a particular system has been guided by an empirical trial-and-error approach, with no systematic research into the structure–activity of DESs. Therefore, the rational design of these solvents for specific purposes is still in its infancy. Data collection on the application properties of DESs and the development of mathematical methods as a tool for the design of novel solvents are imperative for the industrial application of these solvents. The Conductor-like Screening Model for Real Solvents (COSMO-RS) is an ab initio computational method that may be used for the generation of the σ profile of a molecule. The σ profile shows the probability of finding surface segments with σ polarity on the surface of the molecule and contains the most relevant chemical information needed to predict the compound's electrostatic, hydrogen bonding, and dispersion interactions [10]. The distribution of the charge, the width, and the height of the peaks in the σ profile vary with the nature of the molecules. Therefore, any change in the molecular structure can be quantified. By coupling the σ profile of DES-forming compounds with experimental data using model-generating methods such as multiple linear regression (MLR), piecewise linear regression (PLR), or artificial neural networks (ANNs), models for the description of DESs' physicochemical properties can be developed [11–14]. In most studies, good model fitting of the literature viscosity, density, and pH values of the DESs was obtained [12,13]. The results showed that simple linear models such as MLR and more complex ones such as ANN could be used efficiently to predict the physical properties of specific DES groups (e.g., amine or sugar-based DESs), whereas it was difficult to create a single model covering the whole range of possible DES systems [11]. Commonly, simple mathematical models such as MLR were good enough for viscosity and density prediction while in the case of the pH value, more complex ANN models had to be used [11,13,15].

In this work, we report a model for the prediction of the pH values of acidic and basic DESs. For this purpose, the experimental pH values of 38 different DESs were evaluated, described, and mathematically interpreted. For the development of mathematical models, DESs were firstly numerically described using σ profiles estimated by the COSMOtherm software. After the description of DESs, the following models were used: (i) MLR, (ii) PLR, and (iii) ANN to link the experimental values with the descriptors. In the end, the prepared models were statistically verified.

2. Results and Discussion

2.1. DES Characteristics: Experimental pH Values and σ Profiles

This work aimed to develop a simple and robust mathematical model for predicting the pH values of DESs based on S^i_{mix} descriptors. To develop a user-friendly model to predict pH values in the wide range, we selected both acidic and basic DESs from our database. We chose 38 DESs by carefully selecting and varying different HBA, HBD, and water shares (Table 1). Selected HBAs and HBDs can be roughly classified as quaternary ammonium salts (choline chloride, betaine), amino acids (proline), organic acids (citric and malic acid), and sugars (fructose, glucose, sucrose, xylose). In comparison to HBA, there are more HBD candidates from previously mentioned classes and it has been shown that they have an immediate effect on pH values (Table 1). Overall, all synthesized DESs cover a wide range of pH values from 0.36 for Ch:CA containing 30% water (w/w) to 9.31 for Ch:U containing 10% water (w/w). Monitoring the pH values of the same HBA/HBD pair while varying the DES water content shows that water influences the measured pH value. However, this influence is a distinctive characteristic of an individual DES and cannot be extended to all DESs studied in this work.

Table 1. Experimentally measured pH values.

DES	Abbreviation	Molar Ratio	wH_2O [%]	pH (20 °C) ± st.dev.
Betaine:citric acid	B:CA	1:1	30	2.46 ± 0.04
			50	2.46 ± 0.02
Betaine:ethylene glycol	B:EG	1:2	30	6.86 ± 0.00
Betaine:glucose	B:Glc	1:1	10	6.64 ± 0.35
Betaine:glycerol	B:Gly	1:2	30	6.77 ± 0.04
			50	6.38 ± 0.07
Betaine:oxalic acid:glycerol	B:OxA:Gly	1:2:1	30	2.91 ± 0.05
Betaine:malic acid	B:Ma	1:1	30	2.98 ± 0.01
			50	2.92 ± 0.01
Betaine:sucrose	B:Suc	4:1	30	7.85 ± 0.11
Choline chloride:citric acid	Ch:CA	2:1	30	0.34 ± 0.04
			50	0.71 ± 0.00
Choline chloride:ethylene glycol	ChCl:EG	1:2	10	6.19 ± 0.01
			30	6.60 ± 0.57
			50	4.58 ± 0.14
			80	4.41 ± 0.00
Choline chloride:fructose	ChCl:Fru	1:1	30	3.51 ± 0.05
			50	3.35 ± 0.03
Choline chloride:glucose	ChCl:Glc	1:1	30	4.83 ± 0.06
			50	3.56 ± 0.01
Choline chloride:glycerol	ChCl:Gly	1:2	30	3.71 ± 0.06
			50	2.67 ± 0.11
			80	3.06 ± 0.01
Choline chloride:malic acid	ChCl:MA	1:1	30	0.63 ± 0.01
			50	1.03 ± 0.00

Table 1. Cont.

DES	Abbreviation	Molar Ratio	wH_2O [%]	pH (20 °C) ± st.dev.
Choline chloride:proline:malic acid	ChCl:Pro:MA	1:1:1	10	3.23 ± 0.00
			30	2.82 ± 0.01
			50	2.63 ± 0.03
Choline chloride:sorbitol	ChCl:Sol	1:1	50	4.92 ± 0.04
			80	3.80 ± 0.08
Choline chloride:urea	ChCl:U	1:2	10	9.26 ± 0.08
			30	8.85 ± 0.06
			50	8.23 ± 0.04
Choline chloride:urea:ethylene glycol	ChCl:U:EG	1:2:2	10	8.29 ± 0.07
Choline chloride:urea:glycerol	ChCl:U:Gly	1:2:2	10	8.72 ± 0.05
Choline chloride:xylose	ChCl:Xyl	2:1	30	2.86 ± 0.04
			50	3.32 ± 0.03
			80	3.93 ± 0.01
Choline chloride:xylitol	ChCl:Xyol	5:2	30	6.90 ± 0.06
			50	6.50 ± 0.01
			80	6.03 ± 0.06
Choline chloride:fructose	ChCl:Fru	1:1	30	3.51 ± 0.05
			50	3.35 ± 0.03
Citric acid:glucose	CA:Glc	1:1	30	0.53 ± 0.04
Citric acid:sucrose	CA:Suc	1:1	30	0.83 ± 0.00
Fructose:ethylene glycol	Fru:EG	1:2	30	5.31 ± 0.09
Fructose:glucose:ethylene glycol	Fru:Glc:EG	1:1:2	50	3.67 ± 0.06
Fructose:glucose:sucrose	Fru:Glc:Suc	1:1:1	50	2.63 ± 0.03
			80	2.99 ± 0.01
Fructose:glucose:urea	Fru:Glc:U	1:1	30	8.22 ± 0.06
Glucose:ethylene glycol	Glc:EG	1:2	50	4.03 ± 0.02
Glucose:glycerol	Glc:Gly	1:2	50	4.33 ± 0.04
Malic acid:fructose	MA:Fru	1:1	30	0.77 ± 0.01
Malic acid:fructose:glycerol	MA:Fru:Gly	1:1	30	2.77 ± 0.01
Malic acid:glucose	MA:Glc	1:1	30	0.83 ± 0.01
Malic acid:glucose:glycerol	MA:Glc:Gly	1:1:1	10	0.92 ± 0.00
Malic acid:sucrose	MA:Suc	2:1	30	0.66 ± 0.01
Proline:malic acid	Pro:MA	1:1	10	2.63 ± 0.01
			30	2.78 ± 0.02
			50	2.73 ± 0.03
Sucrose:ethylene glycol	Suc:EG	1:2	30	6.05 ± 0.06
Sucrose:glucose:urea	Suc:Glc:U	1:1	30	8.14 ± 0.25
Xylose:ethylene glycol	Xyl:EG	1:2	30	4.57 ± 0.06

Furthermore, DESs were mathematically described using the σ profile defined with the COSMOtherm software. The HBA and HBD molecules were optimized in TmoleX, both from an energy and geometry point of view. The generated COSMO files contain all information necessary for the calculation of the σ profile function and thus for the calculation of the σ profile descriptors. For the preparation of the descriptor set, the DESs were modeled as a molar mixture of HBA and HBD according to Table 1. The σ profile curves for each HBA and HBD were divided into 10 regions, the area under each region was calculated, and their numerical values were correlated with the experimental pH values using mathematical models.

2.2. Multiple Linear Regression and Piecewise Linear Regression

The assessment of the MLR and PLR model applicability to predict the pH values of DESs was based on the correlation coefficient values, R^2, R^2_{adj}, and $RMSE$. The obtained model coefficient values and the basic statistical analysis are presented in Table 2 while a comparison between the experimental and model-estimated pH values is given in Figure 1.

Table 2. MLR and PLR regression coefficients. Statistically significant coefficients are marked in bold.

	MLR		PLR	
	Regression Coeff. ± st. Error	*p*-Value	Regression Coeff. ± st. Error	*p*-Value
Break point			**4.1246 ± 0.3292**	0.0021
b_0	**−13.4623 ± 4.9782**	0.0078	**−1.9449 ± 0.1556 −80.4560 ± 10.6436**	0.0001
$b_1 (S^1_{mix})$	**16.4623 ± 5.1388**	0.0022	**14.8847 ± 2.1908 −23.1982 ± 1.8558**	0.0001
$b_2 (S^2_{mix})$	**9.1349 ± 2.4418**	0.0003	**10.2415 ± 2.3918 27.8095 ± 2.2247**	0.0001
$b_3 (S^3_{mix})$	**9.7560 ± 2.5748**	0.0002	**9.1933 ± 1.7354 35.1992 ± 2.8159**	<0.0001
$b_4 (S^4_{mix})$	**4.2440 ± 1.1602**	0.0004	**4.8581 ± 1.1221 11.2879 ± 1.1902**	<0.0001
$b_5 (S^5_{mix})$	**2.2980 ± 0.6482**	0.0006	**2.5621 ± 0.1188 10.1747 ± 1.3976**	<0.0001
$b_6 (S^6_{mix})$	−0.9176 ± 1.0696	0.3927	−2.4281 ± 0.8779 −14.7126 ± 1.1770	0.2666
$b_7 (S^7_{mix})$	**−4.5381 ± 1.1435**	0.0020	**−4.1497 ± 0.6632 −9.6777 ± 0.7742**	<0.0001
$b_8 (S^8_{mix})$	**−8.9573 ± 1.9634**	<0.0001	**−9.2237 ± 1.6373 −25.6581 ± 2.0526**	<0.0001
$b_9 (S^9_{mix})$	**−10.0312 ± 2.8589**	0.0006	**−11.4736 ± 3.6473 −32.0013 ± 2.5601**	0.0001
$b_{10} (S^{10}_{mix})$	**−12.9604 ± 3.6943**	0.0006	**−13.9250 ± 4.4560 −42.7492 ± 3.4199**	0.0001
R^2	0.7758		0.9654	
R^2_{adj}	0.7564		0.9624	
$RMSE$	1.1865		0.6558	
F value	39.8120		39.8120	
p-value	<0.0001		<0.0001	

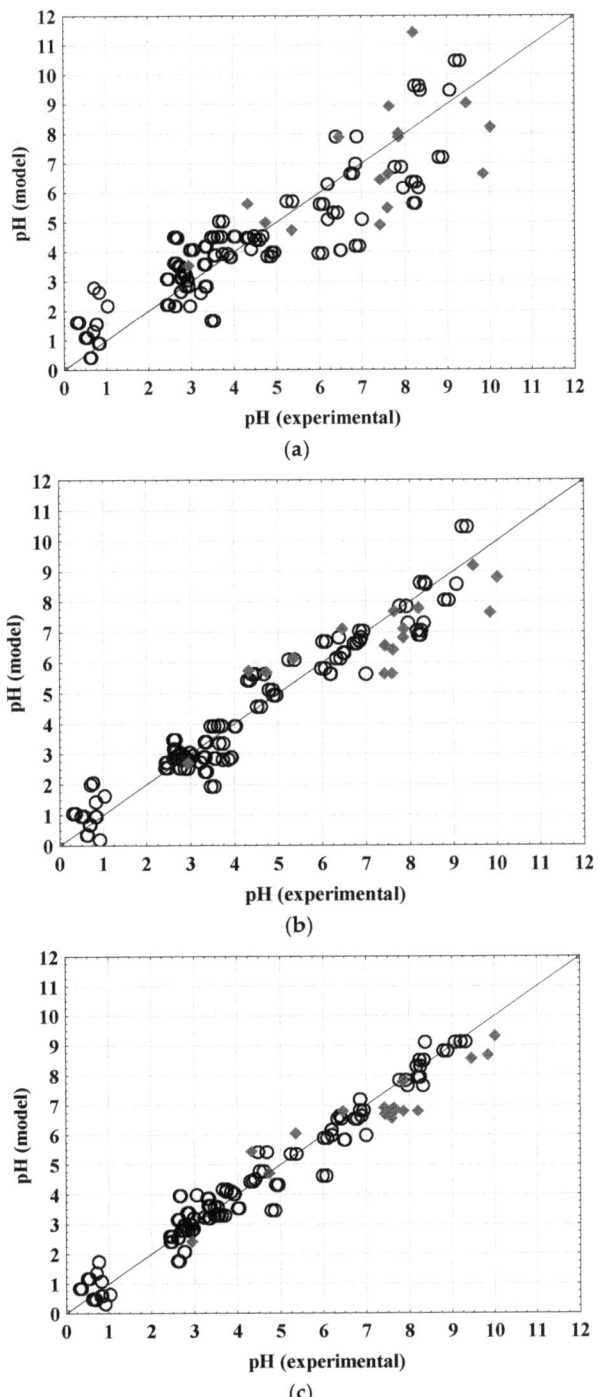

Figure 1. Comparison between experimental data and (**a**) MLR model, (**b**) PLR model, and (**c**) ANN model. (○) data set for model development, (◆) data set for model validation.

As described in the literature, linear regression calculates an equation that minimizes the distance between the fitted line and all data points. In general, a model fits the data well if the discrepancies between the observed and predicted value are minimal and unbiased. According to Cheng et al. (2014) [16], the coefficient of determination and adjusted coefficient of determination can be considered as summary measures for the goodness of fit of any linear regression model. Moreover, Le Mann et al. (2010) stated that the model can be regarded as appropriate if the coefficient of determination is above 0.75 [17]. Based on this, it can be concluded that both the MLR ($R^2 = 0.7758$) and PLR ($R^2 = 0.9654$) models developed in this work are applicable for the description of DESs' pH values based on S^i_{mix} descriptors but not with the same accuracy. When analyzing RMSE errors, it is evident that the PLR model (Figure 1b) ensures significantly smaller data dispersion ($RMSE = 0.6558$) in comparison to the MLR model ($RMSE = 1.1865$) (Figure 1a). As previously described, a high-accuracy model is strongly desired. However, the increase in the accuracy is usually accomplished by the increase in the complexity of the models by increasing the number of model parameters. For practical application, a model with fewer parameters is easier to interpret and, therefore, more suitable for the application.

A high R^2 value alone does not guarantee that the model fits the data well, so the model's goodness of fit was further confirmed by residual analysis. The residuals from a fitted model are the differences between the responses observed and the corresponding prediction of the response computed using the regression function. If the model's fit to the data was correct, the residuals would approximate the random errors that make the relationship between the explanatory variables and the response variable a statistical relationship. Therefore, if the residuals appear to behave randomly, it would suggest that the model fits the data well [18]. Analyzing the results presented in Figure 2, the residuals for the MPLR and PLR models were found to be normally distributed (Figure 2a,b). Furthermore, because the residual plots were gathered roughly along a straight line, the normality condition was met. The bell-shaped histograms that display the measurement distribution also verified the normal distribution of the residuals (Figure 2a,b). The residual vs. predicted value plots (Figure 2a,b) reveal that the residuals have no pattern, implying that the models match the experimental data well. Additionally, the residuals were found to range around the central value (Figure 2a,b) without obvious outliers, which means that the level of randomization was appropriate and that the sequence of testing had no effect on the findings [19].

Analysis of the MLR and PLR model coefficients showed that all coefficients, except b_6 (coefficient multiplying S^6_{mix}), were statistically significant. It can also be noticed that for both models, the coefficients from b_1 to b_5 have a positive influence on the output variable while the coefficients from b_6 to b_{10} have a negative influence on the analyzed model output. The results are easily interpreted in terms of b_1 to b_5, which are associated with the negative potential region and thus with hydrogen bond accepting and basicity properties on the one hand, and b_7 to b_{10}, which are associated with the positive potential region and thus with hydrogen bond donating and acidity properties on the other hand. b_6 turns out to be related to the neutral potential region insignificantly contributing to the pH value. As for the other b coefficient values, the more distant the potential region is from the zero (neutral value), the stronger its influence (whether positive or negative) on the pH value. Thus, the model seems to have a clear and rather simple physical significance. Although statistical analysis showed that the coefficient b_6 was not significant, the variable S_6 was not excluded from the modeling. This result indicates that there is no correlation with the dependent variable at the population level, but this could be changed if a different data set was used.

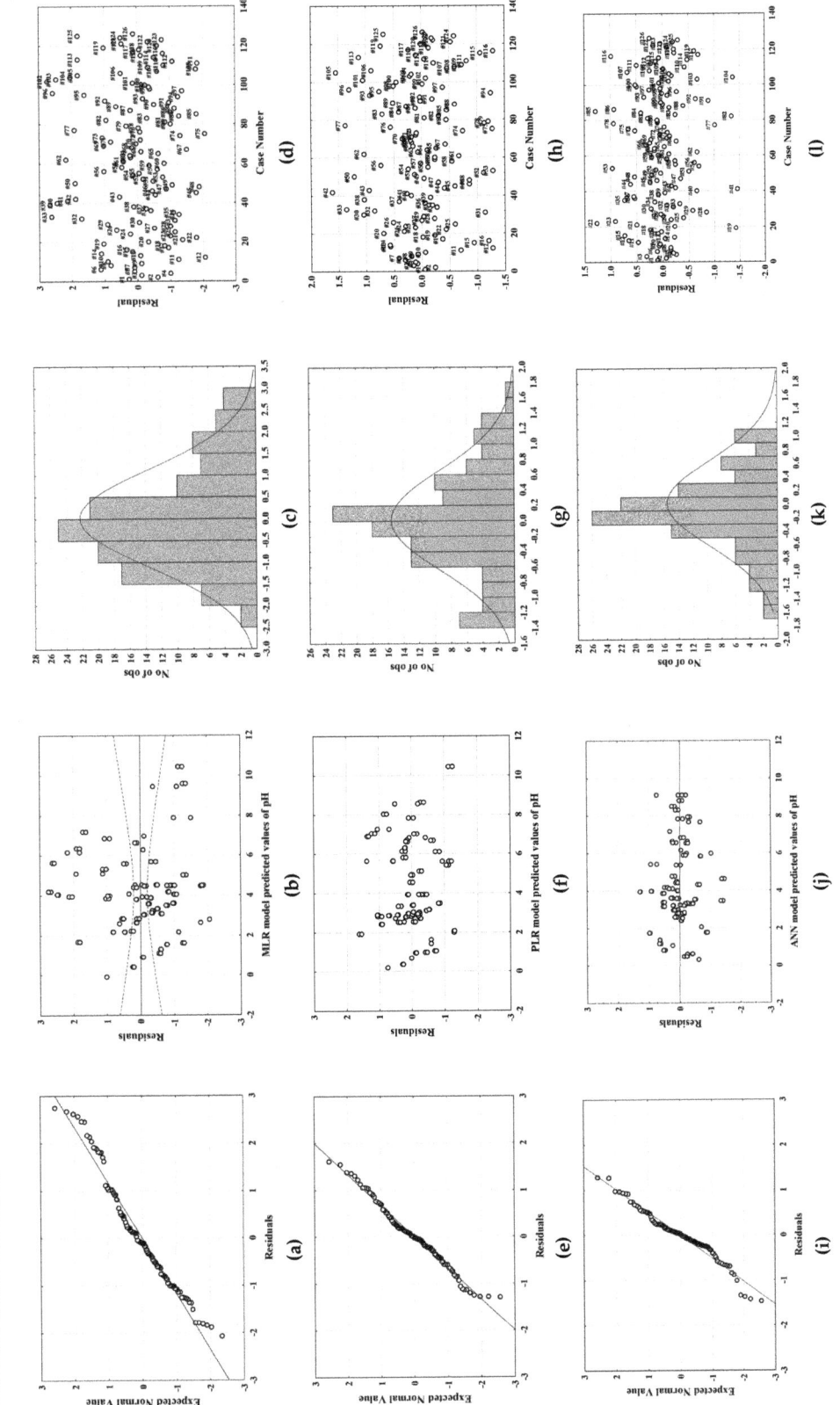

Figure 2. Analysis of the residuals for the MLR model (a–d), PLR model (e–h), and ANN mode (i–l).

The ANOVA revealed that the created MLR and PLR models were statistically significant, with p values < 0.001. Moreover, higher F-test results (F value = 39.8120) and lower p values, according to Greenland et al. (2016) [20], show the relative relevance of the created models. Based on the presented results it can be concluded that the collected findings demonstrate the dependability of the created models throughout the spectrum of variables evaluated.

2.3. Artificial Neural Network Modelling

The applicability of the artificial neural network models for predicting the DES pH values based on the σ profiles was also studied. The best neural network was chosen based on the following criteria: R^2 and $RMSE$ for training, test, and validation sets taking into account the number of neurons in the hidden layer. The properties of the created networks that were chosen are shown in Table 3. Based on the goodness of fit and validation error and considering the number of neurons in the hidden layer, the MLP model 10-5-1 was selected as optimal. Fewer neurons in the hidden layer make the ANN architecture simpler. The selected ANN was characterized by 10 neurons in the input layer, 5 neurons in the hidden layer, and 1 neuron in the output layer. The hidden activation function for the selected ANN was Tanh while the output activation function was Logistic. The described ANN provides a good agreement between the experimental data and the data predicted by the model ($R^2_{validation}$ = 0.9797, $RMSE_{validation}$ = 0.0012). As presented in Figure 1c, it can be observed that the data are distributed around the fitted function and that there are no evident outliers. As for the MLP and PLR models, the residual analysis was also performed for the ANN model (Figure 2c) and confirmed the ANN model's goodness of fit through a normal probability plot of the residuals (Figure 2c), residuals versus the predicted values plot (Figure 2c), histogram of the residuals (Figure 2c), and residuals versus the order of the data plot (Figure 2c).

Table 3. Architecture of the developed ANN (selected network is marked in bold). The numbers in the network name denote the number of neurons in the input, hidden, and output layers, respectively.

Network Name	Training Perf./ Training Error	Test Perf./ Test Error	Validation Perf./ Validation Error	Hidden Activation	Output Activation
MLP 10-13-1	0.9734, 0.0021	0.9751, 0.0031	0.9578, 0.0042	Logistic	Logistic
MLP 10-11-1	0.9812, 0.0013	0.9802, 0.0018	0.9794, 0.0018	Tanh	Exponential
MLP 10-10-1	0.9803, 0.0013	0.9827, 0.0016	0.9788, 0.0019	Tanh	Tanh
MLP 10-10-1	0.9808, 0.0017	0.9806, 0.0021	0.9716, 0.0019	Tanh	Logistic
MLP 10-5-1	**0.9868, 0.0011**	**0.9799, 0.0012**	**0.9797, 0.0012**	**Tanh**	**Logistic**

Based on the presented results, it can be concluded that the σ profiles are good molecular descriptors of DESs since the mathematical correlation of the experimental and predicted values is high. Moreover, based on the obtained R^2 values and the residual analysis, it can be concluded that both the PLR and ANN model can be efficiently applied for the prediction of the DES pH values based on the σ profiles. Due to the simplicity of the PLR model, this model is proposed for the prediction of physicochemical properties.

2.4. MLR, PLR, and ANN Models' Independent Validation

Validation of the MLR, PLR, and ANN models developed for the prediction of the DES pH values based on the σ profiles was performed on the independent set of data. The validation set included the σ profiles of 16 DESs. Comparisons between the experimental data and model-predicted data are shown in Figure 2. The validation performance of the developed models was estimated based on R^2 and $RMSE$ and the obtained values were as follows: (i) for MLR R^2 = 0.7097, $RMSE$ = 1.1140; (ii) for PLR R^2 = 0.8605, $RMSE$ = 0.7652; and (iii) for ANN R^2 = 0.8885, $RMSE$ = 0.82926.

It can be noticed that all three proposed models predict the pH value with high accuracy. As expected, the highest R^2 between the experiment and model-predicted data was obtained for ANN prediction of the analyzed DES pH values while the lowest R^2 between the experiment and model-predicted data was obtained for the MLR model. These findings demonstrate that σ profile ANN modeling is a useful and reliable method for predicting DES pH values based on the σ profiles. Nevertheless, considering RMSE, it can be noticed that the PLR model can efficiently be used for the prediction of pH values based on the σ profiles. As described, the R^2 values are scaled between 0 and 1, whereas the RMSE is not scaled to a specific value and, therefore, provides explicit information about how much the prediction deviates.

As stated before, it was relatively easy to link the parameters of the MLR and PLR models to their physical significance. On the other hand, ANNs, by definition, belong to a class of agnostic models and, thus, it is difficult, if not impossible, to reveal their physical meaning. At the same time, this is the reason why they behave much better in interpolation than in extrapolation. The independent validation presented here may be considered as interpolation since the DES members of the independent validation dataset belong to the same DES classes as those used for constructing the model. However, given the rather simple and rather clear relation between the σ profile and pH as revealed by MLR, there is no true reason to believe that the models would behave poorly in extrapolation, even for ANN, i.e., for DES classes not involved in the development of the models. However, this is yet to be checked, e.g., for DESs based on metal chlorides or DESs containing ionic liquids, etc.

The current literature data refer to the prediction of other physicochemical properties (such as viscosity and density) and only a narrow range of values characteristic for limited groups of structurally related DESs [11–14]. Based on our current knowledge, only one study has investigated the development of a mathematical model for DES pH value prediction [13]. In that study, the pH literature data of 41 DESs were processed in a similar way using the COSMO-RS and mathematical models, MLR and ANN, also covering a variety of cations, anions, and functional groups. The literature study [12] used literature data and included different temperatures (with temperature as an input parameter) while our study used our data obtained at a single temperature. The literature study also showed the potential of MLR and ANN modeling for the prediction of the pH value, however, with more complex models (models with more coefficients) than those developed in this work. Taking into consideration the specific future application of the developed models, it is recommended that they are as simple as possible and as robust as possible. Summing up the presented results, it can be concluded that the PLR model developed in this research can efficiently be used for the prediction of a wide range of DES pH values based on the σ profiles.

3. Materials and Methods

3.1. Materials

Betaine, choline chloride, glucose, L-(−)-proline, oxalic acid, sucrose, sorbitol, and xylitol were all purchased from Acros Organics, USA. Citric acid, D-fructose, D-(+)-xylose, D,L-malic acid, ethylene glycol, glycerol, and urea were all purchased from Sigma-Aldrich, USA. BIOVIA TmoleX19 version 2021 software (Dassault Systèmes, Vélizy-Villacoublay, France) was used for geometry and energy optimization of the HBAs and HBDs used in this study. BIOVIA COSMOtherm 2020 version 20.0.0. software (Dassault Systèmes) was used for the σ profile calculations of the defined DESs.

3.2. Methods

3.2.1. DES Preparation

DESs were prepared by mixing defined molar ratios of HBA to HBD. The two or more components were weighed in a specific ratio in a round-bottomed glass flask, adding 10–50% (w/w) of water. Then, the flasks were sealed, and the mixtures stirred and heated to 50 °C for 2 h until homogeneous transparent colorless liquids formed. The DES abbreviations and corresponding molar ratios are given in Table 1.

3.2.2. pH Value Measurement

The pH values for each DES were determined with a pH/ion meter S220 using an InLab Viscous Pro-ISM pH-electrode (Mettler Toledo, Greifensee, Switzerland), all within the pH measuring range 0.36–9.31 at room temperature. The instrument was calibrated using standard pH buffer solutions. Additionally, the pH values were checked with litmus paper (range 1–14). All measurements were carried out in duplicates and the results were expressed as an average value ± standard deviation.

3.2.3. Calculation of DES Constituents' σ Profiles and Descriptors

All molecules used for DES preparation: HBA, HBD, and water, were geometrically and energetically optimized in the BIOVIA TmoleX19 version 2021 (Dassault Systèmes) software. Quantum chemical calculations were performed by adopting DFT (density functional theory) with the BP86 functional level of theory and def-TZVP basis set [10]. To create a simplified and user-friendly database, for each molecule, the single most abundant non-ionized conformer with the lowest energy was chosen and used for further calculations. Molecules consisting of two or more ions (e.g., choline chloride) were treated as ion pairs and their structures were optimized according to Abranches et al. (2019) [21]. Finally, the software-generated COSMO file for each optimized molecule contained its σ profile curve that provided a quantitative representation of the molecules' polar surface screen charge on the polarity scale. HBAs are characterized by peaks in the negative potential region, HBDs by peaks in the positive potential region, and nonpolar molecules by peaks in the potential region around zero.

To define the molecular descriptors for all DES constituents, the σ profile curve for each HBA, HBD, and water was divided into 10 regions. The width of each region was 0.005 e/Å^2, covering the range from -0.025 to $+0.025$ e/Å^2. The areas under the curve were integrated separately for each defined region. This was achieved by simple summation of the tabulated σ profile data point ordinate values as presented by the BIOVIA COSMOtherm 2020 software. The ordinate values lying on the boundaries of the regions were split into halves and each half was attributed to one of the neighboring regions. Thus, 10 S descriptors (S^1–S^{10}) of the σ profiles were calculated exactly as the numerical values of these 10 areas (Table A1).

3.2.4. Calculation of DES Descriptors

Any change in the DES composition can be described by a change in its σ profile and the associated numerical value of its descriptors. To obtain a unique descriptor set for each particular DES, the σ profiles of its constituents were processed in the following manner. The descriptors of the studied DESs (S^i_{mix}) were calculated from the HBA and HBD component (and in some cases water) descriptors according to Equation (1) proposed by Benguerba et al. (2019) [11]:

$$S^i_{mix} = \sum_{j=1}^{NC} X_j S^i_{\sigma-profile,j} \qquad (1)$$

where i denotes the descriptor number (1–10), j stands for the DES constituent number, X_j is the molar fraction of HBA or HBD or some other constituent such as water if present in the mixture, $S^i_{\sigma-profile,j}$ is the j-th constituent i-th descriptor, and NC is the total number of constituents from which DES is prepared. All the experiments were performed at 20 °C.

3.2.5. Modeling of Correlation between pH and Descriptors

In further calculations, it was assumed that the measured DES pH value can be described as a function of the σ profile of the mixture, expressed by a set of Simix descriptors in Equation (2):

$$pH = f\left(S^1_{mix}, S^2_{mix}, S^3_{mix}, S^4_{mix}, S^5_{mix}, S^6_{mix}, S^7_{mix}, S^8_{mix}, S^9_{mix}, S^{10}_{mix}\right) \qquad (2)$$

Multiple linear regression (MLR) with Equation (3), piecewise linear regression (PLR) with Equation (4), and artificial neural network (ANN) models were attempted to describe the relationship between the input and output variables. The dataset included 142 data points (that included replicates), of which 126 were used for model development and 16 (randomly selected) for independent model validation:

$$pH = b_0 + b_1 \cdot S_{mix}^1 + b_2 \cdot S_{mix}^2 + b_3 \cdot S_{mix}^3 + b_4 \cdot S_{mix}^4 + b_5 \cdot S_{mix}^5 + b_6 \cdot S_{mix}^6 + b_7 \cdot S_{mix}^7 + b_8 \cdot S_{mix}^8 + b_9 \cdot S_{mix}^9 + b_{10} \cdot S_{mix}^{10} \quad (3)$$

$$pH = \left(\begin{cases} b_{01} + \sum_{i=1}^{10} b_{i1} \cdot S_{mix}^i & \forall (pH \leq b_n) \\ b_{02} + \sum_{i=1}^{10} b_{i2} \cdot S_{mix}^i & \forall (pH > b_n) \end{cases} \right) \quad (4)$$

The PLR technique is based on estimating the parameters of two linear regression equations: one for dependent variable values (y) less than or equal to the breakpoint (bn) and the other for dependent variable values (y) higher than the breakpoint.

The MLR parameters in Equation (3) were estimated using least square regression while the PLR parameters in Equation (4) were estimated using the Levenberg–Marquardt algorithm implemented in the software Statistica 13.0 (Tibco Software Inc, Palo Alto, Santa Clara, CA, USA). The algorithm searches for optimal solutions in the function parameter space using the least squares method. The calculations were performed in 50 repetitions with a convergence parameter of 10–6 and a confidence interval of 95% [22].

In addition, multilayer perceptron (MLP) ANNs were used for the prediction of DES pH values based on the Simix descriptors. The ANN models included an input layer, hidden layer, and output layer. The input layer included 10 neurons representing the Simix descriptors, the output layer had only one neuron, and the number of neurons in the hidden layer varied between 4 and 13 and was randomly selected by the algorithm. The hidden activation function and output activation function were selected randomly from the following set: Identity, Logistic, Hyperbolic tangent, and Exponential. The dimension of the data set for ANN modeling was 126 × 11 and was randomly divided into 70% for network training, 15% for network testing, and 15% for model validation. Model training was carried out using a back error propagation algorithm and the error function was a sum of squares implemented in Statistica v.13.0 Automated Neural Networks. The developed model's performance was estimated by calculating the R^2 and root mean squared error (RMSE) values for the training, test, and validation sets.

Validation of the developed MLR, PLR, and ANN models was performed on an independent data set, including the Simix descriptors for 16 randomly selected DESs. The validation performance of the developed models was estimated based on the R^2 and root mean squared error (*RMSE*).

4. Conclusions

The applicability of MLR, PLR, and ANN to predict the pH values of DESs was evaluated. The results indicate that although simple linear regression can be used for the description and prediction, its effectiveness and applicability are limited. On the other hand, PLR and ANN are applicable to predict the pH values of DESs with a very high goodness of fit ($R^2 > 0.8600$). The contribution of this work lies in the development of a user-friendly model to predict pH values in a wide range (from 0.525 to 9.25), indicating that the developed models are good for the prediction of the pH value of newly synthesized DESs. However, due to the simplicity of the developed PLR model, it could be suggested as a model of choice for use in daily work and screening purposes.

Nevertheless, this approach can also be extended to other physicochemical properties since this study confirmed previous findings that showed how the σ profile generated in COSMOtherm is a valuable DES molecular descriptor. It could be a good basis for the evaluation of various mathematical models to develop a simple and applicable prediction model for everyday laboratory or industrial applications.

It is interesting to comment on the influence of the addition of water to a DES. In our previous article [7], based on a limited set of data, it was noticed that the addition of water to extremely acidic DESs increases their pH values, and the addition of water to highly basic DESs decreases their pH values. Thus, it seemed that the addition of water somehow mellowed the pH environments. On the other hand, on a larger set of data, as presented here, this conclusion does not hold any more: there are difficult-to-predict exemptions to the rule. On the other hand, the COSMO-RS calculation results in combination with the non-presumptive numerical models, such as MLR, PLR, and ANN, are perfectly suitable to tackle those difficult-to-predict systems.

Author Contributions: Conceptualization, I.R.R., M.P. and A.J.T.; methodology, M.P. and A.J.T.; software, M.P., M.R. (Mia Radović), M.R. (Marko Rogošić), and J.A.P.C.; validation, M.R. (Mia Radovićand), M.C.B., K.R. and J.A.P.C.; formal analysis, M.P., M.R. (Mia Radović), and M.C.B.; investigation, M.P., M.C.B., K.R. and M.R. (Mia Radović); resources, I.R.R.; data curation, M.P., M.R. (Mia Radović), M.R. (Marko Rogošić), and A.J.T.; writing—original draft preparation, M.P. and A.J.T.; writing—review and editing, M.P., M.R. (Mia Radović), M.C.B., K.R., M.R. (Marko Rogošić), I.R.R., A.J.T. and J.A.P.C.; visualization, M.P., M.R. (Marko Rogošić), and A.J.T.; supervision, I.R.R.; project administration, I.R.R.; funding acquisition, I.R.R. All authors have read and agreed to the published version of the manuscript.

Funding: This work was partly developed within the scope of the project CICECO-Aveiro Institute of Materials, UIDB/50011/2020 & UIDP/50011/2020, financed by national funds through the Portuguese Foundation for Science and Technology/MCTES. This work was also financed by the Croatian science foundation (grant No. 7712).

Institutional Review Board Statement: Not applicable.

Informed Consent Statement: Not applicable.

Data Availability Statement: Not applicable.

Conflicts of Interest: The authors declare no conflict of interest.

Sample Availability: Samples of the compounds are available from the authors.

Appendix A

Table A1. S descriptors (S1–S10) of the σ profiles from compounds from which DESs were prepared.

	Intervals		B Betaine	ChCl Choline Chloride	Pro LD-proline	CA Citric Acid	MA Malic Acid	OxA Oxalic Acid	U Urea	H$_2$O
σ-profile	[−0.025; −0.02]	1	0	0	0.506	4.861	3.5955	0	0	0
	[−0.02; −0.015]	2	0	0	5.186	14.9695	10.5215	7.5105	6.35	6.35
	[−0.015; −0.01]	3	11.869	16.1615	6.9485	13.5665	9.368	20.482	10.027	10.027
	[−0.01; −0.005]	4	59.1185	66.196	17.199	29.212	28.535	9.0145	3.5195	3.5195
	[−0.005; 0.0]	5	36.625	34.4875	60.605	29.3465	23.3925	7.9265	2.1635	2.1635
	[0.0; 0.005]	6	4.5285	5.6435	21.7815	23.467	18.1455	13.051	2.8725	2.8725
	[0.005; 0.01]	7	3.2405	6.6525	10.6	37.877	25.726	7.606	4.055	4.055
	[0.01; 0.015]	8	7.719	18.3	17.614	38.933	30.6435	11.679	5.2285	5.2285
	[0.015; 0.02]	9	22.3525	30.0465	5.2065	1.0135	2.3845	13.8265	8.2765	8.2765
	[0.02; 0.025]	10	8.202	0.0525	1.3475	0	0	0	0.172	0.5775
	Intervals		EG ethylene glycol	Sol sorbitol	Gly glycerol	Xyol xylitol	Fru Dfructose	Glc Dglucose	Suc sucrose	Xyl Dxylose
σ-profile	[−0.025; −0.02]	1	0	0.1725	0.013	0.037	0.1655	0.213	0.108	0.037
	[−0.02; −0.015]	2	3.8055	15.884	7.828	9.216	11.8325	23.022	11.1905	8.7015
	[−0.015; −0.01]	3	7.638	20.8955	11.4065	15.941	16.9895	28.444	14.5935	12.9035
	[−0.01; −0.005]	4	20.2675	41.5705	19.6085	43.0415	36.2095	61.232	35.406	34.6755
	[−0.005; 0.0]	5	28.038	30.5525	34.6465	35.965	39.319	54.567	28.5165	45.0735
	[0.0; 0.005]	6	9.973	18.7645	15.0725	19.083	19.565	29.1605	15.066	17.7475
	[0.005; 0.01]	7	7.9725	21.5775	10.103	20.848	19.4555	26.6145	20.4555	17.2715
	[0.01; 0.015]	8	10.5605	28.283	17.194	28.977	30.9465	47.3685	29.0795	23.517
	[0.015; 0.02]	9	9.6155	20.752	10.7865	10.9745	11.901	26.0425	8.4485	12.688
	[0.02; 0.025]	10	0.0035	0.15	0.0115	0	0	1.082	0	0.005

References

1. Anastas, P.T.; Beach, E.S. Green Chemistry: The Emergence of a Transformative Framework. *Green Chem. Lett. Rev.* **2008**, *1*, 9–24. [CrossRef]
2. Cvjetko Bubalo, M.; Vidović, S.; Radojčić Redovniković, I.; Jokić, S. Green Solvents for Green Technologies. *J. Chem. Technol. Biotechnol.* **2015**, *90*, 1631–1639. [CrossRef]
3. Lanza, V.; Vecchio, G. New Conjugates of Superoxide Dismutase/Catalase Mimetics with Cyclodestrins. *J. Inorg. Biochem.* **2009**, *103*, 381–388. [CrossRef] [PubMed]
4. Abbott, A.P.; Capper, G.; Davies, D.L.; Rasheed, R.K.; Tambyrajah, V. Novel Solvent Properties of Choline Chloride/Urea Mixtures. *Chem. Commun.* **2003**, *10*, 70–71. [CrossRef]
5. Martins, M.A.R.; Pinho, S.P.; Coutinho, J.A.P. Insights into the Nature of Eutectic and Deep Eutectic Mixtures. *J. Solut. Chem.* **2019**, *48*, 962–982. [CrossRef]
6. Paiva, A.; Matias, A.A.; Duarte, A.R.C. How Do We Drive Deep Eutectic Systems towards an Industrial Reality? *Curr. Opin. Green Sustain. Chem.* **2018**, *11*, 81–85. [CrossRef]
7. Mitar, A.; Panić, M.; Prlić Kardum, J.; Halambek, J.; Sander, A.; Zagajski Kučan, K.; Radojčić Redovniković, I.; Radošević, K. Physicochemical Properties, Cytotoxicity, and Antioxidative Activity of Natural Deep Eutectic Solvents Containing Organic Acid. *Chem. Biochem. Eng. Q.* **2019**, *33*, 1–18. [CrossRef]
8. Abbott, A.P.; Alabdullah, S.S.M.; Al-Murshedi, A.Y.M.; Ryder, K.S. Brønsted Acidity in Deep Eutectic Solvents and Ionic Liquids. *Faraday Discuss.* **2017**, *206*, 365–377. [CrossRef]
9. Farias, F.O.; Passos, H.; Coutinho, J.A.P.; Mafra, M.R. PH Effect on the Formation of Deep-Eutectic-Solvent-Based Aqueous Two-Phase Systems. *Ind. Eng. Chem. Res.* **2018**, *57*, 16917–16924. [CrossRef]
10. Klamt, A.; Jonas, V.; Bürger, T.; Lohrenz, J.C.W. Refinement and Parametrization of COSMO-RS. *J. Phys. Chem. A* **1998**, *102*, 5074–5085. [CrossRef]
11. Benguerba, Y.; Alnashef, I.M.; Erto, A.; Balsamo, M.; Ernst, B. A Quantitative Prediction of the Viscosity of Amine Based DESs Using Sσ-Profile Molecular Descriptors. *J. Mol. Struct.* **2019**, *1184*, 357–363. [CrossRef]
12. Lemaoui, T.; Hammoudi, N.E.H.; Alnashef, I.M.; Balsamo, M.; Erto, A.; Ernst, B.; Benguerba, Y. Quantitative Structure Properties Relationship for Deep Eutectic Solvents Using Sσ-Profile as Molecular Descriptors. *J. Mol. Liq.* **2020**, *309*, 113165. [CrossRef]
13. Lemaoui, T.; Abu Hatab, F.; Darwish, A.S.; Attoui, A.; Hammoudi, N.E.H.; Almustafa, G.; Benaicha, M.; Benguerba, Y.; Alnashef, I.M. Molecular-Based Guide to Predict the PH of Eutectic Solvents: Promoting an Efficient Design Approach for New Green Solvents. *ACS Sustain. Chem. Eng.* **2021**, *9*, 5783–5808. [CrossRef]
14. Silva, L.P.; Fernandez, L.; Conceiçao, J.H.F.; Martins, M.A.R.; Sosa, A.; Ortega, J.; Pinho, S.P.; Coutinho, J.A.P. Design and Characterization of Sugar-Based Deep Eutectic Solvents Using Conductor-like Screening Model for Real Solvents. *ACS Sustain. Chem. Eng.* **2018**, *6*, 10724–10734. [CrossRef]
15. Hayyan, A.; Mjalli, F.S.; Alnashef, I.M.; Al-Wahaibi, T.; Al-Wahaibi, Y.M.; Hashim, M.A. Fruit Sugar-Based Deep Eutectic Solvents and Their Physical Properties. *Thermochim. Acta* **2012**, *541*, 70–75. [CrossRef]
16. Cheng, C.L.; Shalabh; Garg, G. Coefficient of Determination for Multiple Measurement Error Models. *J. Multivar. Anal.* **2014**, *126*, 137–152. [CrossRef]
17. Le Man, H.; Behera, S.K.; Park, H.S. Optimization of Operational Parameters for Ethanol Production from Korean Food Waste Leachate. *Int. J. Environ. Sci. Technol.* **2009**, *7*, 157–164. [CrossRef]
18. Feng, C.; Feng, C.; Li, L.; Sadeghpour, A. A Comparison of Residual Diagnosis Tools for Diagnosing Regression Models for Count Data. *BMC Med. Res. Methodol.* **2020**, *20*, 175. [CrossRef]
19. Matešić, N.; Jurina, T.; Benković, M.; Panić, M.; Valinger, D.; Gajdoš Kljusurić, J.; Jurinjak Tušek, A. Microwave-Assisted Extraction of Phenolic Compounds from *Cannabis Sativa* L.: Optimization and Kinetics Study. *Sep. Sci. Technol.* **2020**, *56*, 2047–2060. [CrossRef]
20. Greenland, S.; Senn, S.J.; Rothman, K.J.; Carlin, J.B.; Poole, C.; Goodman, S.N.; Altman, D.G. Statistical Tests, P Values, Confidence Intervals, and Power: A Guide to Misinterpretations. *Eur. J. Epidemiol.* **2016**, *31*, 337–350. [CrossRef]
21. Abranches, D.O.; Larriba, M.; Silva, L.P.; Melle-Franco, M.; Palomar, J.F.; Pinho, S.P.; Coutinho, J.A.P. Using COSMO-RS to Design Choline Chloride Pharmaceutical Eutectic Solvents. *Fluid Phase Equilibria* **2019**, *497*, 71–78. [CrossRef]
22. Jurinjak Tušek, A.; Jurina, T.; Benković, M.; Valinger, D.; Belščak-Cvitanović, A.; Kljusurić, J.G. Application of Multivariate Regression and Artificial Neural Network Modelling for Prediction of Physical and Chemical Properties of Medicinal Plants Aqueous Extracts. *J. Appl. Res. Med. Aromat. Plants* **2020**, *16*, 100229. [CrossRef]

Article

Neutron Total Scattering Investigation of the Dissolution Mechanism of Trehalose in Alkali/Urea Aqueous Solution

Changli Ma [1,2], Taisen Zuo [1,2], Zehua Han [1,2], Yuqing Li [1,2,3], Sabrina Gärtner [4], Huaican Chen [1,2], Wen Yin [1,2], Charles C. Han [5] and He Cheng [1,2,*]

1. Institute of High Energy Physics, Chinese Academy of Sciences (CAS), Beijing 100049, China; machangli@ihep.ac.cn (C.M.); zuots@ihep.ac.cn (T.Z.); hanzh@ihep.ac.cn (Z.H.); liyuqing@ihep.ac.cn (Y.L.); chenhuaican@ihep.ac.cn (H.C.); yinwen@ihep.ac.cn (W.Y.)
2. Spallation Neutron Source Science Center, Dongguan 523803, China
3. University of Chinese Academy of Sciences, Beijing 100049, China
4. STFC ISIS Facility, Rutherford Appleton Laboratory, Didcot OX11 0QX, UK; sabrina.gaertner@stfc.ac.uk
5. Institute for Advanced Study, Shenzhen University, Shenzhen 508060, China; han.polymer@gmail.com
* Correspondence: chenghe@ihep.ac.cn; Tel.: +86-13925802541

Abstract: The atomic picture of cellulose dissolution in alkali/urea aqueous solution is still not clear. To reveal it, we use trehalose as the model molecule and total scattering as the main tool. Three kinds of alkali solution, i.e., LiOH, NaOH and KOH are compared. The most probable all-atom structures of the solution are thus obtained. The hydration shell of trehalose has a layered structure. The smaller alkali ions can penetrate into the glucose rings around oxygen atoms to form the first hydration layer. The larger urea molecules interact with hydroxide groups to form complexations. Then, the electronegative complexation can form the second hydration layer around alkali ions via electrostatic interaction. Therefore, the solubility of alkali aqueous solution for cellulose decreases with the alkali cation radius, i.e., LiOH > NaOH > KOH. Our findings are helpful for designing better green solvents for cellulose.

Keywords: neutron total scattering; cellulose; dissolution mechanism; layered structure; complexation

1. Introduction

Cellulose is the world's most produced natural polymer, and it is a potential candidate to replace petroleum-based materials. To achieve high performance, cellulose has to be dissolved first [1].

Cellulose is difficult to dissolve. Strong inter- and intra-chain interactions prevent its structure from being deconstructed. In industry, the widely used viscose method produces a large amount of alkaline and acidic waste, carbon disulfide and hydrogen sulfide gases. It pollutes the environment. N-methylmorpholine noxide (NMMO) and ionic liquids (ILS) are environment friendly, but they have high costs [1].

Around 2000, Prof. Zhang proposed the use of precooled NaOH/LiOH urea aqueous solution to dissolve cellulose [1,2]. This solvent has the advantages of fast dissolution, low cost and low pollution and, thus, has good application prospects. Unfortunately, the solubility of cellulose in green solvent is still too low to reach the requirements of industrial production. We need to know the dissolution mechanism first to increase its solubility. Lots of methods, such as NMR, FTIR, DSC, TEM, et al., have been used. These studies qualitatively showed that cationic hydrates are more easily adsorbed around cellulose molecules to form a new, stable hydrogen bond network at low temperatures, and hydrates of urea molecules form a sheath-like inclusion complex (IC) around their periphery [3–5]. Bjorn et al. summarized the dissolution mechanism of cellulose and proposed that cellulose is amphiphilic and that hydrophobic interactions are important for its solubility [6–9]. Wolfgang et al. agreed that hydrophobic interactions are the driving forces in an amorphous system, but all types of cellulose are highly crystalline so hydrogen bonds must be broken before hydrophobic interactions can be effective, and this requires a strong alkaline medium [10]. To reveal the dissolution mechanism of cellulose in alkaline/urea aqueous solution, a detailed atomic picture needs to be presented.

The combination of neutron total scattering and empirical potential structure refinement (EPSR) can observe in situ the most probable all-atom structure of the liquids [11,12]. In a previous study, we used the combined methods to study the atomic structure of trehalose in NaOH/urea aqueous solution [13]. We used trehalose as the model molecule for cellulose, because it has similar glucose rings and is one of the disaccharide molecules that has no reducibility (other binary sugars, such as cellobiose, glucose, maltose and lactose, are oxidized in the alkaline solution during neutron total scattering experiments). We found that NaOH, urea and water work cooperatively to dissolve trehalose. Na^+ accumulates around electronegative oxygen atoms in the hydration shell, while urea molecules only participate in the dissolution process via Na^+ bridging. Additionally, we predicted that alkali with smaller ions, such as LiOH, have better solubility for cellulose.

To prove this, we further observed the microscopic dissolution pictures in two different alkali/urea aqueous solutions, i.e., LiOH and KOH. Then, we compared them with previous results in NaOH/urea aqueous solution. The hydration shell of trehalose has a layered structure; cations directly interact with the glucose rings to form the first hydration layer. It destroys their intra- and intermolecular hydrogen bonds. Thus, the smaller the radius of the cation, the easier it approaches the inside of the glucose ring. From K^+ to Na^+ to Li^+, its ability to dissolve cellulose gradually increases. Urea does not directly interact with glucose rings, and it forms strong complexations with hydroxide groups. The urea hydration complexation forms the second hydration layer via electric interaction. It prevents it from re-aggregating. Temperature effect was also investigated. The atomic structure of the solution did not change when it was cooled down to $-10\ °C$ (or $-5\ °C$). Taking into account the fact that cellulose only dissolves in green solvent at lower temperatures, the dissolution had to be a dynamic process.

2. Theory and Methods

2.1. Neutron Scattering Method and SANDALS

The total neutron scattering experiment is an important method for the study of the atomic structure of liquids. In neutron scattering, the observed neutron structure factor ($F(Q)$) and the atomic structure of the sample have the relationship:

$$F(Q) = \sum_{\alpha=1}^{N} c_\alpha b_\alpha^2 + \sum_{\alpha=1, \beta \geq \alpha} (2 - \delta_{\alpha\beta}) c_\alpha c_\beta b_\alpha b_\beta \{4\pi\rho \int_0^\infty r^2 (g_{\alpha\beta}(r) - 1) \frac{\sin(Qr)}{Qr} dr\} \quad (1)$$

where Q and r are the momentum transfer and the distance between the two atoms in the sample; c_i and b_i ($i = \alpha, \beta$) are the quantity ratio and neutron scattering length of i species; $\delta_{\alpha\beta}$ is the Kronecker δ function; ρ is the atomic number density of the sample; and $g_{\alpha\beta}(r)$ is the radial distribution function (hereinafter referred to RDF), which reflects the microscopic atomic structure of amorphous matter.

$$g_{\alpha\beta}(r) = \frac{n_{\alpha\beta}}{4\pi r^2\ dr \rho_\beta} \quad (2)$$

where $n_{\alpha\beta}$ is an average number of β atoms around an α atom contained in a spherical shell with r radius and dr thickness; ρ_β is the average number density of β atoms in the sample; and $g_{\alpha\beta}(r)$ describes the number density change of the β atom as the function of the distance from the α atom, which reflects the interaction between atoms α and β. We used $g_{(1)\alpha\beta}(r)$ to mark the first peak of $g_{\alpha\beta}(r)$. It represents the closest, most probable distance between atoms β and α. The peak height indicates the ratio of the atomic number density of atom β to its average value at this position.

The neutron total scattering experiments were performed at SANDALS in ISIS (Didcot, UK) and Multi-Physics Instrument (MPI) in CSNS (Dongguan, China). The scattering vector (Q) range was from 0.1 Å^{-1} to 50 Å^{-1}. This means that the instrument could only measure the micro-structure of samples from ~0.1 Å to ~30 Å. Therefore, we chose trehalose as the model molecule for cellulose. The size of trehalose is ~11.6 Å; we could, thus, use a neutron total scattering instrument to observe its microscopic atomic structure in alkali/urea aqueous solution.

2.2. EPSR Simulations

The empirical potential structure refinement (EPSR) is a program developed by Prof. Soper to explore the most probable, all-atom structure of an experimental sample based on neutron scattering [11]. EPSR is essentially a Monte Carlo simulation program. In EPSR, there are two kinds of potential energy: reference potential and experimental potential. The reference potential is taken from the molecular dynamics simulation force field, which is used to realize the basic structural constraints of the simulation system, such as molecular structure, the minimum distance between atoms, etc.; the experimental potential is the structural data observed in the neutron scattering experiment [11,12]. After the experiment, we used EPSR simulation to reconstruct the atomic structure of the experimental sample.

2.3. Experiment Samples

The trehalose, LiOH, KOH and urea used in the experiment were purchased from Shanghai Aladdin Reagent Company. The purity of trehalose was over 99%, and the purity of alkali and urea was over 99.9% for both. Both deuterated urea and heavy water were purchased from Sigma-Aldrich China. The deuteration rate of heavy water was over 99.9%, and the deuteration rate of urea was over 98%.

In this study, we used SANDALS and MPI to observe 4 kinds of experimental sample with different chemical components. They were aqueous solutions of trehalose/LiOH/urea, trehalose/KOH/urea, trehalose/LiOH and trehalose/KOH. The molar ratio of each component in the sample was consistent with that in Prof. Zhang's experiment [2]. Each kind of solution included three different deuterated ratio test samples, i.e., full hydrogen, full deuterium and half deuterium. All samples were measured at room temperature (25 °C) and −10 °C. However, the trehalose LiOH/urea aqueous solution froze at −10 °C in the neutron scattering experiment, so we raised the temperature of this sample to −5 °C. Each sample was measured in a flat sample cell of titanium–zirconium alloy with a capacity of 1.3 mL for approximately 6 h. Empty sample cells were measured for approximately 4 h for background subtraction. The sample composition and symbols are shown in Table 1.

Table 1. Sample labels, chemical components, deuterium ratios and molar ratios of the samples.

Sample Labels	Li(K)TrH$_2$O/HDO/D$_2$O	Li(K)TrUrH$_2$O/HDO/D$_2$O
Chemical Component	Li(K)OH Trehalose Water	Li(K)OH Urea Trehalose Water
Deuterium Ratio	0.0/0.5/1.0	0.0/0.5/1.0
Molar Ratio	Li(K)OH:Urea:Trehalose:Water = 222:254:64:5716	

For the EPSR simulation and the following discussion of the results, the symbols of the atoms in trehalose were as shown in Figure 1. Oxygen and hydrogen atoms of H$_2$O were labeled as OW and HW, respectively; alkali cations were labeled as Li, Na and K, respectively; the oxygen and hydrogen atom of hydroxide anion were labeled as OOH and HOH, respectively; and the carbon, oxygen, nitrogen and hydrogen atoms in the urea molecule were labeled as CU, OU, NU and HU, respectively. The label of each atom and the parameters of reference potential used in EPSR simulation are listed in Table 2(1,2). Among them, epsilon and sigma are the parameters of the Lennard-Jones reference potential (L-J), and q is the charge of the atoms [13–17]. The information for the EPSR simulation box is shown in Table 3. The data about NaOH aqueous solutions can be found in our previous study [13].

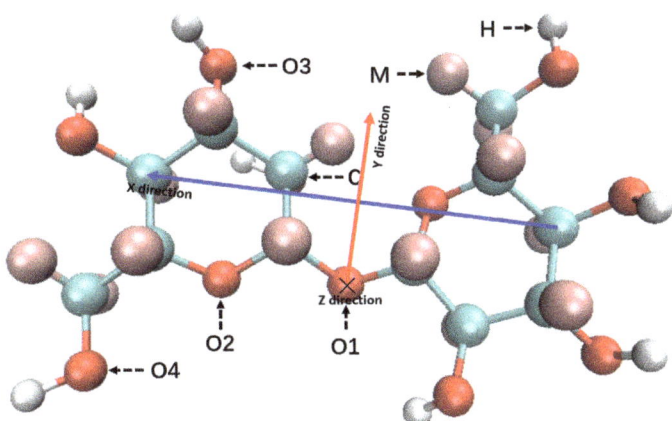

Figure 1. Structure of a trehalose molecule and its atomic labels used in EPSR simulation. All carbon atoms are labeled as C; all hydrogen atoms connected to the oxygen atom are labeled as H; the remaining hydrogen atoms connected to the carbon atom are labeled as M; the oxygen atom linking the two glucose rings is labeled as O1; the oxygen atoms on the glucose ring is labeled as O2; the oxygen atoms on the hydroxyl group connected to the glucose ring are labeled as O3; and the oxygen atom on the methyl group is labeled as O4. In order to describe the distribution of atoms around trehalose with spatial angles, we set up a coordinate system with O1 as the coordinate origin. The blue arrow connecting two C atoms at the glucose rings represents the direction of the X-axis; the red arrow pointing vertically to the blue arrow from O1 is the direction of the Y-axis; the Z-axis is the cross product of the X-axis and the Y-axis, which is approximately perpendicular to the paper.

Table 2. Lennard-Jones (L-J) reference potential parameters and charge (q) of atoms used in EPSR.

	(1)								
Atom Label	OW	HW	Li	Na	K	OOH	HOH	CU	OU
ε [KJ/mole]	0.650	0.000	0.690	0.125	0.500	0.251	0.184	0.439	0.878
σ [Å]	3.166	0.000	1.510	2.500	3.000	2.750	1.443	0.375	2.960
q [e]	−0.848	0.424	0.679	0.679	0.679	−1.103	0.424	0.142	−0.390
	(2)								
Atom Label	NU	HU	C	O1	O2	O3	O4	H	M
ε [KJ/mole]	0.711	0.000	0.276	0.586	0.586	0.711	0.711	0.050	0.121
σ [Å]	3.250	0.000	3.500	3.100	2.900	3.100	3.100	1.700	1.700
q [e]	−0.542	0.333	0.258	−0.500	−0.500	−0.500	−0.500	0.301	0.000

Table 3. Cubic simulation box atomic number density and number of molecules used in EPSR.

Sample Labels	Li(K)OH	Urea	Trehalose	Water	Density (Atoms/Å^3)
Li(K)TrH$_2$O	222	0	64	5716	0.106531 (Li) 0.103323 (K)
Li(K)TrUrH$_2$O	222	254	64	5716	0.106405 (Li) 0.103437 (K)

3. Results

The neutron scattering structure factor and EPSR simulation results of different samples are shown in Figure 2. As shown in the figure, the EPSR simulation results and the neutron scattering profiles were in good agreement. Therefore, the structure reconstructed by EPSR represented the most probable atomic structure of the solutions.

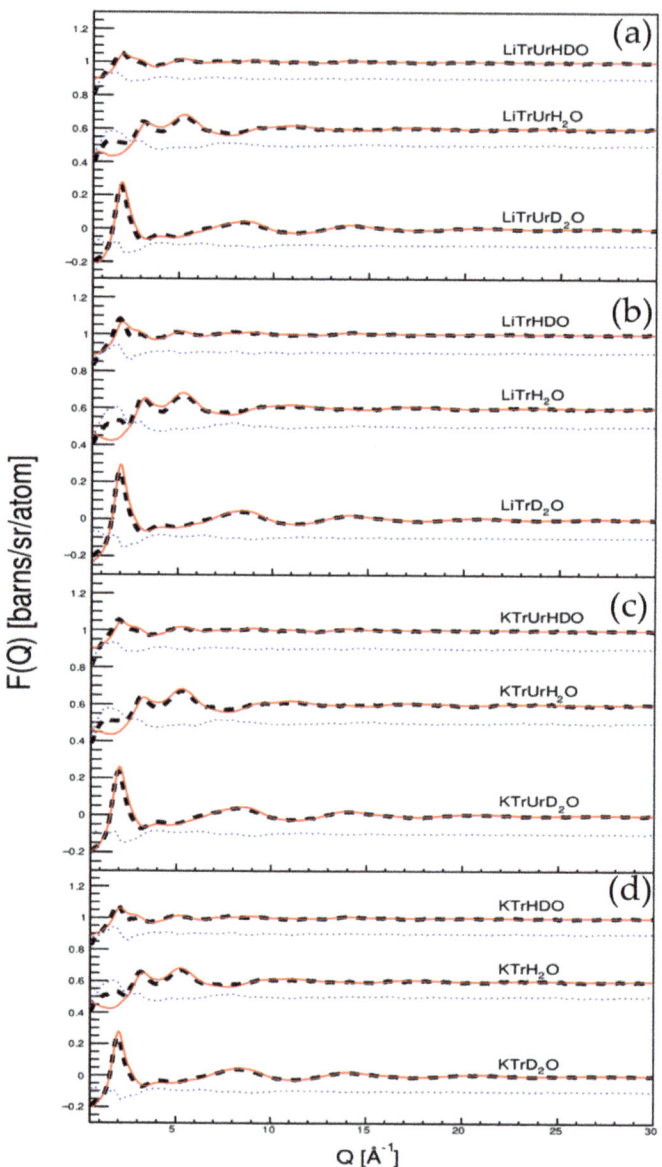

Figure 2. Experimental neutron total scattering profiles (dashed black lines), the EPSR fitted neutron structure factors (red lines) and their differences (blue dots). (**a**) Trehalose in LiOH/urea aqueous solution, (**b**) trehalose in LiOH aqueous solution, (**c**) trehalose in KOH/urea aqueous solution, (**d**) trehalose in KOH aqueous solution. Data have been offset for clarity.

EPSR gave us the most probable, all-atom structures in trehalose alkali/urea aqueous solution. The simulation, as shown in Figure 3, qualitatively showed the distribution diagram of Li^+, Na^+, K^+ ions and urea molecules around trehalose. In order to improve the statistics, the ions and atoms in the figure were the number of atoms gathered in the 100-frame EPSR simulation conformation. Three things could be seen directly. The first thing was that those ions and urea molecules were anisotropically distributed around trehalose because of steric repulsion, so we built a coordinate system around its O1 atom (Figure 1). Thus, RDF could be calculated at different space angles. The second thing was that ions concentrated around the hydrophilic oxygen atoms of trehalose, and smaller ions were much more easily penetrated into the glucose ring, from K$^+$ to Na$^+$ to Li$^+$ (Figure 3a). The third thing was that larger urea molecules could not directly interact with trehalose, and they were further away from trehalose than the ions (Figure 3b).

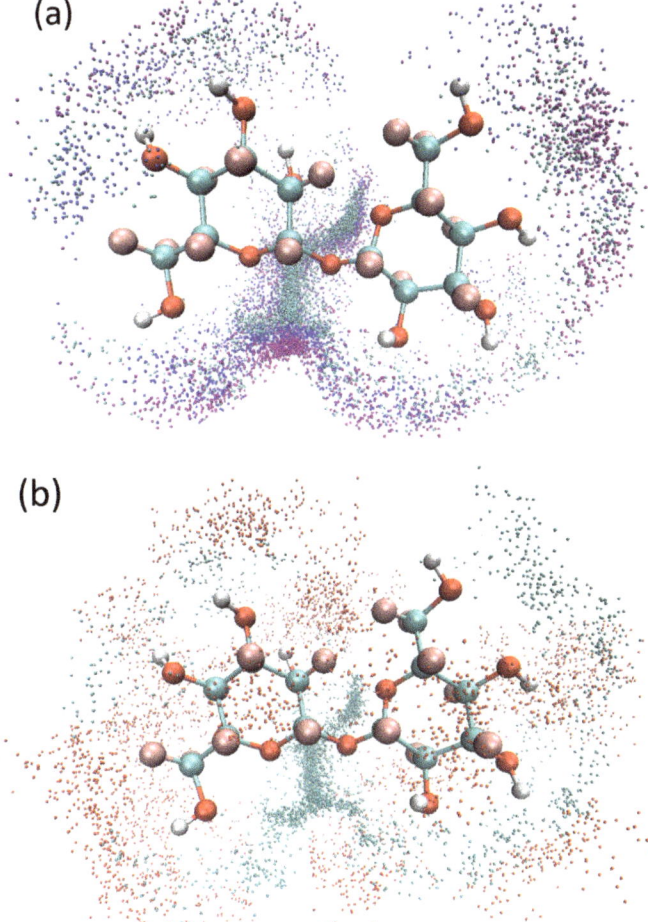

Figure 3. The schematic diagram of the accumulation of alkali metal ions and urea molecules around trehalose. (**a**) Li$^+$ (blue dots), Na$^+$ (dark blue dots), K$^+$ (red dots) ions are distributed around a trehalose molecule. (**b**) Distribution of Li$^+$ and the oxygen atoms of urea (red dots) around a trehalose molecule.

Then, we quantitatively analyzed the distribution of alkali, urea and water, as shown in Figure 3. We paid specific attention to the hydration shell around the oxygen atoms (we call them Os hereafter) of the glucose rings. They included all of the hydrophilic atoms on the glucose rings.

Because Os are electronegative, the first hydration shell had to be electropositive. There are three kinds of electropositive atom in an alkaline system, i.e., hydrogen atoms of urea and water and cations of alkali. Thus, there were 3 × 4 = 12 kinds of g(r) between them and the Os. There was only one $g_{(1)}(r)$ larger than 1, i.e., it was between cation and the Os. From the previous study, we knew that cations play a role in breaking the intra- and inter-molecular hydrogen bonds of cellulose. So, we first examined the RDF distribution between the cations and oxygen atoms of trehalose. Figure 4 compares the RDFs of Li$^+$/Na$^+$/K$^+$ ions and O1~O4. Here, we separated Os into two groups. O1 and O2 were in one group, while O3 and O4 were in the other group. We discussed their RDFs with cations, respectively. In group one, both O1 and O2 were inside the glucose rings. $g_{(1)O1\text{-}K}(r)$~5, centered at r1~2.6 Å. This means that the first K$^+$ shell around O1 was centered about 2.6 Å, where the K$^+$ concentration was five times larger than the average K$^+$ concentration in the solution. $g_{(1)O1\text{-}Na}(r)$~10, located at r1~2.4 Å, and $g_{(1)O1\text{-}Li}(r)$~20 was at r1~2.0 Å. Thus, smaller cations were more close to O1, and their concentrations were 10 and 20 times their averaged concentration in solution. Similar things happened around O2. In group two, both O3 and O4 were in the periphery of the glucose rings. The spaces around them were, thereby, relatively open. $g_{(1)O\text{-}K}(r)$~1.8, centered at r1~2.8 Å; $g_{(1)O\text{-}Na}(r)$~1.9, centered at r1~2.45 Å; and $g_{(1)O3\text{-}Li}(r)$~1.4, centered at r1~2.1 Å. Larger ions more easily concentrated around them.

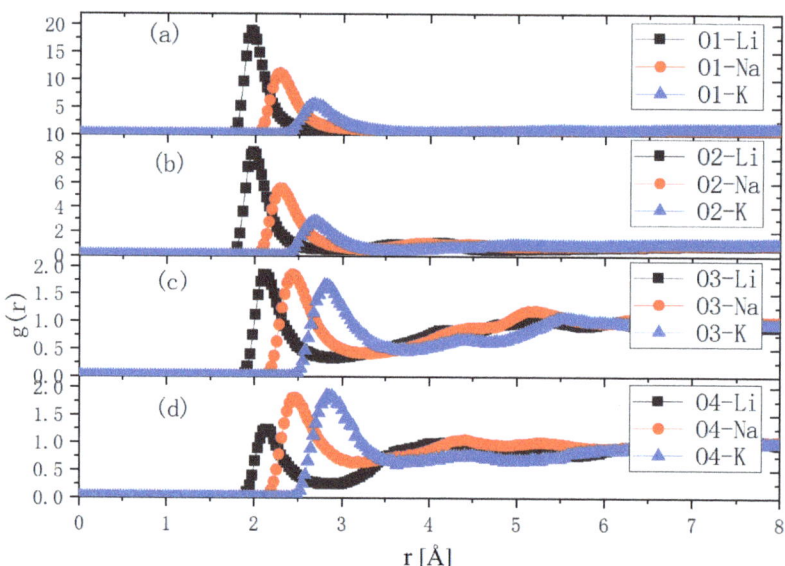

Figure 4. RDF distributions between the alkali metal ions and the solute oxygen atoms. g(r)s between Li$^+$, Na$^+$, K$^+$ and O1 (**a**), O2 (**b**), O3 (**c**), O4 (**d**) of trehalose.

From these comparisons, it was found that the Li$^+$ ions were closest to the two oxygen atoms inside the glucose ring, i.e., O1 and O2, while larger ions, such as Na$^+$ and K$^+$, were more likely to be enriched near O4. Because the LiOH urea aqueous solution had the most powerful solubility, these comparisons confirmed that the cations in the alkaline solution directly interact with the glucose ring. An interesting result was that K$^+$ ions with a relatively larger ionic radius were more concentrated around oxygen atoms away from the center of the glucose ring. This may explain why the KOH/urea aqueous solution had a better dissolving effect on chitin [18]. We suppose that, in the dissolution process of chitin, breaking the association between the side groups may be crucial, while, in the dissolution process of cellulose, breaking the interaction between the glucose backbones is a prerequisite.

Then, we looked for the components in the second layer of the hydration shell. The first hydration shell was electropositive, so the second shell had to be electronegative.

There were four electronegative solvent atoms in the system, i.e., OW, NU, OU and OOH. They themselves or their complexations composed the second hydration layer. We listed all of their g(r) with the Os and compared their structures with $g_{O1-Li}(r)$ in LiOH/urea aqueous solution. Figure 5 assumes that they were isotropically distributed. We added two dash lines to indicate their possible boundary. It is easy to see that all of those g(r), except $g_{(1)OW-O3/O4}(r)$, were smaller than 1. Therefore, three conclusions can be made, i.e., the second hydration layer almost did not exist; it was anisotropically distributed, or it only existed around O3 and O4 on the periphery of the glucose ring. We checked them one by one.

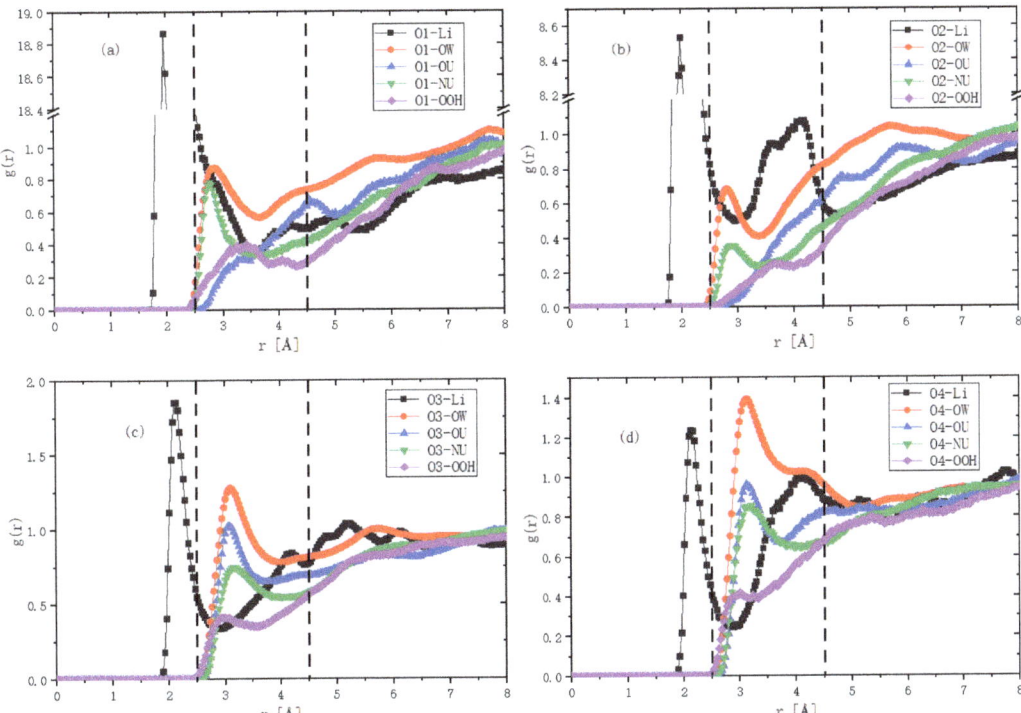

Figure 5. RDF distributions of some solvent atoms, i.e., Li, OW, OU, NU and OOH, around Os atoms in trehalose LiOH urea aqueous solution. (**a**–**d**) is the RDF distribution around O1 to O4, respectively. The dashed lines are used to indicate the boundary of the second hydration shells.

As shown in Figure 3, the distributions of cations and urea around O1 and O2 were very asymmetric. To quantitatively study the asymmetry of cationic and electronegative atoms around O1, we calculated their g(r) as a function of spatial angle. In the calculation, the setting of the coordinate axis was as shown in Figure 1. We first set a cone with an apex angle of 60° and the Y coordinate axis as the axis of symmetry. Then, we placed the apex of this cone at O1. Finally, we let the cone take O1 as the center of rotation and rotated it around the X-axis and calculated the g(r) changes with the rotation angle. The calculation results showed that $g_{(1)O1-Li}(r)$ reached its maximum value between 90° and 180°, which coincided with the Z-axis. Figure 6 shows $g_{O1-Li}(r)$, $g_{O1-OW}(r)$, $g_{O1-OU}(r)$, $g_{O1-NU}(r)$ and $g_{O1-OOH}(r)$ along the Z-axis. $g_{(1)O1-OW}(r)$, $g_{(1)O1-OOH}(r)$ and $g_{(1)O1-NU}(r)$ were larger than 1.0. Therefore, electronegative OW, OOH, NU or their complexation composed the second hydration layer.

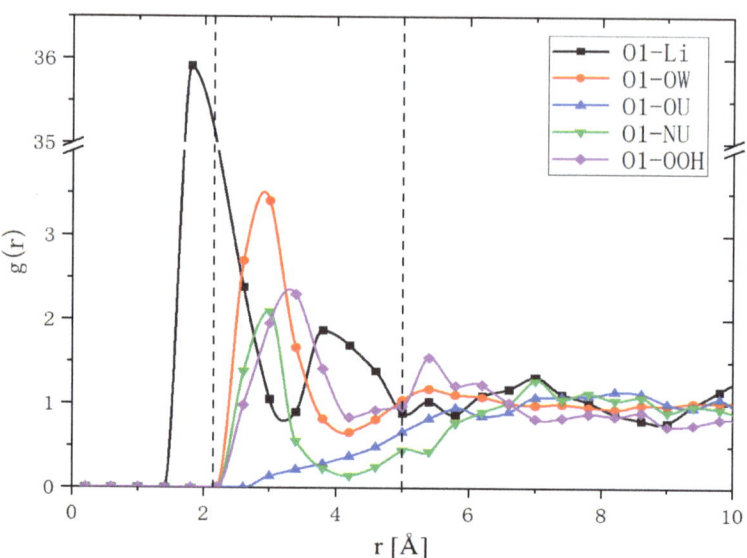

Figure 6. RDF distributions of Li, OW, OU, NU and OOH around O1 atom along Z-axis. The dashed lines are used to indicate the boundary of the second hydration layer.

The existence of urea is crucial for dissolving cellulose in green solvent. Previous NMR observations claimed that the amino group of urea forms an electronegative complexation with OH^- through hydrogen bonding. Such complexations wrap around cations and eventually form cellulose–NaOH–urea–H_2O inclusion complexes (ICs) [5,19]. To clarify this, we listed all of the possible complexations between OH^- and urea (Figure 7a). There were four possibilities, i.e., HU as the proton donor to form a hydrogen bond with OW (HB1) and OOH (HB2), NU as the proton acceptor (HB3) and OU as the proton acceptor (HB4). Figure 7b gives the RDFs between OW and NU, OW and HU and HW and NU. We used 2.9 ± 0.3Å as the donor–acceptor distance constraint and linear bond (±20°) as the angle constraint to define the hydrogen bond. The possibility of forming a hydrogen bond for HB1 and HB3 was 78.5% and 10.9%, respectively. Thus, HU prefers forming a hydrogen bond with OW. Figure 7c shows the RDFs between OOH and NU, OOH and HU and HOH and NU. If we used 2.5 ± 0.4 Å as the donor–acceptor distance constraint and (±20°) as the angle constrain to define the hydrogen bond, the possibility of forming a hydrogen bond for HB2 was 51.9%. Thus, HU can also form a hydrogen bond with OOH. Finally, Figure 7d is the RDF between OW and OU and HW and OU. They could form a hydrogen bond (HB4), but they were away from the second hydration layer, as shown in Figure 6. Therefore, as a proton donor, the amino group of urea can hydrogen bond with hydroxy group to form complexations. The electronegative complexation forms the second hydration layer.

Temperature effect on dissolution was also investigated. Figure 8 shows the neutron scattering profiles of trehalose in LiOH urea aqueous solutions at 25 °C and −5 °C and their differences. It can be seen that there was no obvious difference between the two different temperatures. Therefore, low temperature does not change the atomic structure of the solutions. It only increases the stability of the solution, making it difficult for the dissolved cellulose to re-aggregate.

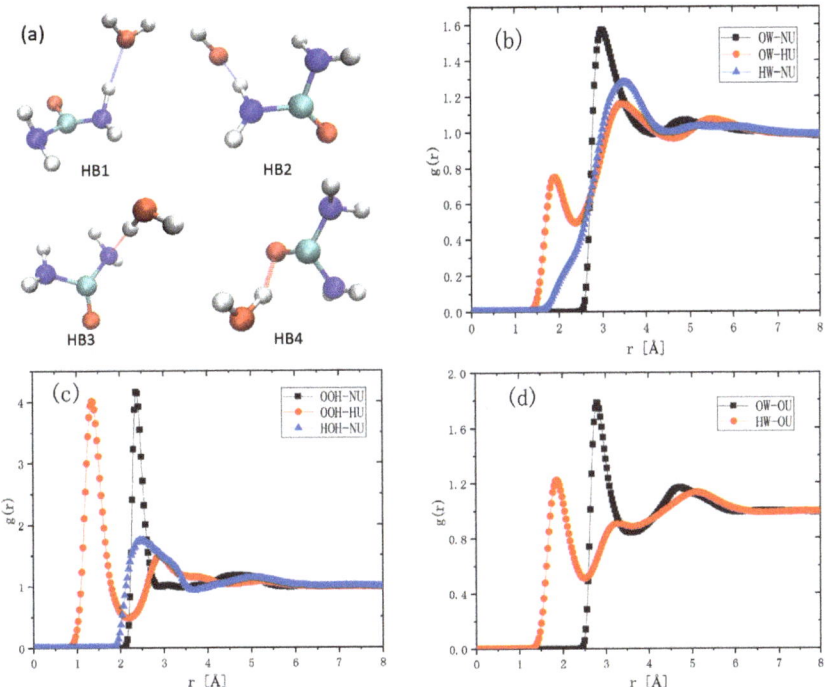

Figure 7. Schematic representation of the four important hydrogen bonds that urea forms with water and OH$^-$ (**a**). The colors of the four atoms C, H, O and N are cyan, white, red and blue, respectively. RDF distributions between the atoms of urea and H$_2$O/OH$^-$ in trehalose LiOH urea aqueous solution (**b**–**d**).

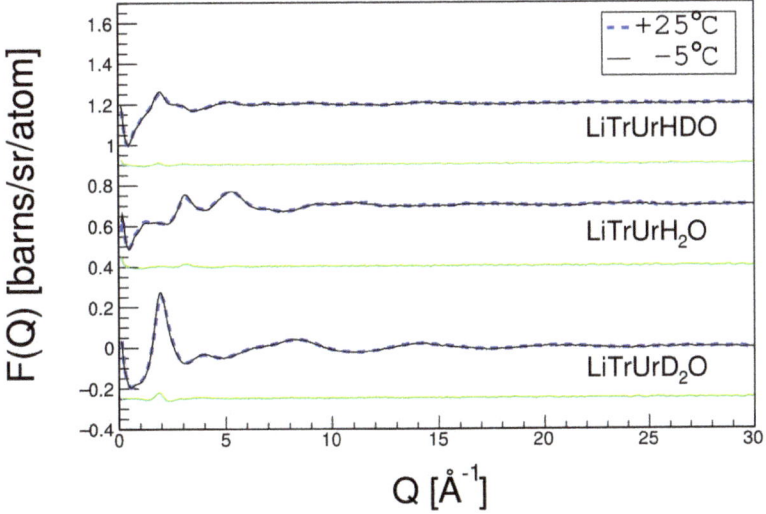

Figure 8. Neutron total scattering profiles of trehalose in LiOH/urea aqueous solution at 25 °C and −5 °C and the difference between them (green lines).

4. Conclusions

In this research work, we used trehalose as a model molecule and neutron scattering and EPSR simulation as the main tools to study the rapid dissolution mechanism of cellulose in alkali/urea aqueous solution. The three-dimensional atomic structures of trehalose in three different alkali/urea aqueous solutions were thus compared. Alkali, urea and water work cooperatively to dissolve trehalose. A layered hydration shell is crucial. Cations directly interact with the Os of the glucose rings. They first break the inter- and intra-hydrogen bonding. The smaller the ions, the more easily they penetrate into the glucose rings. Urea molecules are too large to approach the glucose ring. As proton donor, their amino group can form a hydrogen bond with the hydroxyl group. The resultant electronegative complexation constitutes the second hydration layer via bridging force. They further stabilize trehalose and prevent it from re-aggregating.

We also found an interesting phenomenon, that is, Li^+ ions are more concentrated around O1 and O2, while K^+ ions are more concentrated in the vicinity of O3 and O4, especially near O4. Although the low-temperature aqueous solution of KOH urea could not dissolve cellulose, it could dissolve chitin; at the same time, the low-temperature aqueous solution of LiOH/NaOH urea could dissolve cellulose, but could not dissolve chitin [18]. Based on this experimental evidence, we speculate: in the dissolution of cellulose, breaking the hydrogen bond formed by O1 and O2 is the key factor, while, in the dissolution of chitin, breaking the hydrogen bonds formed by its hydroxyl and amide groups is crucial. These findings are expected to be verified in future experiments.

Author Contributions: Conceptualization, H.C. (He Cheng) and C.M.; methodology, H.C. (He Cheng); software, C.M.; validation, T.Z., Z.H. and Y.L.; formal analysis, C.M.; investigation, H.C. (He Cheng); resources, W.Y.; data curation S.G. and H.C. (Huaican Chen); writing—original draft preparation, C.M.; writing—review and editing, H.C (He Cheng). and C.C.H.; visualization, C.M.; supervision, H.C. (He Cheng); project administration, H.C. (He Cheng); funding acquisition, H.C. (He Cheng) All authors have read and agreed to the published version of the manuscript.

Funding: This research was funded by the National Natural Science Foundation of China (NSFC) (grant numbers U1932161, U1830205 and 11805035); the National High Energy Physics Data Center of China (grant number NHEPSDC-OP-2021-001); and the National Key Research and Development Program of China (grant number 2017YFA-0403703).

Institutional Review Board Statement: Not applicable.

Informed Consent Statement: Not applicable.

Data Availability Statement: The data that support the findings of this study are openly available in ISIS Neutron and Muon Source Data at http://doi.org/10.5286/ISIS.E.RB1820090 (accessed on 27 November 2018) and http://doi.org/10.5286/ISIS.E.RB1920190 (accessed on 15 December 2019).

Acknowledgments: The experiments were conducted using SANDALS at the ISIS Neutron, Muon Source and NOVA of J-PARC and the Multi-Physics Instrument (MPI) of CSNS. The authors thank the staff of the ISIS Disordered Materials Group, Materials and Life Science Experimental Facility of J-PARC and the instrument scientists of SANDALS and NOVA for the support with experiment design, operation and data analysis.

Conflicts of Interest: The authors declare no conflict of interest.

References

1. Wang, S.; Lu, A.; Zhang, L. Recent advances in regenerated cellulose materials. *Prog. Polym. Sci.* **2016**, *53*, 169–206. [CrossRef]
2. Cai, J.; Zhang, L. Rapid Dissolution of Cellulose in LiOH/Urea and NaOH/Urea Aqueous Solutions. *Macromol. Biosci.* **2005**, *5*, 539–548. [CrossRef] [PubMed]
3. Cai, J.; Zhang, L.; Chang, C.; Cheng, G.; Chen, X.; Chu, B. Hydrogen-Bond-Induced Inclusion Complex in Aqueous Cellulose/Lioh/Urea Solution at Low Temperature. *ChemPhysChem* **2007**, *8*, 1572–1579. [CrossRef] [PubMed]
4. Jiang, Z.; Lu, A.; Zhou, J.; Zhang, L. Interaction between –OH groups of methylcellulose and solvent in NaOH/urea aqueous system at low temperature. *Cellulose* **2012**, *19*, 671–678. [CrossRef]

5. Jiang, Z.; Fang, Y.; Xiang, J.; Ma, Y.; Lu, A.; Kang, H.; Huang, Y.; Guo, H.; Liu, R.; Zhang, L. Intermolecular Interactions and 3d Structure in Cellulose-Naoh-Urea Aqueous System. *J. Phys. Chem. B* **2014**, *118*, 10250–10257. [CrossRef] [PubMed]
6. Lindman, B.; Karlström, G.; Stigsson, L. On the mechanism of dissolution of cellulose. *J. Mol. Liq.* **2010**, *156*, 76–81. [CrossRef]
7. Alves, L.; Medronho, B.; Filipe, A.; Antunes, F.E.; Lindman, B.; Topgaard, D.; Davidovich, I.; Talmon, Y. New Insights on the Role of Urea on the Dissolution and Thermally-Induced Gelation of Cellulose in Aqueous Alkali. *Gels* **2018**, *4*, 87. [CrossRef]
8. Lindman, B.; Medronho, B. The Subtleties of Dissolution and Regeneration of Cellulose: Breaking and Making Hydrogen Bonds. *BioResources* **2015**, *10*, 3811–3814. [CrossRef]
9. Medronho, B.; Lindman, B. Brief overview on cellulose dissolution/regeneration interactions and mechanisms. *Adv. Colloid Interface Sci.* **2015**, *222*, 502–508. [CrossRef] [PubMed]
10. Glasser, W.G.; Atalla, R.H.; Blackwell, J.; Brown, R.M.; Burchard, W.; French, A.D.; Klemm, D.O.; Nishiyama, Y. About the Structure of Cellulose: Debating the Lindman Hypothesis. *Cellulose* **2012**, *19*, 589–598. [CrossRef]
11. Soper, A.K. Empirical Potential Monte Carlo Simulation of Fluid Structure. *Chem. Phys.* **1996**, *202*, 12. [CrossRef]
12. Ma, C.L.; Cheng, H.; Zuo, T.S.; Jiao, G.S.; Han, Z.H.; Qin, H. Neudatool: An Open Source Neutron Data Analysis Tools, Supporting Gpu Hardware Acceleration, and across-Computer Cluster Nodes Parallel. *Chin. J. Chem. Phys.* **2020**, *33*, 727–732. [CrossRef]
13. Qin, H.; Ma, C.; Gärtner, S.; Headen, T.F.; Zuo, T.; Jiao, G.; Han, Z.; Imberti, S.; Han, C.C.; Cheng, H. Neutron total scattering investigation on the dissolution mechanism of trehalose in NaOH/urea aqueous solution. *Struct. Dyn.* **2021**, *8*, 014901. [CrossRef] [PubMed]
14. Soper, A.K.; Castner, E.W., Jr.; Luzar, A. Impact of Urea on Water Structure: A Clue to Its Properties as a Denaturant? *Biophys. Chem.* **2003**, *105*, 649–666. [CrossRef]
15. McLain, S.E.; Imberti, S.; Soper, A.K.; Botti, A.; Bruni, F.; Ricci, M.A. Structure of 2 molar NaOH in aqueous solution from neutron diffraction and empirical potential structure refinement. *Phys. Rev. B* **2006**, *74*, 094201. [CrossRef]
16. Imberti, S.; Botti, A.; Bruni, F.; Cappa, G.; Ricci, M.A.; Soper, A.K. Ions in water: The microscopic structure of concentrated hydroxide solutions. *J. Chem. Phys.* **2005**, *122*, 194509. [CrossRef] [PubMed]
17. Soper, A.K.; Ricci, M.A.; Bruni, F.; Rhys, N.H.; McLain, S.E. Trehalose in Water Revisited. *J. Phys. Chem. B* **2018**, *122*, 7365–7374. [CrossRef] [PubMed]
18. Gong, P.; Wang, J.; Liu, B.; Ru, G.; Feng, J. Dissolution of chitin in aqueous KOH. *Cellulose* **2016**, *23*, 1705–1711. [CrossRef]
19. Xiong, B.; Zhao, P.; Hu, K.; Zhang, L.; Cheng, G. Dissolution of Cellulose in Aqueous Naoh/Urea Solution: Role of Urea. *Cellulose* **2014**, *21*, 1183–1192. [CrossRef]

Article

Enhanced Extraction Efficiency of Flavonoids from *Pyrus ussuriensis* Leaves with Deep Eutectic Solvents

Jong Woo Lee [1,†], Hye Yoon Park [2,†] and Junseong Park [1,*]

[1] Department of Engineering Chemistry, Chungbuk National University, Chungbuk, Cheongju 28644, Korea; qoohji@naver.com
[2] Biological and Genetic Resources Assessment Division, National Institute of Biological Resources, Incheon 22689, Korea; rejoice077@korea.kr
[*] Correspondence: jsparkbio@cbnu.ac.kr; Tel.: +82-43-261-3326
[†] There authors contributed equally to this work.

Abstract: In this study, deep eutectic solvents (DESs) were synthesized using different ratios of choline chloride (CC) and dicarboxylic acids, and their eutectic temperatures were determined. The DES synthesized using CC and glutaric acid (GA), which showed a higher extraction efficiency than conventional solvents, was used for the extraction of flavonoid components from *Pyrus ussuriensis* leaves (PUL), and the extraction efficiency was evaluated using the response surface methodology. The flavonoid components rutin, hyperoside, and isoquercitrin were identified through high-performance liquid chromatography (HPLC), equipped with a Waters 2996 PDA detector, and HPLC mass spectrometry (LC-MS/MS) analyses. The optimum extraction was achieved at a temperature of 30 °C using DES in a concentration of 30.85 wt.% at a stirring speed of 1113 rpm and an extraction time of 1 h. The corresponding flavonoid content was 217.56 µg/mL. The results were verified by performing three reproducibility experiments, and a high significance, with a confidence range of 95%, was achieved. In addition, the PUL extracts exhibited appreciable antioxidant activity. The results showed that the extraction process using the DES based on CC and GA in a 1:1 molar ratio could effectively improve the yield of flavonoids from PUL.

Keywords: *Pyrus ussuriensis* leaves; flavonoids; deep eutectic solvent; response surface methodology; green extraction

1. Introduction

Eco-friendly chemistry has recently attracted considerable attention, and "green solvents" based on natural substances are being proposed as useful alternatives to conventional organic solvents [1]. Among them, deep eutectic solvents (DESs) are highly suitable for the extraction of a wide range of natural compounds [2]. Natural DESs (NADESs) occur in a variety of living cells and play an important role as an alternative medium for the biosynthesis, transport, and storage of natural products [3]. DESs are typically composed of nontoxic substances occurring naturally in some plants. These solvents can be used to extract naturally occurring compounds that can be incorporated directly into food formulations without performing additional separation steps. Therefore, DESs have significant advantages over conventional solvents. In addition, several DESs have been found to increase the stability of natural compounds during extraction and storage [4]. Therefore, studies have been performed to determine the potential utility of DESs as extraction media for sparingly soluble bioactive compounds that are not amenable to aqueous extraction [5,6]. DESs consist of mixtures of a wide range of hydrogen bond donor (HBD) and hydrogen bond acceptor (HBA) species. Typical HBAs are nontoxic quaternary ammonium salts or amino acids, such as alanine, proline, and glycine, and HBDs are primarily organic acids, such as oxalic, lactic, and malic acids, or carbohydrate-based substances, such as glucose, fructose, and maltose. Alcohol, amine, aldehyde, ketone, and carboxylate functionalities are highly versatile and can be used to prepare custom solvents of virtually unlimited combinations, as they are capable of both HBA and HBD functions. The nature of the interaction that occurs depends on the type of solid material that forms the liquid [7–9]. The temperature, viscosity, stability, and solvent melting properties are related to the strength of the hydrogen bonds [10].

DESs are typically used to extract hydrophobic components that cannot be efficiently extracted with water, as well as to extract hydrophilic components [11,12]. Flavonoids are important active compounds that exist in more than 5000 natural states, and are known to have excellent antioxidant properties because they effectively eliminate active oxygen species [13–15]. Recently, as the role of active oxygen species in the progression of degenerative diseases has been gradually recognized, research has been underway on the development of antioxidants that can arrest and eliminate active oxygen species [16–18].

Pyrus bretschneideri (genus: *Pyrus*, family: *Rosaceae*), the third most important temperate climate fruit species after grapes and apples, has been cultivated in more than 50 countries since it was first cultivated in China 2000 years ago [19,20]. The genus *Pyrus* contains twenty-two widely recognized primary species, including at least six wild species and three interspecific hybrids [21]. Pear trees are a deciduous tree plant species belonging to the genus *Pyrus*, and are one of the representative cultivated varieties that have been introduced and grown nationwide in Korea [22]. *Pyrus ussuriensis* is a deciduous tree belonging to the family *Rosaceae*. Parts of this tree have been used in the private sector as medicinal products to treat vomiting, diarrhea, and fever. The fruits have been historically enjoyed by our ancestors [23]. PUL are used as a traditional herbal medicine for treating asthma, coughs, and fever, and although they are effective in treating atopic dermatitis, reports on the efficacy of certain components are insufficient [24]. The separation of isoprenoid alcohols and polyphenols from the leaves of *Pyrus ussuriensis* has been reported, along with the separation of flavonoids from the stems. However, more studies have been focused on the well-known Korean tree *Pyrifolia Nakai* [25,26].

The main purpose of this study is to synthesize DESs using choline chloride (CC) and dicarboxylic acids, and apply them to extracting flavonoids from PUL. In the process of extracting flavonoid components from PUL, the conditions that can maximize the extraction efficiency have been determined using the response surface methodology (RSM).

2. Results and Discussion

2.1. Chemical Profiling of PUL Extracts

A total of 30.02 g of dried PUL was extracted to obtain 4.24 g of freeze-dried extract powder (yield: 14.12%). To analyze the active compounds present in the PUL, the extract was subjected to liquid chromatography–tandem mass spectrometry (LC-MS/MS) analysis. On performing LC-MS/MS analysis, valid peaks corresponding to the flavonoid components were detected in the range of 35–37 min at a wavelength of 270 nm. As can be observed in Figure S1A, the m/z value of [M-H]$^-$ was 609.15 at 35.0 min. The m/z value of [M-H]$^-$ was 463.09 at 35.2 min and 35.6 min (Figure S1B,C). From the results of the MS analysis, rutin, hyperoside, and isoquercitrin were determined to be the flavonoid components of PUL. In this study, the HPLC-MS/MS results show that rutin, hyperoside, and isoquercitrin are the active flavonoid components present in the PUL extracts. To confirm the presence of the flavonoids in the PUL extracts, standard samples of rutin, hyperoside, and isoquercitrin were subjected to HPLC analysis equipped with a Waters 2996 PDA detector, and the resulting chromatograms were compared with that obtained for the PUL extract (Figure 1). The retention times of the standard rutin, hyperoside, and isoquercitrin samples are consistent with those of the peaks 1, 2, and 3 of the PUL extract.

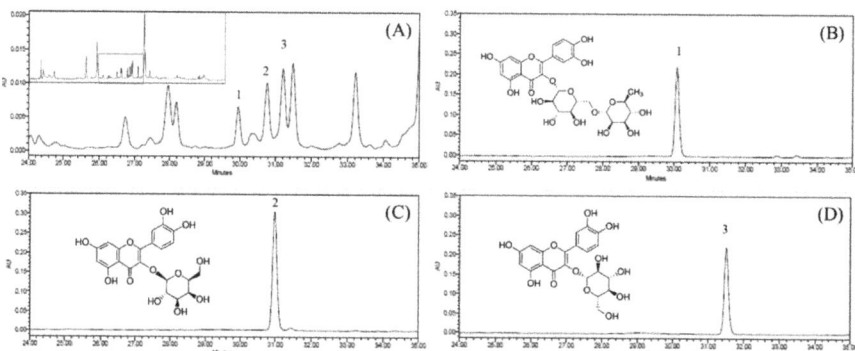

Figure 1. HPLC chromatograms of PUL extract and standard samples. Detection interval of (**A**) extract, (**B**) rutin, (**C**) hyperoside, and (**D**) isoquercitrin.

2.2. Preparation of DES and Selection of DES by Extraction Efficiency

Five types of DESs were synthesized, using CC as the HBA. Different dicarboxylic acids were employed as the HBDs. All the HBDs contain two carboxyl groups, but differ in the total number of carbon atoms. The efficiencies of the solvent extractions of the flavonoids from PUL, using the various DESs, are listed in Table 1. The extraction efficiency of the flavonoids is expressed as the sum of the extraction contents of three components: rutin, hyperoside, and isoquercitrin. Three standard samples of each extract were evaluated using HPLC with a correlation coefficient of 0.99. The contents of the flavonoids in the extract were determined using this standard curve. All the extraction efficiencies with the DESs were higher than those obtained with water (used for comparison), except for that pertaining to the DES synthesized using adipic acid. The extraction efficiencies of PUL with the malonic- and glutaric acid (GA)-based DESs were significantly higher than those with the other DESs. Therefore, the DES prepared using GA and CC was selected for the extraction of PUL prior to the optimization of the extraction condition. The HBDs and HBAs in the DES can interact with the cellulose and lignin in the plant cell walls to promote loosening of the cell wall structure and enable easy extraction of the intracellular components. A higher maximum extraction efficiency was achieved with the DES than without it, which was consistent with previous reports [27]. In addition, the current results show that when CC is used as the HBA, the extraction efficiency varies with the types of HBD used. Higher extraction efficiencies were achieved using DESs based on structurally similar dicarboxylic acids with odd numbers of carbon atoms. It has been confirmed that the structures of the HBDs affect the extraction efficiency, and it can be observed in Table 1 that the molar ratios vary depending on the structures. The current results also indicate that the eutectic temperature is determined by the structure of the HBD. Field emission scanning electron microscopy (FE-SEM) studies indicated the penetration of trace amounts of the solvents on the PUL surface during the extraction process. The FE-SEM images of the pre- and post-extraction PUL samples are shown in Figure 2. As can be observed in Figure 2A, there was no solvent penetration on the surface of the pre-extraction PUL, while traces of solvent penetration were observed when the extraction was performed with water (Figure 2B). The FE-SEM images of the PUL surfaces after DES extraction (Figure 2C,D) indicate the occurrence of sufficient solvent penetration. The high solubility of the plant cell wall components, such as lignin, cellulose, and flavonoids, in the DES facilitated the penetration of the solvent, which led to structural changes in the overall leaf surface. The efficient penetration of plant tissues by the DES prepared using GA and CC proves that it can serve as an environmentally friendly solvent for isolating natural active components. Based on the penetration, it can be concluded that the synthesized covalent solvent is sufficiently functional.

Table 1. Extraction efficiency of flavonoids from PUL.

HBD, Carbon Number	Choline Chloride/HBD (Molar Ratio)	Melting Point (°C)	Flavonoid Contents * (µg/mL)
Control (H_2O)	Control	-	105.926 ± 1.652
Oxalic acid (C_2)	2:1	75	150.586 ± 3.072
Malonic acid (C_3)	1:1	69	162.169 ± 3.016
Succinic acid (C_4)	2:1	80	155.945 ± 2.070
Glutaric acid (C_5)	1:1	60	171.326 ± 3.615
Adipic acid (C_6)	2:1	84	32.113 ± 0.353

* Flavonoid contents are the sum of contents of the three components (rutin, hyperoside, and isoquercitrin).

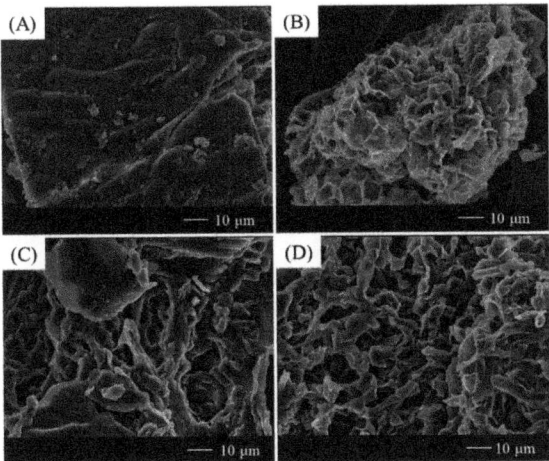

Figure 2. Configuration of PUL surface after extraction using different solvents. (**A**) PUL surface before extraction; (**B**) PUL surface after extraction with water; (**C**) PUL surface after extraction with DES synthesized using CC and malonic acid; (**D**) PUL surface after extraction with DES synthesized using CC and GA.

2.3. Optimization of Extraction Conditions for Flavonoids by RSM

Four factors were considered for optimizing the extraction conditions of the DES prepared using GA and CC: temperature, extraction time, DES content, and stirring speed. According to the results of the single-factor experiment (Figure S2), the four factors were subjected to the Box–Behnken design (BBD) analysis. The three levels of extraction temperature were 30, 60, and 90 °C, while those of the extraction time were 1, 24.5, and 48 h. The three levels of DES content were 10, 50, and 90%, and those of the stirring speed were 450, 850, and 1250 rpm. The design parameters were determined using the Design-Expert software and the results are listed in Table 2. The Design-Expert software was used to simulate the results of the BBD analysis. The obtained quadratic polynomial regression equation for the extraction efficiency Y, as the sum of the contents of rutin, hyperoside, and isoquercitrin, versus the temperature (A), extraction time (B), DES content (C), and stirring speed (D) was as follows:

Y (content of flavonoids) = $147.43 - 16.44A - 10.20B - 28.63C + 25.72D - 32.66AB + 1.18AC - 15.59AD + 10.62BC - 0.6583BD - 9.24CD + 20.68A^2 + 10.77B^2 - 48.72C^2 + 13.77D^2$

Table 2. Design and experimental response values based on BBD analysis.

Run	Factor				Flavonoid Content (μg/mL)
	Temperature (A, °C)	Extraction Time (B, h)	DES Content (C, %)	Stirring Speed (D, rpm)	
1	30	24.5	50	450	126.1
2	30	24.5	10	850	118.6
3	60	1	90	850	79.0
4	60	24.5	90	450	93.4
5	90	24.5	50	1250	223.0
6	90	24.5	10	850	95.8
7	60	24.5	50	850	149.6
8	60	48	50	450	147.2
9	60	24.5	90	1250	83.9
10	60	24.5	50	850	139.9
11	60	24.5	50	850	147.4
12	90	24.5	90	850	76.4
13	90	48	50	850	114.8
14	30	24.5	90	850	94.5
15	60	48	50	1250	157.3
16	30	48	50	850	221.3
17	30	24.5	50	1250	291.1
18	60	1	10	850	173.1
19	60	1	50	1250	151.9
20	60	48	10	850	135.2
21	60	24.5	50	850	147.6
22	60	24.5	50	850	152.6
23	60	48	90	850	83.6
24	90	1	50	850	231.3
25	30	1	50	850	207.2
26	90	24.5	50	450	120.3
27	60	1	50	450	139.2
28	60	24.5	10	450	152.1
29	60	24.5	10	1250	179.7

The results of the variance analysis of the above equation are listed in Table 3. According to the analysis of variance (ANOVA), the p-value (0.0441) of the four-factor global model was significant. The model F-value of 2.57 implied that the model was significant. The quadratic coefficients (A^2, B^2, C^2, and D^2) also had a significant effect, while the linear coefficients (A, B, C, and D) and terms representing the interaction between two factors had a negligible effect. Therefore, the above model can be employed to predict the extraction efficiency of the DES toward PUL. In the DES extraction, the content of DES was found to be the variable with the highest influence on the content of flavonoids. It also affected the order of the stirring speed, extraction temperature, and extraction time. The results confirmed that all the interactions, except that between the extraction temperature and extraction time, were insignificant. Figure 3 depicts a three-dimensional (3D) response surface curve based on the regression polynomial. It represents the range of three independent variables, while the fourth variable is held at level 0. As can be observed in Figure 3A, the extraction efficiency increased as the extraction temperature and time increased. On the other hand, it tended to increase initially and then decrease, depending on the DES content. The maximum extraction efficiency was achieved at a DES content of 30.85%. At low stirring speeds, the extraction efficiency was found to increase with temperature. However, when the stirring speed is fast, the extraction efficiency decreases as the temperature increases, and then increases again when the temperature reaches 60 °C. The optimal extraction conditions predicted from the three-dimensional response surface curves were as follows: extraction temperature: 30°C, extraction time: 1 h, DES content: 30.85%, and stirring speed: 1113.1 rpm. The optimized conditions corresponded to a predicted flavonoid content of 217.515 μg/mL. Three confirmatory tests, performed under the predicted conditions, showed high significance, with error rates within 5%, and flavonoid yields of 223.37, 219.10, and 227.11 μg/mL.

Table 3. ANOVA statistical results.

Source	Sum of Squares	Degree of Freedom	Mean of Square	F-Value	p-Value
model	53,142.77	14	3795.912143	2.57	0.0441
A-A	3241.96	1	3241.96	2.2	0.1606
B-B	1248.81	1	1248.81	0.8457	0.3733
C-C	9834.27	1	9834.27	6.66	0.0218
D-D	7936.66	1	7936.66	5.38	0.0361
AB	4266.84	1	4266.84	2.89	0.1112
AC	5.53	1	5.53	0.0037	0.9521
AD	972.46	1	972.46	0.6586	0.4306
BC	451.55	1	451.55	0.3058	0.589
BD	1.73	1	1.73	0.0012	0.9732
CD	341.66	1	341.66	0.2314	0.6379
A^2	2775.34	1	2775.34	1.88	0.192
B^2	751.98	1	751.98	0.5093	0.4872
C^2	15,394.76	1	15,394.76	10.43	0.0061
D^2	1229.1	1	1229.1	0.8324	0.377
Residual	20,672.24	14	1476.588571		
Lack of Fit	20,584.39	10	2058.439	93.73	0.0003
Pure Error	87.85	4	21.9625		
Cor. Total	73,815.01	28			

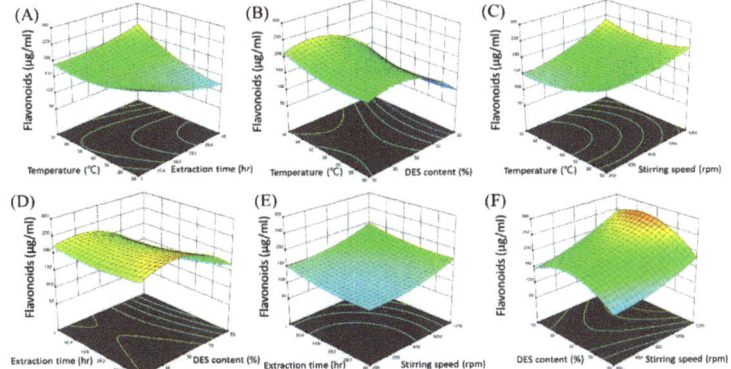

Figure 3. Response surface in Box–Behnken experimental design for the four factors influencing the extraction of flavonoids. (**A**) Effect of temperature and extraction time; (**B**) effect of temperature and DES content; (**C**) effect of temperature and stirring speed; (**D**) effect of extraction time and DES content; (**E**) effect of extraction time and stirring speed; and (**F**) effect of DES content and stirring speed.

2.4. Evaluation of Antioxidant Activity

The antioxidant activity of the flavonoids extracted from the PUL under the optimum conditions was evaluated at various concentrations. The flavonoid concentration corresponding to 50% free radical scavenging activity was calculated based on L-ascorbic acid, which constituted the positive control group. As can be observed in Figure 4, the 2,2-diphenyl-1-picrylhydrazyl (DPPH) free radical scavenging activity increases as the concentration of the extract is increased. However, it does not increase when the maximum concentration is reached. The concentrations of L-ascorbic acid (positive group) and the eluted extract corresponding to 50% DPPH radical scavenging activity were 59.54 and 299.68 ppm, respectively. The antioxidant activity of the active component of the PUL extracted using the DES was approximately 1/6 that of L-ascorbic acid. It was, thus, confirmed that the active components of the PUL extracted using the DES had potential utility as antioxidants.

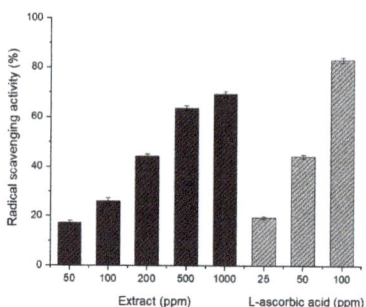

Figure 4. DPPH radical scavenging activity of *PUL* extract under optimal extraction conditions.

2.5. Evaluation of Anti-Inflammatory Effectiveness

Inflammatory reactions are defense mechanisms against external invaders, which produce a variety of inflammatory factors. Among them, nitric oxide (NO) is a highly reactive biomolecule. When cells are stimulated by lipopolysaccharides (LPSs), they produce NO in the presence of inducible nitric oxide synthase (iNOS) [28,29]. The inhibition of NO production in RAW 264.7 cells stimulated by LPSs is illustrated in Figure 5. Both the DES and PUL extract obtained using the DES showed cytotoxicity at concentrations exceeding 0.3%. As can be observed in Figure 5A, the DES did not inhibit the generation of NO at a concentration of 0.1%, while the PUL extracts obtained using the DES significantly inhibited NO generation (up to 26.1%) at the same concentration (Figure 5B). The anti-inflammatory properties of the active components of the PUL extracted using the DES were, thus, demonstrated.

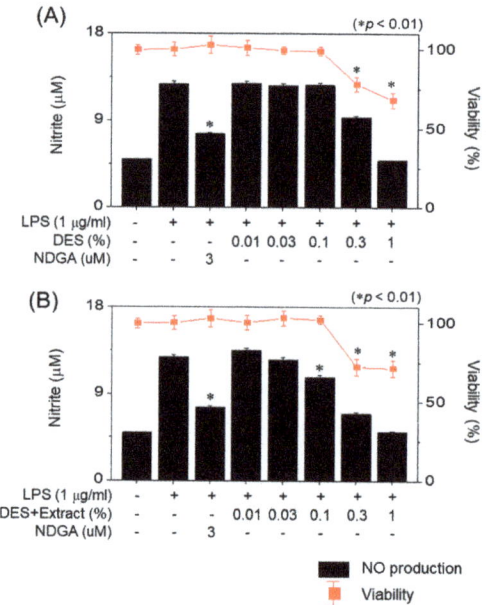

Figure 5. Effect of DES and PUL extract on the production of NO in LPS-stimulated RAW 264.7 cells. (**A**) DES synthesized using CC and GA; (**B**) PUL extract obtained using DES synthesized from CC and GA.

3. Materials and Methods

3.1. Materials

Dried PUL were provided by the National Institute of Biological Resources. Choline chloride ($\geq$99.0%), oxalic acid, anhydrous ($\geq$98.0%), and malonic acid ($\geq$99.0%) were supplied by Samchun Chemical Co., Seoul, Korea. Succinic acid ($\geq$99.5%) was purchased from Junsei Chemical Co., Japan, Glutaric acid ($\geq$99.0%) was supplied by Daejung Chemicals and Metals, Gyunggi-do, Korea. Adipic acid ($\geq$99.0%) was procured from Sigma Aldrich, USA. Acetonitrile (ACN, CH_3CN, 99.9%, Fisher Chemical, Loughborough, UK) was used as a solvent for the mobile phase of HPLC, and acetic acid (CH_3COOH, ACS reagent, Sigma Aldrich, St. Louis, MO, USA) was used as an additive.

3.2. Extraction and Determination of Flavonoids in PUL

PUL were dried and crushed up to 50 mesh. The crushed leaves (30 g) were extracted by stirring for 2 h at 30 °C with 600 mL of 70% ethanol, followed by filtration. The extract was obtained in powder form by concentrating and freeze-drying. The extract was subjected to HPLC analysis on a Waters 2695 separation module equipped with a Mightysil RP-18GP column (KANTO CHEMICAL, Japan). The sample was examined at a wavelength of 270 nm (UV) using a detector (Tunable Absorbance Detector, Waters, USA). Two solution systems were used in the HPLC mobile phase: the A solution (acetonitrile containing 0.1 wt.% acetic acid) and B solution (distilled water containing 0.1 wt.% acetic acid). The analysis was performed at a flow rate of 1.0 mL/min. The mobile phase B was eluted from 0 min to 100% and to be 70% to 40 min, 40% from 40 to 45 min, 10% to 50 min, and 100% to 60 min.

3.3. Preparation of DES

CC, a naturally occurring quaternary ammonium salt, was used as the HBA, and a dicarboxylic acid with two carboxy groups was used as the HBD to prepare the DES. The dicarboxylic acids were classified according to their carbon numbers, and the ratios were determined experimentally from oxalic (C2) to adipic acid (C6). The corresponding DESs were synthesized and used for extraction. In the ratio determination experiment, a thermomixer (Eppendorf thermomixer comfort, Eppendorf, Germany) was used to mix the HBA and HBD at ratios ranging from 3:1 to 1:3 at room temperature. The HBA and HBD were allowed to react under stirring at 950 rpm for 30 min. The temperature was raised by 2 °C/30 min until the formation of a transparent liquid, and the process was recorded. After determining the ratio, the mixture was heated at 80 °C for 2 h to obtain a uniform and stable transparent solution, which was aged in an oven for 24 h at 60 °C prior to use in extraction. The characteristic properties of the synthesized DESs are listed in Table 1. The molar ratio of the HBA and HBD used to synthesize each DES is included in Table 1.

3.4. Extraction with DES

The extractions with the DESs and purified water as extraction solvents were performed as follows: 30 mg of powdered PUL extract was mixed with the solvent and stirred at 40 °C at 950 rpm for 60 min. Each extract was subsequently centrifuged at 13,000 rpm to separate the supernatant and filtered through a syringe filter prior to analysis. The extract was subjected to HPLC analysis to determine the content of flavonoids.

3.5. Optimization of the Extraction Condition using the Box–Behnken Design

Based on the results of the screening test, the BBD experimental design of the RSM was performed using the Design-Expert 12.0 software to optimize the four selected factors, extraction temperature (°C), extraction time (h), DES content (%), and stirring speed (rpm), that influenced the extraction of PUL and could enhance the extraction efficiency of the flavonoids. The four independent factors were investigated at three different levels (−1, 0, and +1). The complete experimental design consisted of 29 runs. Each run was repeated three times.

3.6. SEM Analysis

The native PUL samples (prior to extraction) and those subjected to extraction with water, the DES synthesized from CC and malonic acid, and the DES synthesized from CC and GA were dried at 60 °C for 48 h. The surface characteristics of the samples were observed through a field emission scanning electron microscope (ULTRAPLUS, Carl Zeiss NTS GmbH, Oberkochen, Germany). The effects of different extraction solvents on the surface structure were analyzed.

3.7. Antioxidant Activity

The antioxidant activity of the PUL extracts obtained using the DESs synthesized from CC and different dicarboxylic acids was evaluated using the DPPH assay. DPPH is a stable radical that exists as a purple solution, which turns bile yellow as the free radicals are eliminated by hydrogen or electron donors. The radical scavenging activity was estimated by measuring the change in absorbance [30,31]. L-ascorbic acid was used as a positive control to compare the scavenging activity. DPPH was dissolved in ethanol at a concentration of 100 µM. L-ascorbic acid solutions of 50, 100, and 200 ppm concentrations were prepared, and the samples to be tested were diluted to the desired concentrations. Each test solution (10 µL) was mixed with the same amount of ethanol in a 96-well plate. To each of these solutions, 190 µL of DPPH solution (100 µM) was added. After 30 min of reaction at room temperature, the absorbance was measured at 530 nm using a microplate reader. The DPPH free radical scavenging activity was calculated according to the following equation:

$$AA\% = 100 - \left[\frac{\left(Abs_{sample} - Abs_{blank}\right) \times 100}{Abs_{control}}\right]$$

Abs_{sample} is the absorbance of the sample in which the extract and the DPPH solution are mixed, Abs_{blank} is the absorbance inherent in the extract, and $Abs_{control}$ is the absorbance of the DPPH solution.

3.8. Anti-Inflammatory Effectiveness

RAW 264.7 cells were loaded on a 96-well plate at a concentration of 1×10^5 cells/well and incubated for 1 d in a 37 °C incubator. Lipopolysaccharide (LPS) was added at a concentration of 500 ng/mL to induce NO production, and the samples were allowed to react for 24 h. After the reaction, the upper solution layer was reacted with Griess reagent, and the absorbance was measured at 540 nm to estimate the NO production. The cells were treated with 5 mg/mL of 3-(4,5-dimethylthiazol-2-yl)-2,5-diphenyl tetrazolium bromide (MTT) reagent and incubated for an additional hour. The cytotoxicity was determined by removing all the upper fluid and treating it with 100 µL DMSO to completely dissolve the cells. Subsequently, the absorbance was measured at 550 nm. In addition, the significance of the test substance was determined by the two-sample test to meet the biological statistical criterion of (α): 1%. It treated nordihydroguaiaretic acid (NDGA) as a benign control group [28,29].

4. Conclusions

In this study, the flavonoid contents of PUL were investigated, and the PUL were discovered to be rich in flavonoids, such as rutin, hyperoside, and isoquercitrin. Extraction using different DESs based on CC and dicarboxylic acids was found to effectively improve the extraction efficiency of the three flavonoids from the PUL. Among the different DESs tested, the DES based on CC and GA exhibited the highest extraction efficiency. The extraction conditions were optimized using the BBD analysis. According to these conditions, the highest extraction efficiency (217.515 µg/mL) could be achieved by performing the extraction of PUL for 1 h at a temperature of 30 °C and stirring speed of 1113.1 rpm, using the DES based on CC and GA at a concentration of 30.85 wt.%. The extraction efficiency achieved using the DES was more than twice as high as that of aqueous extraction. In addition, the PUL extract obtained using the DES synthesized from CC and GA exhibited appreciable antioxidant activity, which was approximately 1/6 that of L-ascorbic acid. The potential utility of the PUL extracts as antioxidants was, thus, demonstrated. The PUL extract obtained using the DES synthesized from CC and GA significantly inhibited NO generation (up to 26.1%) at a concentration of 0.1%. These results indicated the anti-inflammatory properties of the active components of the PUL extracted using the DES.

Supplementary Materials: The following supporting information can be downloaded at: https://www.mdpi.com/article/10.3390/molecules27092798/s1: Figure S1: liquid chromatogram of PUL extract and mass spectra of flavonoid peaks; Figure S2: effect of extraction conditions on extraction efficiency of DES.

Author Contributions: Conceptualization, H.Y.P. and J.P.; methodology, J.W.L.; validation, J.W.L.; resources, H.Y.P.; writing—original draft preparation, J.P.; writing—review and editing, J.P. All authors have read and agreed to the published version of the manuscript.

Funding: This research was funded by the National Institute of Biological Resources (NIBR), funded by the Ministry of Environment (MOE) of the Republic of Korea, grant number NIBR202219101, and the National Research Foundation of Korea (NRF) grant funded by the Korean government (MSIT), grant number 2019R1F1A1058645.

Institutional Review Board Statement: Not applicable.

Informed Consent Statement: Not applicable.

Data Availability Statement: Not applicable.

Acknowledgments: This work was supported by a grant from the National Institute of Biological Resources (NIBR), funded by the Ministry of Environment (MOE) of the Republic of Korea (NIBR202219101) and a National Research Foundation of Korea (NRF) grant funded by the Korean government (MSIT) (2019R1F1A1058645).

Conflicts of Interest: The authors declare no conflict of interest.

Sample Availability: Not applicable.

References

1. Boyko, N.; Zhilyakova, E.; Malyutina, A.; Novikov, O.; Pisarev, D.; Abramovich, R.; Potanina, O.; Lazar, S.; Mizina, P.; Sahaidak-Nikitiuk, R. Studying and Modeling of the Extraction Properties of the Natural Deep Eutectic Solvent and Sorbitol-Based Solvents in Regard to Biologically Active Substances from *Glycyrrhizae* Roots. *Molecules* **2020**, *25*, 1482. [CrossRef]
2. Mišan, A.; Nađpal, J.; Stupar, A.; Pojić, M.; Mandić, A.; Verpoorte, R.; Choi, Y.H. The perspectives of natural deep eutectic solvents in agri-food sector. *Crit. Rev. Food Sci. Nutr.* **2020**, *60*, 2564–2592. [CrossRef]
3. Choi, Y.H.; van Spronsen, J.; Dai, Y.; Verberne, M.; Hollmann, F.; Arends, I.W.C.E.; Witkamp, G.-J.; Verpoorte, R. Are Natural Deep Eutectic Solvents the Missing Link in Understanding Cellular Metabolism and Physiology? *Plant Physiol.* **2011**, *156*, 1701–1705. [CrossRef]
4. Savi, L.K.; Carpiné, D.; Waszczynskyj, N.; Ribani, R.H.; Haminiuk, C.W.I. Influence of temperature, water content and type of organic acid on the formation, stability and properties of functional natural deep eutectic solvents. *Fluid Phase Equilibria* **2019**, *488*, 40–47. [CrossRef]
5. Jakovljević, M.; Vladić, J.; Vidović, S.; Pastor, K.; Jokić, S.; Molnar, M.; Jerković, I. Application of Deep Eutectic Solvents for the Extraction of Rutin and Rosmarinic Acid from *Satureja montana* L. and Evaluation of the Extracts Antiradical Activity. *Plants* **2020**, *9*, 153. [CrossRef]
6. Sut, S.; Faggian, M.; Baldan, V.; Poloniato, G.; Castagliuolo, I.; Grabnar, I.; Perissutti, B.; Brun, P.; Maggi, F.; Voinovich, D.; et al. Natural Deep Eutectic Solvents (NADES) to Enhance Berberine Absorption: An In Vivo Pharmacokinetic Study. *Molecules* **2017**, *22*, 1921. [CrossRef]
7. de los Ángeles Fernández, M.; Boiteux, J.; Espino, M.; Gomez, F.J.; Silva, M.F. Natural deep eutectic solvents-mediated extractions: The way forward for sustainable analytical developments. *Anal. Chim. Acta* **2018**, *1038*, 1–10. [CrossRef]
8. Abbott, A.P.; Boothby, D.; Capper, G.; Davies, D.L.; Rasheed, R.K. Deep Eutectic Solvents Formed between Choline Chloride and Carboxylic Acids: Versatile Alternatives to Ionic Liquids. *J. Am. Chem. Soc.* **2004**, *126*, 9142–9147. [CrossRef]
9. Abbott, A.P.; Capper, G.; Davies, D.L.; Rasheed, R.K.; Tambyrajah, V. Novel solvent properties of choline chloride/urea mixtures. *Chem. Commun.* **2003**, *1*, 70–71. [CrossRef]

10. Jablonský, M.; Škulcová, A.; Malvis, A.; Šima, J. Extraction of value-added components from food industry based and agro-forest biowastes by deep eutectic solvents. *J. Biotechnol.* **2018**, *282*, 46–66. [CrossRef]
11. Liu, Y.; Garzon, J.; Friesen, J.B.; Zhang, Y.; McAlpine, J.B.; Lankin, D.C.; Chen, S.-N.; Pauli, G.F. Countercurrent assisted quantitative recovery of metabolites from plant-associated natural deep eutectic solvents. *Fitoterapia* **2016**, *112*, 30–37. [CrossRef]
12. Dai, Y.; Witkamp, G.-J.; Verpoorte, R.; Choi, Y.H. Natural Deep Eutectic Solvents as a New Extraction Media for Phenolic Metabolites in *Carthamus tinctorius* L. *Anal. Chem.* **2013**, *85*, 6272–6278. [CrossRef]
13. Tsao, R. Chemistry and Biochemistry of Dietary Polyphenols. *Nutrients* **2010**, *2*, 1231–1246. [CrossRef]
14. Heim, K.E.; Tagliaferro, A.R.; Bobilya, D.J. Flavonoid antioxidants: Chemistry, metabolism and structure-activity relationships. *J. Nutr. Biochem.* **2002**, *13*, 572–584. [CrossRef]
15. Williams, R.J.; Spencer, J.P.E.; Rice-Evans, C. Flavonoids: Antioxidants or signalling molecules? *Free. Radic. Biol. Med.* **2004**, *36*, 838–849. [CrossRef]
16. Kitahara, K.; Matsumoto, Y.; Ueda, H.; Ueoka, R. A Remarkable Antioxidation Effect of Natural Phenol Derivatives on the Autoxidation of γ-Irradiated Methyl Linoleate. *Chem. Pharm. Bull.* **1992**, *40*, 2208–2209. [CrossRef]
17. Hatano, T. Constituents of natural medicines with scavenging effects on active oxygen species-tannins and related polyphenols. *Nat. Med.* **1995**, *49*, 357–363.
18. Masaki, H.; Sakaki, S.; Atsumi, T.; Sakurai, H. Active-oxygen scavenging activity of plant extracts. *Biol. Pharm. Bull.* **1995**, *18*, 162–166. [CrossRef]
19. Moazedi, R.; Nahandi, F.Z.; Mahdavi, Y.; Ebrahemi, M.A. Assessment of genetic relationships of some cultivars of Asian pears (*Pyrus pyrifolia* Nakai) with some native pears of Northern Iran using SSR markers. *Int. J. Farming Allied Sci.* **2014**, *3*, 923–929.
20. Bell, R. Chapter 14. Pears (Pyrus). Genetic resources of temperate fruit and nut crops. International Society for Horticultural Science, Wageningen, Netherlands. *Acta Hortic.* **1990**, *290*, 657–700.
21. Bell, R.; Quamme, H.; Layne, R.; Skirvin, R. Pears. In *Fruit Breeding, Volume I: Tree and Tropical Fruits*; Janick, J., Moore, J.N., Eds.; Wiley: New York, NY, USA, 1996.
22. Kim, M.; Lee, S.; Lee, H.; Lee, S. Phenological Response in the Trophic Levels to Climate Change in Korea. *Int. J. Environ. Res. Public Health* **2021**, *18*, 1086. [CrossRef]
23. Choi, H.-J.; Park, J.-H.; Han, H.-S.; Son, J.-H.; Son, K.-M.; Bae, J.-H.; Choi, C. Effect of polyphenol compound from Korean pear (Pyrus pyrifolia Nakai) on lipid metabolism. *J. Korean Soc. Food Sci. Nutr.* **2004**, *33*, 299–304.
24. Banerjee, S.; Mazumdar, S. Electrospray Ionization Mass Spectrometry: A Technique to Access the Information beyond the Molecular Weight of the Analyte. *Int. J. Anal. Chem.* **2012**, *2012*, 282574. [CrossRef] [PubMed]
25. Chojnacki, T.; Vogtman, T. The occurrence and seasonal distribution of C50-C60-polyprenols and of C100-and similar long-chain polyprenols in leaves of plants. *Acta Biochim. Pol.* **1984**, *31*, 115–126.
26. Park, D.E.; Adhikari, D.; Pangeni, R.; Panthi, V.K.; Kim, H.J.; Park, J.W. Preparation and Characterization of Callus Extract from Pyrus pyrifolia and Investigation of Its Effects on Skin Regeneration. *Cosmetics* **2018**, *5*, 71. [CrossRef]
27. Lu, W.; Alam, M.A.; Pan, Y.; Wu, J.; Wang, Z.; Yuan, Z. A new approach of microalgal biomass pretreatment using deep eutectic solvents for enhanced lipid recovery for biodiesel production. *Bioresour. Technol.* **2016**, *218*, 123–128. [CrossRef]
28. Kundu, J.K.; Surh, Y.-J. Inflammation: Gearing the journey to cancer. *Mutat. Res./Rev. Mutat. Res.* **2008**, *659*, 15–30. [CrossRef]
29. Nathan, C. Nitric oxide as a secretory product of mammalian cells. *The FASEB journal* **1992**, *6*, 3051–3064. [CrossRef]
30. Ratty, A.; Sunamoto, J.; Das, N.P. Interaction of flavonoids with 1, 1-diphenyl-2-picrylhydrazyl free radical, liposomal membranes and soybean lipoxygenase-1. *Biochem. Pharmacol.* **1988**, *37*, 989–995. [CrossRef]
31. Garcia, E.J.; Oldoni, T.L.C.; de Alencar, S.M.; Reis, A.; Loguercio, A.D.; Grande, R.H.M. Antioxidant activity by DPPH assay of potential solutions to be applied on bleached teeth. *Braz. Dent. J.* **2012**, *23*, 22–27. [CrossRef]

Article

Copper-Free Halodediazoniation of Arenediazonium Tetrafluoroborates in Deep Eutectic Solvents-like Mixtures

Giovanni Ghigo [1,*], Matteo Bonomo [1,2,*], Achille Antenucci [1,2], Chiara Reviglio [1] and Stefano Dughera [1,*]

1. Department of Chemistry, University of Turin, Via Pietro Giuria 7, 10125 Turin, Italy; achille.antenucci@unito.it (A.A.); chiara.reviglio@edu.unito.it (C.R.)
2. NIS Interdepartmental Centre and INSTM Reference Centre, Universiy of Turin, Via Gioacchino Quarello 15/a, 10125 Turin, Italy
* Correspondence: giovanni.ghigo@unito.it (G.G.); matteo.bonomo@unito.it (M.B.); stefano.dughera@unito.it (S.D.)

Abstract: Deep Eutectic Solvent (DES)-like mixtures, based on glycerol and different halide organic and inorganic salts, are successfully exploited as new media in copper-free halodediazoniation of arenediazonium salts. The reactions are carried out in absence of metal-based catalysts, at room temperature and in a short time. Pure target products are obtained without the need for chromatographic separation. The solvents are fully characterized, and a computational study is presented aiming to understand the reaction mechanism.

Keywords: Sandmeyer reactions; deep eutectic solvents; reaction mechanism

1. Introduction

In recent years, the interest in Deep Eutectic Solvents (DESs) has been growing since they are a potentially environmentally benign and sustainable alternative to conventional organic solvents [1–3]. Despite the increasing use of DESs in organic synthesis [4–6], they have scarcely been utilized in the reactions of diazonium salts. In fact, the literature shows only three very recent examples, where DESs are employed as innocent solvents [7–9].

In our previous paper [10] we studied the behaviour of arenediazonium tetrafluoroborates in a new DES formed by KF as a hydrogen bond acceptor (HBA) and glycerol as a hydrogen bond donor (HBD) as reaction media. A controlled decomposition of diazonium salts took place, with the formation of arenes as hydrodediazoniation products. We proposed a plausible mechanism where a relatively fast (strictly depending on the electronic effects of the substituents bound to the aromatic ring) reduction reaction occurs initiated by the formation of a glycerolate-like species. Interestingly, in some preliminary tests we found that the behaviour of 4-nitrobenzenediazonium tetrafluoroborate (**1a**) in a mixture formed by glycerol (HBD) and KBr or KCl (HBA) was completely different. In fact, a certain amount of halogenation product was obtained, still alongside the reduction product. Therefore, in the present paper, we decided to further investigate the reactivity of these mixtures.

The transformations of arenediazonium salts into several functional groups, such as halogen, hydroxyl, and cyano, are known as Sandmeyer or Sandmeyer-type reactions (Scheme 1) and have been widely used both in the fields of research and in industrial production [11–14].

In particular, halodediazoniation [15–18] represents an important organic transformation that converts arylamines to aryl halides via a diazonium salt intermediate and in the presence of Cu(I) as a catalyst. It must be stressed that aryl halides have always been drawing notable attention as beneficial reagents for arylation or for further aromatic core modification [19–21]. On these grounds, they have found a wide application in the synthesis of compound libraries, first of all because of smoothly running arylation reactions, as well as metal-catalyzed couplings. Moreover, they are an important structural motif in molecules, usually of marine origin, that exhibit interesting biological and pharmacological activity [22,23].

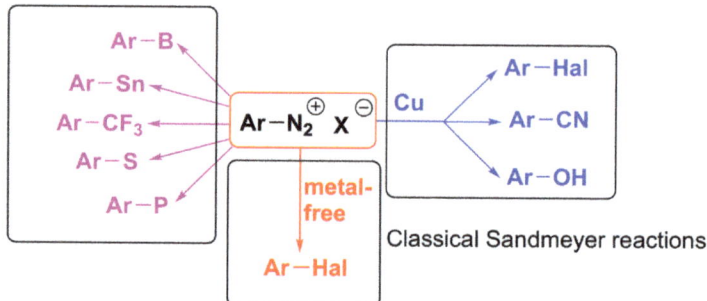

Scheme 1. History of Sandmeyer reactions.

In recent years, this "old" reaction (proposed by Sandmeyer at the end of the 19th century [14]) has been revisited and has become an interesting research topic (Scheme 1) [24,25]. In light of these, a conversion of aromatic amines via arenediazonium salts in aromatic compounds bearing a C-CF$_3$ [26,27], C-B [28,29], C-Sn [30], C-S [31–34], and C-P [35,36] bonds have been proposed. Furthermore, some research groups, aiming at increasing the sustainability of the reaction, have developed interesting synthetic procedures that allow aryl halide to be obtained without the use of copper as a catalyst [37–41].

On this basis, in this paper we report a thorough experimental and theoretical study of halodediazoniation reactions carried out in various types of DES-like systems and in total absence of a Cu catalyst.

2. Results and Discussion

As mentioned above, in one of our recent papers [10], we proposed a new DES based on glycerol and KF as an effective solvent for the green reduction of arenediazonium salts with almost quantitative yields. In that context, glycerol/KF 6:1 was unambiguously categorized as a DES system due to an unexpectedly high conductivity (31 S m^2 mol^{-1}), much higher than the values predicted by the Walden plot. Similar behaviour was found when K$^+$ was replaced by Cs$^+$, indicating the main role of the anion in the formation of a DES. Indeed, when F$^-$ was replaced with Cl$^-$ or Br$^-$, the resulting systems presented, from an electrochemical point of view, a conventional behaviour. Additionally, they seemed to be ineffective in the reduction reaction whose mechanism is based on the formation of a glycerolate-like species (i.e., a glycerol molecule involved in a hydrogen bond network). This evidence pushed us to preliminarily classify glycerol/KBr 6:1 and glycerol/KCl 6:1 as "simple" salts in solvent mixtures. Similar findings were evidenced when the cation was tetrabutylammonium (TBA), characterized by a slight decrease in the conductivity of the system ascribable to the increase in the viscosity due to higher steric hindrance of TBA$^+$ with respect to K$^+$. Indeed, in the case of TBABr-Glycerol systems, Andruch et al. [42] reported on the formulation of three mixtures in which the molar ratio of TBABr decreased from 0.5 to 0.20. The latter, having a TBABr:Glycerol ratio of 1:4, presented a DES-like behaviour with a homogeneous aspect and a relatively low melting point (233 K). However, our attempt in the formulation of the same mixture led to unstable solutions (some salt is still present as suspension), proving the needs of a higher amount of HBD (i.e., glycerol).

In this research, we propose a more detailed analysis, also taking into account the thermal behaviour of the proposed mixtures. Conventionally, Differential Scan Calorimetry (DSC) is the technique of election to monitor the freezing/melting process, which is clearly evidenced by a positive/negative peak due to the exothermic/endothermic nature of the transition. Unfortunately, in the case of glycerol, these peaks are not detectable due to the almost heat-neutrality of the freezing process [43], and the same applies to its mixture with the investigated salts (see Supporting Information). These findings prevented us from reliably employing DSC to determine the freezing point of our mixtures. Indeed, the glycerol freezing point is tabulated, but, on the other hand, no values are available for its mixture with unconventional salts. As a matter of fact, the freezing temperature of a (glycerol-based) mixture can be theoretically calculated by means of the freezing point depletion rule ($\Delta T = K_{cr} \times i \times m$, where K_{cr} is the cryoscopic constant, 3.5 K Kg mol^{-1} for glycerol [44], i is maximum number of ions in which the solute could be divided, and m is the molality of the solution). Having this in mind, an ideal mixture should present a ΔT of 13 K for all systems being independent on the nature of the solute. This should give a freezing point of 278 K (T_f = 291 K for pure glycerol) for the mixture that is investigated here. Once ruling out the possibility of effectively employed DSC to detect the freezing point of the mixtures, we resolved to a different approach (see Experimental part for further details).

The results point out that all investigated mixtures present a freezing point below 260 K (see Table 1), proving an unconventional behaviour. This allows us to unambiguously classify the analysed solvent as a low-melting system. However, a more detailed study is required to clarify if the 6:1 ratio effectively corresponds to the exact eutectic mixture or if a lower freezing point can be obtained by varying the hydrogen bond donor (HBD): hydrogen bond acceptor (HBA) ratio. Therefore, hereafter we will refer to them as DES-like systems, also according to what was reported by Abbott and co-workers for similar mixtures [45]. To shed more light on this aspect but limiting the analyses to the case of Gly/KBr systems, we prepared two other mixtures having a lower (Gly/KBr 4:1) or higher (Gly/KBr 10:1) amount of HBD, leading to a variation of the magnitude of HBD-HBA interaction. In both the cases, compared to Gly/KBr 6:1 (T_m = 225 K), higher melting temperature values (i.e., 238 K and 247 K, respectively) were measured. Therefore, in the following, we focused our analyses on systems having an HBD/HBA ratio equal to 6.

Table 1. Melting temperature of the different DES-like systems investigated within the paper.

Code	HBD	HBA	HBD/HBA Ratio	Melting Temperature/K
Gly/KBr 4:1	Glycerol	KBr	4:1	238
Gly/KBr 6:1	Glycerol	KBr	6:1	225
Gly/KBr 10:1	Glycerol	KBr	10:1	247
Gly/KCl 6:1	Glycerol	KCl	6:1	239
Gly/KI 6:1	Glycerol	KI	6:1	251
Gly/TBABr 6:1	Glycerol	TBABr	6:1	245
Gly/TBACl 6:1	Glycerol	TBACl	6:1	241
Glycerol [1]	Glycerol	Glycerol	-	291 [1]
Glycerol [2]	Glycerol	Glycerol	-	269 [2]

[1] Tabulated value for pure glycerol. [2] Measured value due to the vitrification process.

Comparing mixtures in which the halogen atom of the HBD is changed, the mixture containing the Br$^-$ demonstrated the lowest melting temperature, whereas Cl$^-$ and I$^-$-based systems melt at higher temperatures (i.e., 239 K and 251 K, respectively). This evidence proves that the melting temperature of the mixtures cannot be directly correlated with the electronegativity of the halogen (Cl$^-$ > Br$^-$ > I$^-$), proving the significance of the interplay between the different interactions and thus allowing the formation of a stable DES (-like) system.

Once the nature of the proposed solvent had been elucidated, a model reaction was preliminarily studied (Table 2), consisting in the reaction, at room temperature, of 4-nitrobenzenediazonium tetrafluoroborate (**1a**) in both mixtures of glycerol/KBr and conventional organic solvents. The choice of privileging the bromination reactions arises from the reason that, from the synthetic point of view, bromides, compared to chlorides, can be more useful. For example, in the cross-coupling reactions, C–Cl bonds are often too inert, and bromides (or iodides) leaving groups are required for acceptable yields [46].

Table 2. Bromodediazotation: trial reactions.

$$O_2N-C_6H_4-N_2^+ \; BF_4^- \longrightarrow O_2N-C_6H_4-Br + O_2N-C_6H_5$$

$$\text{1a} \qquad\qquad\qquad \text{2a} \qquad\quad \text{3a}$$

Entry	Solvent	Amount	Time (h)	Yields of 2a [1] (%)
1	Gly and KBr [2]	2.5 mL	24	- [3]
2	Gly and KBr [4]	2.5 mL	24	- [3]
3	EtOH and KBr [2]	2.5 mL	24	- [3]
4	DMSO and KBr [2]	2.5 mL	24	- [3]
5	MeCN and KBr [2]	2.5 mL	24	- [3]
6	Gly/KBr 4:1 [5]	2.5 mL	24	Traces [3]
7	Gly/KBr 6:1 [5]	2.5 mL	24	Traces [3]
8	Gly/KBr 10:1 [5]	2.5 mL	24	Traces [3]
9	Gly/KBr 4:1 [5]	3.5 mL	12	34 [6]
10	Gly/KBr 6:1 [5]	3.5 mL	12	32 [6]
11	Gly/KBr 10:1 [5]	3.5 mL	12	29 [6]
12	Gly/KBr 6:1 [5]	5 mL	4	72 [6]
13	Gly/KBr 4:1 [5]	5 mL	4	62 [6]
14	Gly/KBr 6:1 [5]	10 mL	4	72 [6,7,8]
15	Gly/KBr 4:1 [5]	10 mL	4	70 [6]

[1] All reactions were carried out at room temperature. Yields refer to pure and isolate 2a. [2] A suspension of KBr (4 mmol for 2 mmol of 1a) in 2.5 mL of suitable solvent was prepared. [3] GC and MS analyses showed the presence of nitrobenzene 3a as the only or main product. [4] A suspension of KBr (8 mmol for 2 mmol of 1a) in 2.5 mL of glycerol was prepared. [5] The reactions were carried out with 2 mmol of 1a. [6] GC-MS analyses also showed the presence of nitrobenzene 3a. It was removed by evaporation under vacuum. [7] Heating to 40 °C, no increase in yield was observed. [8] After adding 0.1 mmol of Cu(0), same results were obtained. However, the reaction finished in 30 min.

Firstly, it must be stressed that, at room temperature, when adding salt 1a (2 mmol) to a suspension of KBr in glycerol, target 1-bromo-4-nitrobenzene (2a) was not formed (Table 2; entry 1), not even in strong excess of KBr (Table 2; entry 2). Furthermore, the same negative result was obtained with a KBr suspension in common organic solvents (Table 2; entries 3–5). This evidence shows that a close interplay between glycerol and salt is required to activate the target reaction. Thus, homogeneous mixtures of glycerol and KBr (in various ratios, 10:1; 6:1 and 4:1) were tested. When dissolving 1a (2 mmol) in 2.5 mL of this solution (Table 1; entries 6–8), only traces of 2a were detected. We obtained the target 2a (Table 1; entries 9–11), albeit still in unsatisfactory yields, using 3.5 mL of solvent systems. Consequently, we decided to double their amount (5 mL). The reaction time was 4 h; in these conditions, to our delight, a dramatic increase in yield occurred (72%. Table 1; entries 12,13). The by-product nitrobenzene (3a) was easily removed under vacuum. Alternatively, it could be recovered (20% yield) by chromatographing the crude residue on a short column (see Experimental). It must be stressed that the increase in the amount of solvent or concentration of KBr (Table 2, entries 14,15) does not lead to higher yields, further confirming the likely necessity of the instauration of intermolecular interactions.

After this optimization and choosing glycerol/KBr 6:1 (method A) as privileged solvent systems, the scope of this reaction was extended to a wider library of arenediazonium salts 1 containing both electron-donating and electron-withdrawing groups. The results are summarized in Figure 1.

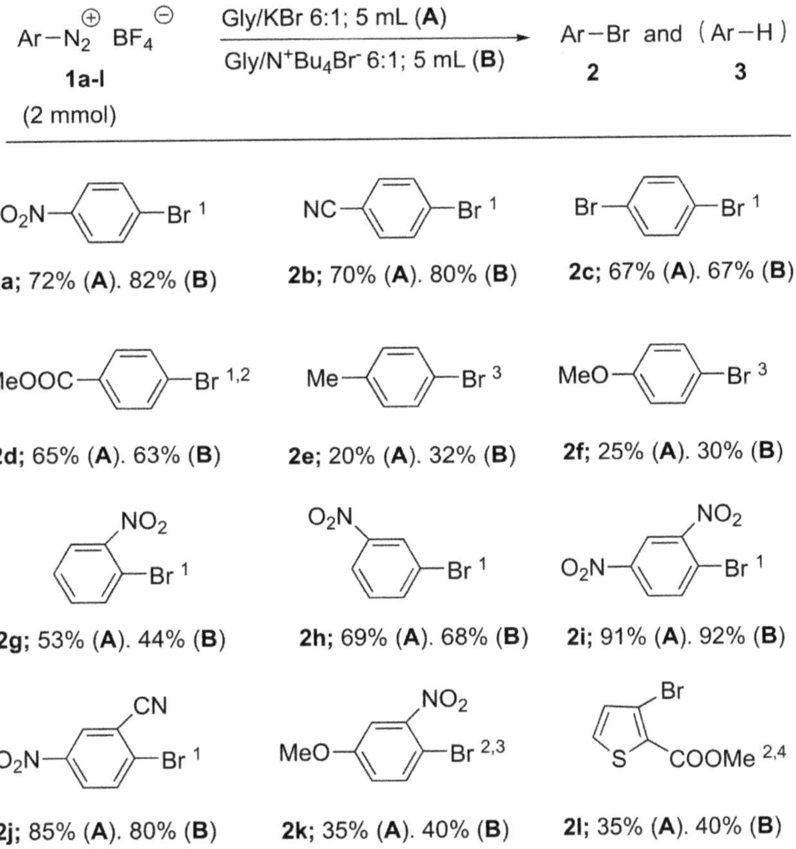

$$\text{Ar-N}_2^+ \text{ BF}_4^- \xrightarrow[\text{Gly/N}^+\text{Bu}_4\text{Br}^- \text{ 6:1; 5 mL (B)}]{\text{Gly/KBr 6:1; 5 mL (A)}} \text{Ar-Br and (Ar-H)}$$

1a-l (2 mmol) → 2 and 3

2a; 72% (A). 82% (B) — O$_2$N-C$_6$H$_4$-Br [1]
2b; 70% (A). 80% (B) — NC-C$_6$H$_4$-Br [1]
2c; 67% (A). 67% (B) — Br-C$_6$H$_4$-Br [1]
2d; 65% (A). 63% (B) — MeOOC-C$_6$H$_4$-Br [1,2]
2e; 20% (A). 32% (B) — Me-C$_6$H$_4$-Br [3]
2f; 25% (A). 30% (B) — MeO-C$_6$H$_4$-Br [3]
2g; 53% (A). 44% (B) — 2-NO$_2$-C$_6$H$_4$-Br [1]
2h; 69% (A). 68% (B) — 3-NO$_2$-C$_6$H$_4$-Br [1]
2i; 91% (A). 92% (B) — 2,4-(NO$_2$)$_2$-C$_6$H$_3$-Br [1]
2j; 85% (A). 80% (B) — 2-CN-4-O$_2$N-C$_6$H$_3$-Br [1]
2k; 35% (A). 40% (B) — 2-NO$_2$-4-MeO-C$_6$H$_3$-Br [2,3]
2l; 35% (A). 40% (B) — 3-Br-thiophene-2-COOMe [2,4]

Figure 1. Bromodediazotation scope. [1] Reactions were carried out at room temperature for 4 h: 2 mmol of salts **1** and 5 mL of Gly/KBr (method A) or 5 mL of Gly/TBAB (method B). Longer reaction times or higher amounts of solvent systems did not lead to increased yields. Yields refer to pure and isolated **2**. [2] In order to remove **3**, the crude residue was purified in a chromatography column (eluent: petroleum ether). The yield of **3d** was 26%, and the yield of **3l** was 42%. [3] Reactions were carried out at room temperature for 24 h: 2 mmol of salts **1** and 5 mL of Gly/KBr (method A) or 5 mL of Gly/TBAB (method B). Longer reaction times as well as higher amounts of solvent systems or higher temperatures did not lead to increased yields. Yields refer to pure and isolated **2**. [4] The reaction was carried out at 35 °C suspension of KBr (4 mmol for 2 mmol of **1a**) in 2.5 mL of suitable solvent that was prepared.

Good yields of target aryl bromides **2** were achieved in the presence of electron-withdrawing groups. On the other hand, a decreasing of the yields was observed in the presence of electron-donating groups. In any case, longer reaction times or a greater amount of Glycerol/KBr 6:1 did not lead to sizeable improvements. It must be stressed that all aryl bromides **2** were obtained in adequate purity, making further chromatographic purification usually unnecessary since the by-products arenes **3** were usually removed under vacuum. However, from a GC-MS analysis of the crude residues, it is possible, albeit in a completely qualitative way, to have indications on the amount of **3** that were formed. The data are shown in Table 3.

Table 3. GC-MS analyses of crude residues of bromodediazotation and chlorodediazotation.

Entry	Salt	Estimated Yield (%) of 2 or 4	Estimated Yield (%) of 3	Yields (%) of Isolated 2a	Method
1	1a	2a; 75	3a; 24 [1]	2a; 72	A
2	1a	2a; 88	3a; 10	2a; 82	B
3	1b	2b; 78	3b; 22	2b; 70	A
4	1b	2b; 82	3b; 17	2b; 80	B
5	1c	2c; 65	3c; 34	2c; 67	A
6	1c	2c; 71	3c; 28	2c; 80	B
7	1d	2d; 69	3d; 30	2d; 65	A
8	1d	2d; 67	3d; 33 [2]	2d; 63	B
9	1e	2e; 28	3e; 72	2e; 20	A
11	1f	2f; 32	3f; 68	2f; 25	A
12	1f	2f; 36	3f; 63	2f; 30	B
13	1g	2g; 60	3g; 38	2g; 53	A
14	1g	2g; 36	3g; 63	2g; 30	B
15	1h	2h; 74	3h; 25	2h; 69	A
16	1h	2h; 77	3h; 22	2h; 68	B
17	1i	2i; 98	3i; traces	2i; 91	A
18	1i	2i; 98	3i; traces	2i; 92	B
19	1j	2j; 94	3j; traces	2j; 85	A
20	1j	2j; 93	3j; traces	2j; 80	B
21	1k	2k; 37	3k; 60	2k; 35	A
22	1k	2k; 42	3k; 55	2k; 40	B
23	1l	2l; 39	3l; 58 [3]	2l; 35	A
24	1l	2l; 44	3l; 56	2l; 40	B
25	1a	4a; 27	3a; 65	4a; 21	A
26	1a	4a; 25	3a; 67	4a; 18	B
27	1e	4e; -	3e; 74	4e; -	A
28	1e	4e; -	3e; 76	4e; -	B
29	1f	4f; -	3f; 81	4f; -	A
30	1f	4f; -	3f; 81	4f; -	B
31	1i	4i; 100	3i; -	4i; 100	A
32	1i	4i; 100	3i; -	4i; 100	B
33	1j	4j; 100	3j; -	4j; 94	A
34	1j	4j; 100	3j;-	4j; 92	B
35	1m	4m; 42	3m; 55	4m; 35	A
36	1m	4m; 43	3m; 53	4m; 33	B

[1] The yield of isolated 3a, after purification in chromatography column (eluent: petroleum ether) was 20%. [2] The yield of isolated 3d, after purification in chromatography column (eluent: petroleum ether) was 26%. [3] The yield of isolated 3l, after purification in chromatography column (eluent: petroleum ether) was 42%.

In light of the data described above, a second DES-like mixture formed by glycerol and tetrabutylammonium bromide (TBAB) in ratio 6:1 was used as an alternative solvent to glycerol/KBr (method B). As can be seen from Figure 1, the results achieved with method A or, alternatively, with method B, are almost identical. At the end of the reactions, the solvent system glycerol/KBr was easily recovered (for details see Experimental) and was used in four consecutive runs (Table 4), without observing a decrease in the yield of 2a. In fact, potassium tetrafluoroborate, formed in the reaction as a by-product, remained trapped in the solvent system upon evaporation of the aqueous layer of the extraction, without interfering in the following runs.

Table 4. Recovery and reuse of solvent system Gly/KBr 6:1.

Entry	Time (h)	Yield of 2a (%) [1,2]	Recovery of Solvent
1	4	72	4.9 mL [3]
2	4	70	4.9 mL [4]
3	4	67	4.7 mL [5]
4	6	70	4.6 mL [6]
5	6	68	4.5 mL

[1] Yields refer to the pure and isolated product. [2] The reaction was carried out at room temperature with 2 mmol of **1a** and 5 mL of Gly/KBr 6:1. [3] Was used as a solvent in entry 2. [4] Was used as a solvent in entry 3. [5] Was used as a solvent in entry 4. [6] Was used as a solvent in entry 5.

In order to further expand the scope of the reaction, chloro and iododediazotation reactions were also studied. Regarding chlorodediazotation, in order to find optimal reaction conditions, a model reaction was preliminarily studied with 2,4-nitrobenzenediazonium tetrafluoroborate (**2i**) reacting at room temperature in various solvents (Table 5).

Table 5. Chlorodediazotation: trial reactions.

Entry	Solvent	Amount	Time (h)	Yields of 4i [1] (%)
1	Gly and KCl [2]	2.5 mL	24	- [3]
2	Gly and KCl [4]	2.5 mL	24	- [3]
3	EtOH and KCl [2]	2.5 mL	24	- [3]
4	DMSO and KCl [2]	2.5 mL	24	- [3]
5	MeCN and KCl [2]	2.5 mL	24	- [3]
6	Gly/KCl 4:1 [5]	5 mL	12	65 [6,7]
7	Gly/KCl 6:1 [5]	5 mL	12	63 [6,7]
8	Gly/KCl 10:1 [5]	5 mL	12	49 [6,7]
9	Gly/KCl 6:1 [5]	5 mL	4	87 [6,7]
10	Gly/KCl 6:1 [5]	10 mL	4	100 [6]

[1] All reactions were carried out at room temperature; yields refer to pure and isolate **4i**. [2] A suspension of KBr (4 mmol for 2 mmol of **1i**) in 2.5 mL of suitable solvent was prepared. [3] GC and MS analyses showed the presence of 1,3-dinitrobenzene **3i** as the only product. [4] A suspension of KBr (8 mmol for 2 mmol of **1i**) in 2.5 mL of glycerol was prepared. [5] The reactions were carried out with 2 mmol of **1i**. [6] GC-MS analyses also showed the presence of 1,3-dinitrobenzene **3i**. [7] GC calculated yields.

The optimal reaction conditions (Table 5; entry 10) were found by reacting salt **1i** (2 mmol) with 10 mL of solvent system Glycerol/KCl 6:1 for 4 h. In fact, target 1-chloro-2,4-dinitrobenzene (**4i**) was obtained in a quantitative yield. After this optimization, the scope was extended to arenediazonium salts **1**, containing both electron-donating and electron-withdrawing groups. The results are reported in Figure 2. In the best conditions, salt **1j** furnished **2j** in excellent yield. It is necessary to point out that, for the success of chlorodediazotation, it is essential to have strong electron-withdrawing groups linked to the aromatic ring. In the case of the salt **1a** with a single electron-withdrawing group, a low amount of **4a** was obtained and only with an excess of Gly/KCl 6:1 (20 mL) and heating to 35 °C for 4 h. In the presence of electron-donating groups (salts **1e** and **1f**), the reaction did not occur. Similar results were obtained using a solvent system formed by glycerol and tetrabutylammonium chloride (6:1 ratio).

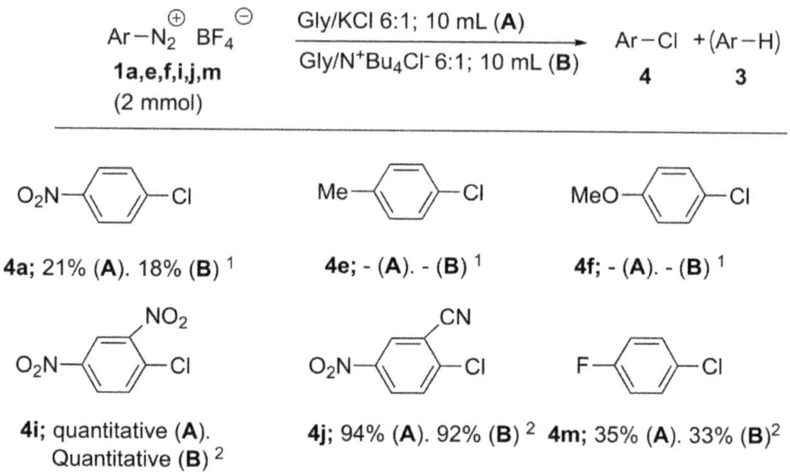

Ar—N₂⁺ BF₄⁻   Gly/KCl 6:1; 10 mL (**A**)
1a,e,f,i,j,m ────────────────────→   Ar—Cl + (Ar—H)
(2 mmol)   Gly/N⁺Bu₄Cl⁻ 6:1; 10 mL (**B**)   **4**   **3**

O₂N—C₆H₄—Cl
4a; 21% (A). 18% (B) [1]

Me—C₆H₄—Cl
4e; - (A). - (B) [1]

MeO—C₆H₄—Cl
4f; - (A). - (B) [1]

O₂N—C₆H₃(NO₂)—Cl
4i; quantitative (A). Quantitative (B) [2]

O₂N—C₆H₃(CN)—Cl
4j; 94% (A). 92% (B) [2]

F—C₆H₄—Cl
4m; 35% (A). 33% (B) [2]

Figure 2. Chlorodediazotation scope.[1] Reaction was carried out at 35 °C for 4 h with 20 mL of solvent. Longer reaction times or higher amounts of solvent systems did not lead to increased yields. Yields refer to pure and isolated **4**. GC and GC-MS analyses showed the presence of **3a** as main product or **3e** and **3f** as only products. [2] Reactions were carried out at room temperature for 4 h: 2 mmol of salts **1** and 10 mL of Gly/KCl (method A) or 10 mL of Gly/TBAC (method B). Yields refer to pure and isolated **4**.

Finally, although the iododediazotization reaction is normally carried out in the absence of Cu,11,12, Figure 3 reports the preparation of some iododerivatives **5** reacting some arenediazonium salts **1** in Gly/KI solvent system. The yields are always virtually quantitative, independently on the nature of the substituent.

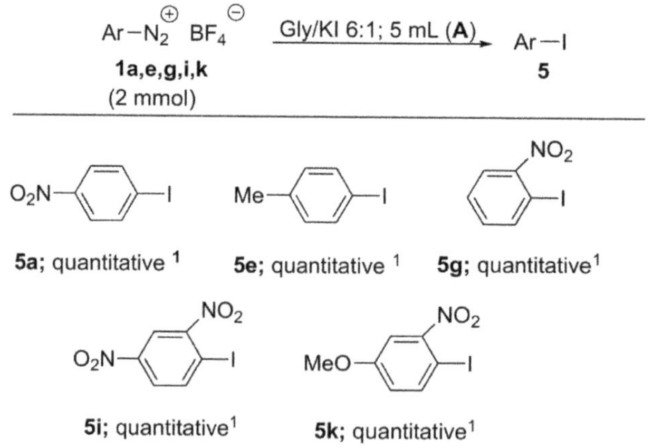

Ar—N₂⁺ BF₄⁻   Gly/KI 6:1; 5 mL (**A**)
1a,e,g,i,k ────────────────────→   Ar—I
(2 mmol)   **5**

O₂N—C₆H₄—I
5a; quantitative [1]

Me—C₆H₄—I
5e; quantitative [1]

C₆H₄(NO₂)—I
5g; quantitative [1]

O₂N—C₆H₃(NO₂)—I
5i; quantitative [1]

MeO—C₆H₃(NO₂)—I
5k; quantitative [1]

Figure 3. Iododediazotation scope. [1] Reactions were carried out at room temperature for 2 h: 2 mmol of salts **1** and 5 mL of Gly/KI. Yields refer to pure and isolated **5**.

As mentioned above, the Sandmeyer reaction has been rediscovered and "modernized" in recent years. In particular, new methodologies that allow the target products to be obtained in the absence of copper catalysts have been proposed [37–41]. As a result, in the light of the above experiments, we have proposed here a sustainable version of the Sandmeyer halodediazotation reaction. In fact, in our conditions, we were able to achieve important benefits, including the reaction's use at room temperature in mild conditions and without any metal catalyst and a recoverable and reusable sustainable reaction medium, which plays the role not only of solvent but also of reagent. Moreover, the target halides are usually obtained in adequate purity, making further chromatographic purification unnecessary, and the only solid waste product is potassium tetrafluoroborate, which is soluble in water.

Beside these aspects, we found it interesting to study this reaction from a mechanistic point of view. The mechanisms for the reactions in glycerol with fluoride [10], chloride, bromide, iodide, and pure glycerol were calculated for the electron-poor **1a** and for the electron-rich **1f** *para*-substituted benzenediazonium tetrafluoroborates. The preliminary calculations indicated that the dissociations of **1a** and **1f** to BF_4^- and to 4-nitrobenzenediazonium (**1a′**) and 4-methoxybenzenediazonium (**1f′**), respectively, are thermodynamically favoured by 0.5 and 2.2 kcal mol^{-1} in terms of free energies, as expected by the ion solvation properties of glycerol. The tests for the reaction of **1a′** with fluoride suggested that the appropriate model for the computational study requires the presence of some explicit glycerol molecules.

In the textbooks [47–49], the nucleophilic substitution in arenediazonium compounds is supposed to take place by the S_N1 mechanism where the key intermediate is the phenyl cation (i.e., phenylium). However, this cation is so reactive that it can recapture the nitrogen generated in the decomposition (N_2-scrambling) [50].

Additionally, attempts to observe the formation of phenyl cations by ionization of aryl triflates have only succeeded when especially stabilizing groups, such as trimethylsilyl groups, are present at the 2- and 6-positions of the aromatic ring [51]. Moreover, the experiments that prove the existence of the phenylium have all been performed in an ionizing but almost non-nucleophilic solvent such as 2,2,2-trifluoroethanol. The transition structure (TS) that we optimized for the N_2-scrabling in 4-nitrobenzenediazonium in glycerol is located 30.6 kcal mol^{-1} above **1a′**. However, in all the attempts made to optimize this TS in the presence of explicit glycerol molecules, we found that the C-N bond breaking was immediately followed by the solvent capture (solvolysis). Therefore, the free 4-phenylium is not generated, but, after a proton transfer, the final products are the mixed ether **6a** and the protonated glycerol (Scheme 2).

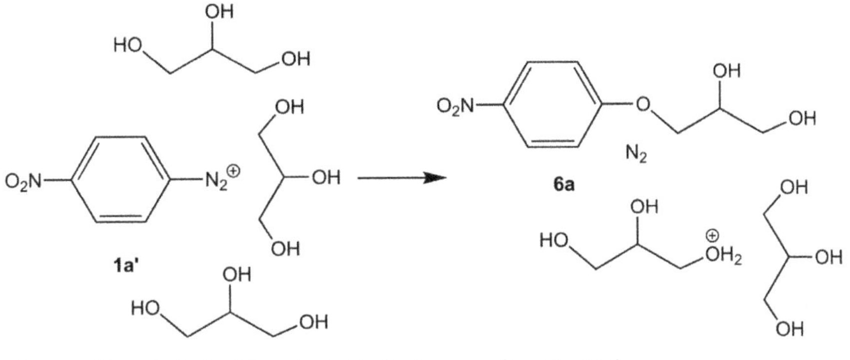

Scheme 2. The solvolysis of the solvated 4-nitrobenezenediazonium **1a′**.

Thus, the solvolysis takes place through a concerted TS corresponding to a *front*-S_N2 as illustrated in Figure 4 (up). Indeed, such a mechanism was already proposed in the literature [52,53], and it is operative also for the halogenations (Scheme 3; an example of the TS is shown in Figure 4 (down), and the pictures of all structures are reported in the Supporting Information). All halogenations as well as the solvolysis are irreversible, being strongly exoergic by 55–60 kcal mol^{-1} for **1a′** and by 40–56 kcal mol^{-1} for **1f′** in terms of free energies. From the activation free energies, we calculated the reaction rate constants **k** (Table 6) by means of the Eyring equation [54]. For the halogenations in Gly/potassium halide 6:1 DES-like systems, we calculated a concentration for the halogenides of 22.8 M, and from this value as well as from **k** we estimated the lifetimes **k,** which are reported in Table 5. Although underestimated because of the difficulty in modelling the solvent effects, we observed that the values of τ for the halogenations were in qualitative agreement with the experimental findings; the reactions of **1a'** are sensibly faster than those of **1f'**. However, in the case of fluorination, we should remember that in the Gly/KF DES (6:1 ratio) the reductions are the fastest processes [10]. For the other halogenations, because of the limitation in the calculation accuracy, the lifetimes are too close to perform any reliable comparison among the halogens. For iodination we could not exclude radical mechanism that we did not analyse, being out of the scope of the present paper. The solvolyses, which are treated as monomolecular processes as a result of the nucleophile being the solvent, show low **k** and long lifetimes, in agreement with the fact that mixed ethers **6a** and **6f** are never found in the experiments because the other reactions always prevail.

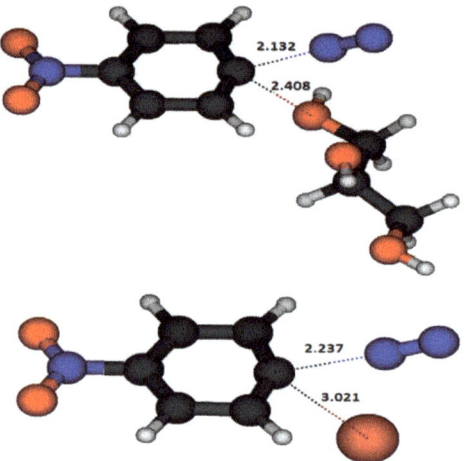

Figure 4. Transition structures for the solvolysis (**up**) and for the bromination (**down**) of **1a'** in glycerol. Unreactive explicit glycerol solvent molecules have been removed for clarity.

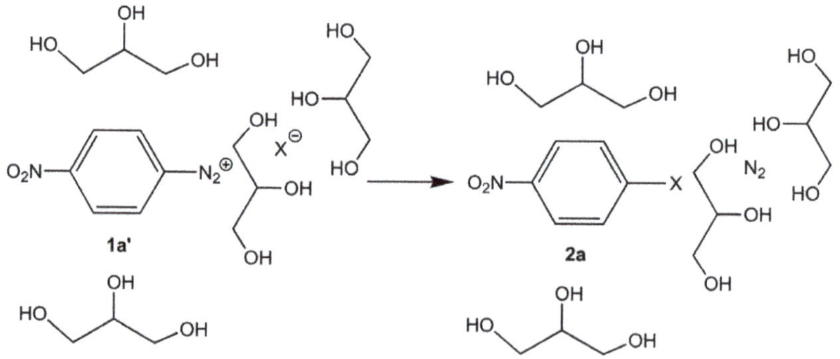

Scheme 3. The halogenation of the solvated 4-nitrobenezenediazonium **1a'**.

Table 6. Calculated rate constants and lifetimes for the halogenations and solvolysis of **1a'** and **1f'**.

	1a'		**1f'**	
	K [1]	τ	K [1]	τ
Fluorination	2.3×10^1	0.002 s	1.2×10^{-5}	1.1 h
Chlorination	1.2×10^{-1}	0.4 s	2.7×10^{-5}	0.5 h
Bromination	3.6×10^{-2}	1.2 s	1.0×10^{-5}	1.1 h
Iodination	4.6×10^{-1}	0.1 s	-[2]	-[2]
Solvolysis	4.1×10^{-6}	2.8 days	5.3×10^{-10}	60 years

[1] In $M^{-1} s^{-1}$ for the halogenation and in sec-1 for the solvolysis. [2] Non-calculated.

We also stress that the observation of a secondary deuterium kinetic isotope effect (KIE) of 1.49 in the dediazotation of benzenediazonium [55], interpreted as evidence of the formation of the phenylium, is also compatible with the present *front*-$S_N 2$ mechanism. In fact, the calculated KIE values for the solvolysis of the 2,6-dideuterated **1a'** and **1f'** are, respectively, 1.40 and 1.43. This is because the electronic and molecular structures of the aryl moieties in the transition structures are similar to that of the arylium. For the solvolysis of **1a'**, for example, the natural charges on C^1 (0.034 in the 4-nitrobenzenediazonium) are 0.551 in the TS and 0.649 in the 4-nitrophenilium. The $C^2 C^1 C^6$ angles (which is 125.4 degree in the 4-nitrobenzenediazonium) are 140.0 degrees in the transition structure and 149.3 in the 4-nitrophenilium.

The reduction of the arenediazonium in DES (see [10]) involves the generation of a phenyl radical Ar• that abstracts a hydrogen from the solvent by a Hydrogen Atom Transfer (HAT). As an alternative, the radical can bind a halide ion X^-, starting a radical chain that yields to the aryl halide Ar-X (Scheme 4):

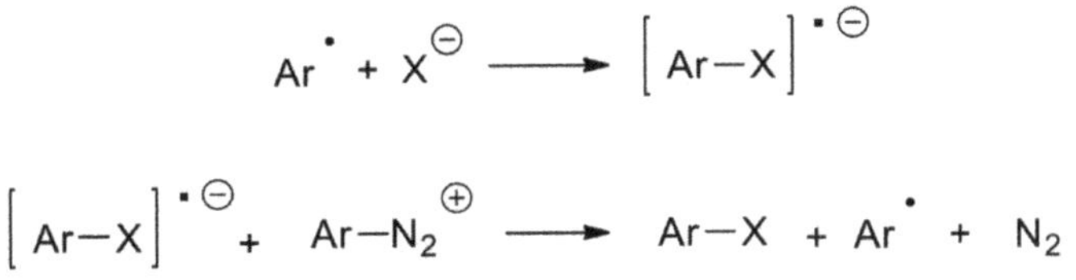

$$Ar^\bullet + X^{\ominus} \longrightarrow [Ar-X]^{\bullet \ominus}$$

$$[Ar-X]^{\bullet \ominus} + Ar-N_2^{\oplus} \longrightarrow Ar-X + Ar^\bullet + N_2$$

Scheme 4. Possible radical mechanism.

The first step of the radical halogenation is in competition with the HAT that yields to the reduction products. The calculated pseudo-first order rate constants for the binding of the 4-nitrophenyl radical with fluoride, chloride, bromide, and iodide are, respectively, 10^2, 10^5, 10^6, and 10^8 s^{-1}, to be compared with 10^9 s^{-1} for the HAT [10]. Clearly, the halogenation by a radical mechanism is not feasible for the three lighter halogens, but it is a possible alternative for the iodination where the radical chain can be initiated by the reduction of the arenediazonium by iodide, as proposed in the literature [11,12]. For the electron-rich 4-methoxybenzenediazonium, the chlorination and the bromination (which are the main interest of this work) by the radical-chain pathway appears to be even less feasible. The reaction of the 4-methoxyphenyl radical with chloride and bromide are strongly endoergic (in term of free energies, respectively 30.3 and 29.5 kcal mol^{-1} compared to -11.9 and -10.4 kcal mol^{-1} for the reaction with the 4-nitrophenyl radical). Therefore, the radical chain for these halogenations cannot proceed.

Finally, we studied the mechanism for the direct reaction of **1a'** with BF_4^- (an intramolecular reaction from the point of view of **1a**). The rate constant for this exoergic reaction ($\Delta G = -34.4$ kcal mol^{-1}) is 2.0×10^{-7} s^{-1}, which corresponds to a lifetime of 57 days. This value is in agreement with the experiments where this reaction was performed for 4-*t*butybenzenediazonium tetrafluoroborate in CH_2Cl_2, yielding the corresponding fluoride with a yield of 50% ca. after 10 days at room temperature [56].

3. Materials and Methods

General. All the reactions were carried out in open air glassware. Analytical-grade reagents and solvents (purchased from Merck or Alfa Aesar) were used, and reactions were monitored by GC, GC-MS, and TLC. Column chromatography and TLC were performed on Merck silica gel 60 (70–230 mesh ASTM) and GF 254, respectively. Petroleum ether refers to the fraction boiling in the range 40–70 °C. Room temperature is 22 °C. Mass spectra were recorded on an HP 5989B mass selective detector connected to an HP 5890 GC with a methyl silicone capillary column. GC analyses were performed on a Perkin Elmer AutoSystem XL GC with a methyl silicone capillary column. ^{1}H NMR and ^{13}C NMR spectra were recorded on a Jeol ECZR spectrometer at 600 and 150 MHz, respectively. Arenediazonium tetrafluoroborates were prepared as reported in the literature [57]. Structures and purity of bromoarenes **2**, chloroarenes **4**, and iodoarenes **5** were confirmed by their spectral (NMR, MS) and physical data, substantially identical to those reported in the literature. Their NMR spectra are reported in the Supporting Information. Differential Scanning Calorimetry (DSC) experiments were performed on Q200 DSC TA instruments with a ramp of 5 or 20 °C/min under N_2 atmosphere. After a stabilization at −85 °C, the sample was heated up to 80 °C and then cooled down to -80 °C, and the cycle was repeated two times. As a result of the DSC being unsuitable for the determination of its melting point in glycerol-based mixtures, we resolved to an unconventional approach: we inserted a thermometer into a vial containing the mixture, and then the vial was moved to a freezer (at −79 °C) for 15 min. After this time, the vial was removed from the fridge and was allowed to warm up to room temperature. The T_m value was selected as the temperature signed by thermometer when the latter was extracted (with no solid residues on it) from the vial.

Preparation of solvent systems: typical procedure for glycerol/KBr 6:1. KBr (11.9 g, 0.1 mol) was added at room temperature to glycerol (55.2 g, 0.6 mol). The suspension was stirred at 80 °C for about 2 h. It was cooled to room temperature, and a clear solution was obtained, which was used without any further purification.

Bromodediazotation of 4-nitrobenzenediazonium tetrafluoroborate (1a): typical procedure. Preparation of 1-bromo-4-nitrobenzene (2a). 4-Nitrobenzenediazonium tetrafluoroborate (**1a**, 0.44 g, 2 mmol) was added at room temperature to glycerol/KBr 6:1 (5 mL). The mixture was stirred at room temperature for 4 h, and the completion of the reaction was confirmed by the absence of azo coupling with 2-naphthol. Then, the reaction mixture was poured into Et_2O/H_2O (10 mL, 1:1). The aqueous layer was separated and extracted with Et_2O (5 mL). The combined organic extracts were washed with H_2O (5 mL), dried with Na_2SO_4, and evaporated under reduced pressure. GC-MS analyses of the crude residue showed a mixture of **2a**, as the major product, MS (EI, 70 eV): m/z (%) = 201 (100) [M]$^+$, 203 (100) [M +2]$^+$ and nitrobenzene **3a** MS (EI, 70 eV): m/z (%) = 123 (100) [M]$^+$. Further evaporation at reduced pressure allowed **3a** to be completely removed and pure **2a** to be obtained (GC, GC-MS, TLC and NMR; 290 mg, 72%). Alternatively, in order to quantify **3a**, the crude residue was chromatographed on a short column (silica gel; eluent: PE). The first eluted product was **3a** (50 mg, 20%). The second one was **2a** (284 mg, 70%).

Recovery and reuse of solvent system glycerol/KBr 6:1 The aqueous layers (about 15 mL) were collected and gathered. In order to remove solid residues, they were filtered on a funnel. H_2O was evaporated under reduced pressure. The recovered glycerol/KBr (4.9 mL), which showed NMR and IR spectra virtually identical to the initial one, was reused in four consecutive reactions. The average yield of **2a** was 69%, and the recovered solvent system (at the end of fifth run) continued to show NMR and IR spectra almost identical to the initial one.

Computational method.

The details and the references related to the computational method are all reported in the Supplementary Materials.

4. Conclusions

We have proposed here a synthetic and computational study of halodediazotation reactions (a Sandmeyer's class of reactions) carried out in deep eutectic solvent-like mixtures. The computational results are coherent with the experimental findings; the reactions of the control electron-poor **1a** are faster than that of the control electron-rich **1f**, and the halogenations are always much faster than the solvolysis that would yield the mixed ethers **6a** and **6f**. Both halogenations and the solvolysis take place through a mechanism similar to that of front-S_N2. Diazonium salts are resourceful building blocks in organic synthesis, and the last few years have seen a dramatic resurgence of their chemistry, as new functionalities have been introduced on their aromatic ring. The marriage between diazonium salts and deep eutectic solvents, which have also been defined as "the solvents of the century" [5], can become not only a valid tool to make their chemistry more sustainable but can also allow the exploration of new types of hitherto unknown reactivity.

Supplementary Materials: The following supporting information can be downloaded at: https://www.mdpi.com/article/10.3390/molecules27061909/s1, Computational method, Tables of calculated absolute and relative energies, Pictures and Cartesian Coordinates of all optimized structures, IRCs from the TSs for the chlorination and solvolysis of **1a'**, physical and NMR data of aryl halides **2, 4, 5** and NMR of glycerol/KBr solvent system can be downloaded at: https://www.mdpi.com/article/10.3390/molecules27061909/s1. References [58–82] are cited in the supplementary materials.

Author Contributions: G.G.: general organization, DFT calculations; M.B.: general organization, analysis of solvent systems; A.A.: halodediazotation reactions; C.R.: halodediazotation reactions; S.D.: general organization, halodediazotation reactions. All authors have read and agreed to the published version of the manuscript.

Funding: This research received no external funding.

Data Availability Statement: The data reported in this article can be obtained from the authors upon reasonable request.

Acknowledgments: This work was Presented at VI International Symposium «The Chemistry of Diazo Compounds and Related Systems» Saint Petersburg, Russia, 6-10/9/2021. Dughera, S.; Ghigo, G.; Bonomo, M.; Antenucci, A.; Barolo, C.; Gontrani, L.; Reviglio, C. How do diazonium salts behave in Deep Eutectic solvents? VI International Symposium "The Chemistry of Diazo Compounds and Related Systems"-Book of Abstracts. P. 41 (ISBN 978-5-9651-1365-1).

Conflicts of Interest: The authors declare no conflict of interest.

Sample Availability: Samples of the compounds **2, 4, 5** are available from the authors.

References

1. Abbott, A.P.; Boothby, D.; Capper, G.; Davies, D.L.; Rasheed, R.K. Deep Eutectic Solvents formed between choline chloride and carboxylic acids: Versatile alternatives to ionic liquids. *J. Am. Chem. Soc.* **2004**, *126*, 9142–9147. [CrossRef] [PubMed]
2. Marcus, Y. *Deep Eutectic Solvents*; Springer International Publishing: Berlin/Heidelberg, Germany, 2019.
3. Hansen, B.B.; Spittle, S.; Chen, B.; Poe, D.; Zhang, Y.; Klein, J.M.; Horton, A.; Adhikari, L.; Zelovich, T.; Doherty, B.W.; et al. Deep Eutectic Solvents: A Review of Fundamentals and Applications. *Chem. Rev.* **2020**, *121*, 1232–1285. [CrossRef] [PubMed]
4. Khandelwal, S.; Tailor, Y.K.; Kumar, M. Deep eutectic solvents (DESs) as eco-friendly and sustainable solvent/catalyst systems in organic transformations. *J. Mol. Liq.* **2016**, *215*, 345–386. [CrossRef]
5. Alonso, D.A.; Baeza, A.; Chinchilla, R.; Guillena, G.; Pastor, I.M.; Ramón, D.J. Deep Eutectic Solvents: The Organic Reaction Medium of the Century. *Eur. J. Org. Chem.* **2016**, *2016*, 612–632. [CrossRef]
6. Hooshmand, S.E.; Afshari, R.; Ramón, D.J.; Varma, R.S. Deep eutectic solvents: Cutting-edge applications in cross-coupling reactions. *Green Chem.* **2020**, *22*, 3668–3692. [CrossRef]
7. Kamble, S.S.; Shankarling, G.S. Room temperature diazotization and coupling reaction using a DES-ethanol system: A green approach towards the synthesis of monoazo pigments. *Chem. Commun.* **2019**, *55*, 5970–5973. [CrossRef]
8. Lenne, Q.; Andrieux, V.; Levanen, G.; Bergamini, J.F.; Nicolas, P.; Paquin, L.; Lagrost, C.; Leroux, Y.R. Electrochemical grafting of aryl diazonium salts in deep eutectic solvents. *Electrochim. Acta* **2021**, *369*, 137672. [CrossRef]
9. Ma, X.; Li, Z. Synthesis of Diarylethynes from Aryldiazonium Salts by Using Calcium Carbide as an Alkyne Source in a Deep Eutectic Solvent. *Synlett* **2021**, *32*, 631–635. [CrossRef]

10. Antenucci, A.; Bonomo, M.; Ghigo, G.; Gontrani, L.; Barolo, C.; Dughera, S. How do arenediazonium salts behave in deep eutectic solvents? A combined experimental and computational approach. *J. Mol. Liq.* **2021**, *339*, 116743. [CrossRef]
11. Saunders, K.H.; Allen, R.L.M. *Aromatic Diazo Compounds*; Edward Arnold: London, UK, 1985.
12. Zollinger, H. *Diazochemistry: Aromatic and Heteroaromatic Compounds*; Wiley-VCH: Weinheim, Germany, 1994.
13. Schareina, T.; Bellen, M. Copper-Catalyzed Cyanations of Aryl Halides and Related Compounds. In *Copper-Mediated Cross-Coupling Reactions*; Evano, G., Blanchard, N., Eds.; Wiley-VCH: Weinheim, Germany, 2013; pp. 313–334.
14. Sandmeyer, T. Ueber die Ersetzung der Amid-gruppe durch Chlor, Brom und Cyan in den aromatischen Substanzen. *Ber. Dtsch. Chem. Ges.* **1884**, *17*, 2650–2653. [CrossRef]
15. Krasnokutskaya, E.A.; Semenischeva, N.I.; Filimonov, V.D.; Knochel, P. A new, one-step, effective protocol for the iodination of aromatic and heterocyclic compounds via aprotic diazotization of amines. *Synthesis* **2007**, *2007*, 81–84. [CrossRef]
16. Beletskaya, I.P.; Sigeev, A.S.; Peregudov, A.S.; Petrovskii, P.V. Catalytic Sandmeyer bromination. *Synthesis* **2007**, *2007*, 2534–2538. [CrossRef]
17. Hubbard, A.; Okazaki, T.; Laali, K.K. Halo- and azidodediazoniation of arenediazonium tetrafluoroborates with trimethylsilyl halides and trimethylsilyl azide and Sandmeyer-type bromodediazoniation with Cu(I)Br in [BMIM][PF6] ionic liquid. *J. Org. Chem.* **2008**, *73*, 316–319. [CrossRef] [PubMed]
18. Filimonov, V.D.; Semenischeva, N.I.; Krasnokutskaya, E.A.; Tretyakov, A.N.; Ho, Y.H.; Chi, K.W. Sulfonic acid based cation-exchange resin: A novel proton source for one-pot diazotization-iodination of aromatic amines in water. *Synthesis* **2008**, *2008*, 185–187. [CrossRef]
19. Biffis, A.; Centomo, P.; Del Zotto, A.; Zecca, M. Pd Metal Catalysts for Cross-Couplings and Related Reactions in the 21st Century: A Critical Review. *Chem. Rev.* **2018**, *118*, 2249–2295. [CrossRef]
20. Ayogu, J.I.; Onoabedje, E.A. Recent advances in transition metal-catalysed cross-coupling of (hetero)aryl halides and analogues under ligand-free conditions. *Catal. Sci. Technol.* **2019**, *9*, 5233–5255. [CrossRef]
21. Nicolaou, K.C.; Bulger, P.G.; Sarlah, D. Palladium-catalyzed cross-coupling reactions in total synthesis. *Angew. Chem. Int. Ed.* **2005**, *44*, 4442–4489. [CrossRef]
22. Cabrita, M.T.; Vale, C.; Rauter, A.P. Halogenated compounds from marine algae. *Mar. Drugs* **2010**, *8*, 2301–2317. [CrossRef]
23. Wilcken, R.; Zimmermann, M.O.; Lange, A.; Joerger, A.C.; Boeckler, F.M. Principles and applications of halogen bonding in medicinal chemistry and chemical biology. *J. Med. Chem.* **2013**, *56*, 1363–1388. [CrossRef]
24. Mo, F.; Qiu, D.; Zhang, Y.; Wang, J. Renaissance of Sandmeyer-Type Reactions: Conversion of Aromatic C-N Bonds into C-X Bonds (X = B, Sn, P, or CF_3). *Acc. Chem. Res.* **2018**, *51*, 496–506. [CrossRef]
25. Mo, F.; Qiu, D.; Zhang, L.; Wang, J. Recent Development of Aryl Diazonium Chemistry for the Derivatization of Aromatic Compounds. *Chem. Rev.* **2021**, *121*, 5741–5829. [CrossRef] [PubMed]
26. Dai, J.J.; Fang, C.; Xiao, B.; Yi, J.; Xu, J.; Liu, Z.J.; Lu, X.; Liu, L.; Fu, Y. Copper-promoted sandmeyer trifluoromethylation reaction. *J. Am. Chem. Soc.* **2013**, *135*, 8436–8439. [CrossRef] [PubMed]
27. Danoun, G.; Bayarmagnai, B.; Grünberg, M.F.; Matheis, C.; Risto, E.; Gooßen, L.J. Sandmeyer trifluoromethylation. *Synthesis* **2014**, *46*, 2283–2286. [CrossRef]
28. Zhao, C.J.; Xue, D.; Jia, Z.H.; Wang, C.; Xiao, J. Methanol-promoted borylation of arylamines: A simple and green synthetic method to arylboronic acids and arylboronates. *Synlett* **2014**, *25*, 1577–1584. [CrossRef]
29. Qi, X.; Li, H.P.; Peng, J.B.; Wu, X.F. Borylation of aryldiazonium salts at room temperature in an aqueous solution under catalyst-free conditions. *Tetrahedron Lett.* **2017**, *58*, 3851–3853. [CrossRef]
30. Qiu, D.; Wang, S.; Tang, S.; Meng, H.; Jin, L.; Mo, F.; Zhang, Y.; Wang, J. Synthesis of trimethylstannyl arylboronate compounds by sandmeyer-type transformations and their applications in chemoselective cross-coupling reactions. *J. Org. Chem.* **2014**, *79*, 1979–1988. [CrossRef]
31. Zhong, T.; Pang, M.K.; Chen, Z.D.; Zhang, B.; Weng, J.; Lu, G. Copper-free Sandmeyer-type Reaction for the Synthesis of Sulfonyl Fluorides. *Org. Lett.* **2020**, *22*, 3072–3078. [CrossRef]
32. Li, Y.; Pu, J.; Jiang, X. A highly efficient Cu-catalyzed S-transfer reaction: From amine to sulfide. *Org. Lett.* **2014**, *16*, 2692–2695. [CrossRef]
33. Matheis, C.; Bayarmagnai, B.; Jouvin, K.; Goossen, L.J. Convenient synthesis of pentafluoroethyl thioethers: Via catalytic Sandmeyer reaction with a stable fluoroalkylthiolation reagent. *Org. Chem. Front.* **2016**, *3*, 949–952. [CrossRef]
34. Koziakov, D.; Majek, M.; Jacobi Von Wangelin, A. Metal-free radical thiolations mediated by very weak bases. *Org. Biomol. Chem.* **2016**, *14*, 11347–11352. [CrossRef]
35. Wang, S.; Qiu, D.; Mo, F.; Zhang, Y.; Wang, J. Metal-Free Aromatic Carbon-Phosphorus Bond Formation via a Sandmeyer-Type Reaction. *J. Org. Chem.* **2016**, *81*, 11603–11611. [CrossRef] [PubMed]
36. Estruch-Blasco, M.; Felipe-Blanco, D.; Bosque, I.; Gonzalez-Gomez, J.C. Radical arylation of triphenyl phosphite catalyzed by salicylic acid: Mechanistic investigations and synthetic applications. *J. Org. Chem.* **2020**, *85*, 14473–14485. [CrossRef] [PubMed]
37. Kutonova, K.V.; Trusova, M.E.; Postnikov, P.S.; Filimonov, V.D. The first example of the copper-free chloro- and hydrodediazoniation of aromatic amines using sodium nitrite, CCl_4, and $CHCl_3$. *Russ. Chem. Bull.* **2012**, *61*, 206–208. [CrossRef]
38. Leas, D.A.; Dong, Y.; Vennerstrom, J.L.; Stack, D.E. One-Pot, Metal-Free Conversion of Anilines to Aryl Bromides and Iodides. *Org. Lett.* **2017**, *19*, 2518–2521. [CrossRef] [PubMed]

39. Mukhopadhyay, S.; Batra, S. Direct Transformation of Arylamines to Aryl Halides via Sodium Nitrite and N-Halosuccinimide. *Chem. Eur. J.* **2018**, *24*, 14622–14626. [CrossRef]
40. Liu, Q.; Sun, B.; Liu, Z.; Kao, Y.; Dong, B.W.; Jiang, S.D.; Li, F.; Liu, G.; Yang, Y.; Mo, F. A general electrochemical strategy for the Sandmeyer reaction. *Chem. Sci.* **2018**, *9*, 8731–8737. [CrossRef]
41. Filimonov, V.D.; Trusova, M.; Postnikov, P.; Krasnokutskaya, E.A.; Lee, Y.M.; Hwang, H.Y.; Kim, H.; Chi, K.W. Unusually stable, versatile, and pure arenediazonium tosylates: Their preparation, structures, and synthetic applicability. *Org. Lett.* **2008**, *10*, 3961–3964. [CrossRef]
42. Chromá, R.; Vilková, M.; Shepa, I.; Makoś-Chełstowska, P.; Andruch, V. Investigation of tetrabutylammonium bromide-glycerol-based deep eutectic solvents and their mixtures with water by spectroscopic techniques. *J. Mol. Liq.* **2021**, *330*, 115617. [CrossRef]
43. Claudy, P.; Commerçon, J.C.; Lètoffé, J.M. Quasi-static study of the glass transition of glycerol by DSC. *Thermochim. Acta* **1988**, *128*, 251–260. [CrossRef]
44. Lane, L.B. Freezing Points of Glycerol and Its Aqueous Solutions. *Ind. Eng. Chem.* **1925**, *17*, 924. [CrossRef]
45. Abbott, A.P.; D'Agostino, C.; Davis, S.J.; Gladden, L.F.; Mantle, M.D. Do group 1 metal salts form deep eutectic solvents? *Phys. Chem. Chem. Phys.* **2016**, *18*, 25528–25537. [CrossRef] [PubMed]
46. Littke, A.F.; Fu, G.C. Palladium-Catalyzed Coupling Reactions of Aryl Chlorides. *Angew. Chem. Int. Ed.* **2002**, *41*, 4176–4211. [CrossRef]
47. Clayden, J.; Greeves, N.; Warren, S. *Organic Chemistry and Solution Manual*, 2nd ed.; Oxford University Press: Oxford, UK, 2013.
48. Smith, M.B.; March, J. *March's Advanced Organic Chemistry*, 6th ed.; Wiley: Hoboken, NJ, USA, 2007.
49. Carey, F.A.; Sundberg, R.J. *Advanced Organic Chemistry—Part A*; Springer: Berlin/Heidelberg, Germany, 2007.
50. Bergstrom, R.G.; Landells, R.G.M.; Wahl, G.H.; Zollinger, H. Dediazoniation of arenediazonium ions in homogeneous solution. 7. Intermediacy of the phenyl cation. *J. Am. Chem. Soc.* **1976**, *98*, 3301–3305. [CrossRef]
51. Himeshima, Y.; Kobayashi, H.; Sonoda, T. A first example of generating aryl cations in the solvolysis of aryl triflates in trifluoroethanol. *J. Am. Chem. Soc.* **1985**, *107*, 5286–5288. [CrossRef]
52. Martínez, A.G.; Cerero, S.d.; Barcina, J.O.; Jiménez, F.M.; Maroto, B.L. The mechanism of hydrolysis of aryldiazonium ions revisited: Marcus theory vs. canonical Variational transition state theory. *Eur. J. Org. Chem.* **2013**, *2013*, 6098–6107. [CrossRef]
53. Wu, Z.; Glaser, R. Ab Initio Study of the S_N1Ar and S_N2Ar Reactions of Benzenediazonium Ion with Water. On the Conception of "Unimolecular Dediazoniation" in Solvolysis Reactions. *J. Am. Chem. Soc.* **2004**, *126*, 10632–10639. [CrossRef]
54. Laidler, K.J.; King, M.C. The development of transition-state theory. *J. Phys. Chem.* **1983**, *87*, 2657–2664. [CrossRef]
55. Swain, C.G.; Sheats, J.E.; Gorenstein, D.G.; Harbison, K.G. Aromatic hydrogen isotope effects in reactions of benzenediazonium salts. *J. Am. Chem. Soc.* **1975**, *97*, 791–795. [CrossRef]
56. Swain, C.G.; Rogers, R.J. Mechanism of formation of aryl fluorides from arenediazonium fluoborates. *J. Am. Chem. Soc.* **1975**, *97*, 799–800. [CrossRef]
57. Roe, A. Preparation of Aromatic Fluorine Compounds from Diazonium Fluoborates. In *Organic Reactions*; John Wiley & Sons, Inc.: Hoboken, NJ, USA, 2011; pp. 193–228. [CrossRef]
58. Parr, R.G. Density Functional Theory of Atoms and Molecules. In *Horizons Quantum Chem*; Springer: Berlin/Heidelberg, Germany, 1980; pp. 5–15. [CrossRef]
59. Zhao, Y.; Truhlar, D.G. The M06 suite of density functionals for main group thermochemistry, thermochemical kinetics, noncovalent interactions, excited states, and transition elements: Two new functionals and systematic testing of four M06-class functionals and 12 other functionals. *Theor. Chem. Acc.* **2008**, *120*, 215–224.
60. Zhao, Y.; Truhlar, D.G. Density Functionals with Broad Applicability in Chemistry. *Acc. Chem. Res.* **2008**, *41*, 157–167. [CrossRef] [PubMed]
61. McLean, A.D.; Chandler, G.S. Contracted Gaussian basis sets for molecular calculations. I. Second row atoms, Z=11–18. *J. Chem. Phys.* **1980**, *72*, 5639–5648. [CrossRef]
62. Clark, T.; Chandrasekhar, J.; Spitznagel, G.W.; Schleyer, P.V.R. Efficient diffuse function-augmented basis sets for anion calculations. III. The 3-21+G basis set for first-row elements, Li–F. *J. Comput. Chem.* **1983**, *4*, 294–301. [CrossRef]
63. Frisch, M.J.; Pople, J.A.; Binkley, J.S. Self-consistent molecular orbital methods 25. Supplementary functions for Gaussian basis sets. *J. Chem. Phys.* **1984**, *80*, 3265–3269. [CrossRef]
64. Foresman, J.; Frisch, A. *Exploring Chemistry with Electronic Structure Methods*; Gaussian Inc.: Pittsburgh, PA, USA, 1996; Available online: http://gaussian.com/expchem3/ (accessed on 4 June 2021).
65. Ribeiro, R.F.; Marenich, A.V.; Cramer, C.J.; Truhlar, D.G. Use of Solution-Phase Vibrational Frequencies in Continuum Models for the Free Energy of Solvation. *J. Phys. Chem. B* **2011**, *115*, 14556–14562. [CrossRef] [PubMed]
66. Tomasi, J.; Mennucci, B.; Cammi, R. Quantum Mechanical Continuum Solvation Models. *Chem. Rev.* **2005**, *105*, 2999–3093. [CrossRef]
67. Frisch, D.J.; Trucks, M.J.; Schlegel, G.W.; Scuseria, H.B.; Robb, G.E.; Cheeseman, M.A.; Scalmani, R.J.; Barone, V.G.; Petersson, G.A.; Nakatsuji, H.; et al. *Gaussian 16, Revision A.03*; Gaussian, Inc.: Wallingford, CT, USA, 2016.
68. Schaftenaar, G.; Noordik, J.H. Molden: A pre-and post-processing program for molecular and electronic structures. *J. Comput. Aided. Mol. Des.* **2000**, *14*, 123–134. [CrossRef]
69. Zarei, M.; Noroozizadeh, E.; Moosavi-Zare, A.R.; Zolfigol, M.A. Synthesis of Nitroolefins and Nitroarenes under Mild Conditions. *J. Org. Chem.* **2018**, *83*, 3645–3650. [CrossRef]

70. Zhu, C.; Chen, F.; Liu, C.; Zeng, H.; Yang, Z.; Wu, W.; Jiang, H. Copper-Catalyzed Unstrained C–C Single Bond Cleavage of Acyclic Oxime Acetates Using Air: An Internal Oxidant-Triggered Strategy toward Nitriles and Ketones. *J Org. Chem.* **2018**, *83*, 14713–14722. [CrossRef]
71. Xin, H.-L.; Pang, B.; Choi, J.; Akkad, W.; Morimoto, H.; Ohshima, T. C–C Bond Cleavage of Unactivated 2-Acylimidazoles. *J. Org. Chem.* **2020**, *85*, 11592–11606. [CrossRef]
72. Li, T.; Cui, X.; Sun, L.; Li, C. Economical and efficient aqueous reductions of high melting-point imines and nitroarenes to amines: Promotion effects of granular PTFE. *RSC Adv.* **2014**, *4*, 33599–33606. [CrossRef]
73. Shao, H.; Foley, D.W.; Huang, S.; Abbas, A.Y.; Lam, F.; Gershkovich, P.; Bradshaw, T.D.; Pepper, C.; Fischer, P.M.; Wang, S. Structure-based design of highly selective 2,4,5-trisubstituted pyrimidine CDK9 inhibitors as anti-cancer agents. *Eur. J. Med. Chem.* **2021**, *214*, 113244. [CrossRef] [PubMed]
74. Lerch, U.; Moffatt, J.G. Carbodiimide-Sulfoxide Reactions. XIII. Reactions of Amines and Hydrazine Derivatives. *J. Org. Chem.* **1971**, *36*, 3861–3869.
75. Baudet, H.P. The replaceability of the halogen atom in 1-chloro- and 1-bromo-2-cyano-4-nitrobenzene. *Recl. Trav. Chim. Pays-Bas* **1924**, *43*, 707–726. [CrossRef]
76. Gianni, J.; Pirovano, V.; Abbiati, G. Silver triflate/p-TSA co-catalysed synthesis of 3-substituted isocoumarins from 2-alkynylbenzoates. *Org. Biomol. Chem.* **2018**, *16*, 3213–3219. [CrossRef] [PubMed]
77. Liu, J.; Li, J.; Ren, J.; Zeng, B.-B. Oxidation of aromatic amines into nitroarenes with m-CPBA. *Tetrahedron Lett.* **2014**, *55*, 1581–1584. [CrossRef]
78. Ramananda, D.; Uchil, J. ^{15}Cl NQR study of charge-transfer complexes. *J. Mol. Struct.* **1994**, *319*, 193–196. [CrossRef]
79. Wilshire, J.F.K. The preaparation of 2-Fluoro-5-nitrobenzonitrile and the proton magnetic resonance spectra of some compounds containing the N-(2-Cyano-4-nitrophenyl)group. *Aust. J. Chem.* **1967**, *20*, 1663–1670. [CrossRef]
80. Sloan, N.; Luthra, S.K.; McRobbie, G.; Pimlott, S.L.; Sutherland, A. A one-pot radioiodination of aryl amines via stable diazonium salts: Preparation of 125I-imaging agents. *Chem. Commun.* **2017**, *53*, 11008–11011. [CrossRef]
81. Gupta, S.; Ansari, A.; Sashidhara, K.V. Base promoted peroxide systems for the efficient synthesis of nitroarenes and benzamides. *Tetrahedron Lett.* **2019**, *60*, 151076. [CrossRef]
82. Fu, Z.; Li, Z.; Song, Y.; Yang, R.; Liu, Y.; Cai, H. Decarboxylative Halogenation and Cyanation of Electron-Deficient Aryl Carboxylic Acids via Cu Mediator as Well as Electron-Rich Ones through Pd Catalyst under Aerobic Conditions. *J. Org. Chem.* **2016**, *81*, 2794–2803. [CrossRef] [PubMed]

Article

Hydrothermal CO₂ Reduction by Glucose as Reducing Agent and Metals and Metal Oxides as Catalysts

Maira I. Chinchilla, Fidel A. Mato, Ángel Martín and María D. Bermejo *

High Pressure Process Group, Department of Chemical Engineering and Environmental Technology, BioEcoUva Research Institute on Bioeconomy, Universidad de Valladolid, 47011 Valladolid, Spain; mairaivette.chinchilla@alumnos.uva.es (M.I.C.); fidel@iq.uva.es (F.A.M.); mamaan@iq.uva.es (Á.M.)
* Correspondence: mdbermejo@iq.uva.es

Abstract: High-temperature water reactions to reduce carbon dioxide were carried out by using an organic reductant and a series of metals and metal oxides as catalysts, as well as activated carbon (C). As CO_2 source, sodium bicarbonate and ammonium carbamate were used. Glucose was the reductant. Cu, Ni, Pd/C 5%, Ru/C 5%, C, Fe_2O_3 and Fe_3O_4 were the catalysts tested. The products of CO_2 reduction were formic acid and other subproducts from sugar hydrolysis such as acetic acid and lactic acid. Reactions with sodium bicarbonate reached higher yields of formic acid in comparison to ammonium carbamate reactions. Higher yields of formic acid (53% and 52%) were obtained by using C and Fe_3O_4 as catalysts and sodium bicarbonate as carbon source. Reactions with ammonium carbamate achieved a yield of formic acid up to 25% by using Fe_3O_4 as catalyst. The origin of the carbon that forms formic acid was investigated by using $NaH^{13}CO_3$ as carbon source. Depending on the catalyst, the fraction of formic acid coming from the reduction of the isotope of sodium bicarbonate varied from 32 to 81%. This fraction decreased in the following order: Pd/C 5% > Ru/C 5% > Ni > Cu > C ≈ Fe_2O_3 > Fe_3O_4.

Keywords: hydrothermal reaction; CO_2 conversion; glucose; metal catalysts; metal oxide catalysts

1. Introduction

Global warming is still one of the main worldwide concerns in the present time [1,2]. Increasing the efficiency of processes, implementing renewable sources of energy and fuel switching are some of the alternatives for the reduction of greenhouse emissions [3]. However, in the transition to a decarbonated economy, oil and gas still play a relevant role in energy generation for electricity and transportation; in this context, technological initiatives as carbon capture and utilization become of high interest to mitigate these emissions by using CO_2 to synthesize high value-added products [4].

CO_2 is attractive as a raw material for industry because it is cheap, has very low toxicity, is available in great quantity [5] and can be used as feedstock for different processes. Physical methods are widely used to revalorize CO_2 as refrigerant, solvent, dry ice, etc. Chemical methods are also used to convert it into valuable compounds such as urea and DME [6,7] One of the main difficulties faced in the chemical conversion of CO_2 is the high thermodynamic stability of the compound [8]. Electrochemical and photochemical reduction for CO_2 hydrogenation have shown favorable results in overcoming this issue [9]. However, the high costs and low yields of these techniques [10] have led to the study of other alternatives such as the hydrothermal treatment in which CO_2 reduction takes places in water media at high pressures and temperatures [11–13]. In this process, water acts as hydrogen donor instead of H_2, which is flammable and complex to store [14].

In these processes, CO_2 is captured in the aqueous media in basic conditions. In most works, it is captured as $NaHCO_3$, but it can be also in the shape of carbamates that are formed when CO_2 is captured by ammonia or amines. This process opens the possibility to connect carbon capture in basic solutions directly with the CO_2 conversion process, avoiding costly intermediate separation steps [12,15].

In the hydrothermal reduction of CO_2, the most frequently obtained product is formic acid. This is a compound of great interest for the energy sector because it is an alternative source of hydrogen and can be used directly as an energy-dense carrier for fuel cells [16]; besides, it is biodegradable and less flammable than other fuels at room temperature [17].

In order to reduce CO_2, several metals have been suggested as CO_2 reductants: Zn [18], Fe [19], Mn [20], Mg [20] and Al [20] (efficiency: Al > Zn > Mn > Fe) [10,21] are mentioned as favorable for the process. Metallic catalysts metals such as Ni-ferrite [22], Ni nanoparticles [23], Ni [24], Raney Ni [25], Cu [20,21], Fe_2O_3 [26], Ru/C [27] and Pd/C [12,27] can be used as-is or coupled with "zero-valent" metals to improve the reaction; however, the reduction of the metal after the process in order to recycle the material should be considered.

Organic compounds containing alcohol groups, such as isopropanol [8], glycerol [28], glucose, C2 and C3 alcohols, saccharides and lignin derivatives [29], are often used as reductants as well. It is known that many of these molecules can be obtained from the hydrolysis of lignocellulosic biomass in hydrothermal media.

Subcritical water can act as a basic or acidic catalyst; it has a higher ion product and lower dielectric constant than room-temperature water. In these conditions, the cellulose, hemicellulose and lignin from biomass can be isolated and depolymerized into monomeric units (mainly sugars or phenols). Alongside the decarbonization approach, the usage of biomass is of great interest mainly because of the valorization of lignocellulosic residues that can be converted into several intermediate products such as lactic acid, acetic acid and vanillin [30,31]. Carrying out the hydrolysis of biomass simultaneously with the reduction of CO_2 captured as a basic solution (i.e., bicarbonate, amine carbamates of ammonia) could be interesting because many of the hydrolysis products of biomass contain alcohol groups that can act as reductants of CO_2 in hydrothermal media.

So far, there has been a number of studies in which the use of catalysts (Cu, Ni, Pd/C) in hydrothermal CO_2 reduction was carried out using metals as reductants (Zn, Mg, Al, etc.) [12,18,20–22,25]. This work studies, for the first time, the influence of different catalysts in the hydrothermal reduction of CO_2 by using an organic (glucose) as a reducing agent. As CO_2 source, sodium bicarbonate ($NaHCO_3$) and ammonium carbamate ($NH_4[H_2NCO_2]$) were used. There are literature studies stating that ammonium carbamate is reduced by using metals or hydrogen [12], but this is the first time that the reduction is performed using organics containing an alcohol group. The main objective of the present work is to develop batch screening reactions to find the best catalyst that can improve the formic acid production, as well as lowering the temperature for the reduction reactions normally fixed at 300 °C in other works [29]. In addition, as formic acid can be also derived by sugar hydrolysis [32–35], experiments to study the origin of the carbon forming formic acid by using $NaH^{13}CO_3$ as carbon source were performed.

2. Results

To perform the reduction of CO_2, several experiments were carried out by using glucose as organic reductant, sodium bicarbonate (SB) and ammonium carbamate (AC) as sources of carbon and several metals and metal oxides as catalysts (Cu, Ni, Pd/C 5%, Ru/C 5%, Fe_2O_3 and Fe_3O_4). Activated carbon (C) was used in some experiments in order to compare its performance with the palladium and ruthenium supported catalysts.

It should be noted that only the products of the liquid phase were analyzed. Gas products were not formed or were produced in so small amounts that they could not be collected. It is not excluded that gases such as CH_4 [23,24] could be produced in a very small amount in the case of Ni catalyst.

2.1. Particle Size of the Catalysts

As seen in Figure 1, SEM images of the unreacted catalysts were taken in order to measure the average particle size. The approximate diameters of the particles of each the catalysts before the hydrothermal reaction were as follows: Cu: 400 µm; Ni: 9 µm; Pd/C: 25 µm; C: 55 µm; Ru/C: 175 µm; Fe_2O_3; Fe_3O_4: 93 µm.

Figure 1. SEM images of the catalyst particles.

2.2. Results of Hydrothermal Reactions with Sodium Bicarbonate as Carbon Source

In the reaction of SB with glucose, typical products derived from the hydrothermal reduction of glucose were observed [29,33–37]. Formic acid (FA), acetic acid (AA) and lactic acid (LA) were the main compounds produced in the reactions. In minor amounts, glyceraldehyde, glycolaldehyde, formaldehyde, ethylene glycol, acetone, pyruvaldehyde, galacturonic acid and 5-HMF were obtained. The yields of the three main products of the catalyzed reactions are shown in Tables 1–3. Each experiment was repeated at least twice, the average error being around 5%.

After carrying out the reduction of CO_2 captured as SB, it was found that the highest yields of FA were obtained by using C and Fe_3O_4 as catalysts, reaching yields of 53% and 52%, respectively (Table 1). The conditions at which the maximum values were obtained were 200 °C and 30 min of reaction for C and 250 °C and 30 min of reaction for Fe_3O_4.

The highest yield for AA was obtained in the sample without catalyst: 45% at 250 °C and 30 min. This was followed by Ni and Cu catalysts, which achieved yields of 45% and 44%, at 250 °C and 30 min and 250 °C and 120 min (Table 2).

For LA, the maximum yield was achieved with Fe_3O_4, 43% at 250 °C and 30 min of reaction (Table 3).

It is remarkable that most of the catalysts promoted similar or less yield of FA in comparison to the sample without catalyst; in fact, only C and Fe_3O_4 improved the yield of FA over AA and LA over the sample with no catalyst.

Table 1. Yields of formic acid obtained after the hydrothermal reaction of NaHCO$_3$ with glucose in the presence of catalysts. The higher yields obtained after the hydrothermal reactions are marked with an asterisk (*).

Reaction Time (min)	Cu	Ni	Pd/C	C	Ru/C	Fe$_3$O$_4$	Fe$_2$O$_3$	No Catalyst
Reaction Temperature: 200 °C								
30	44%	44%	20%	*53%	31%	49%	45%	44%
60	43%	40%	16%	46%	30%	51%	49%	48%
90	43%	35%	18%	51%	25%	48%	45%	-
120	39%	34%	20%	51%	27%	49%	48%	51%
180	39%	32%	20%	37%	23%	46%	41%	40%
Reaction Temperature: 250 °C								
30	41%	41%	29%	50%	38%	*52%	40%	47%
60	36%	-	31%	47%	33%	44%	42%	41%
90	37%	35%	36%	47%	31%	44%	39%	38%
120	37%	40%	35%	46%	32%	49%	40%	40%
180	32%	34%	38%	46%	26%	45%	40%	39%

Table 2. Yields of acetic acid obtained after the hydrothermal reaction of NaHCO$_3$ with glucose in the presence of catalysts. The higher yields obtained after the hydrothermal reactions are marked with an asterisk (*).

Reaction Time (min)	Cu	Ni	Pd/C	C	Ru/C	Fe$_3$O$_4$	Fe$_2$O$_3$	No Catalyst
Reaction Temperature: 200 °C								
30	39%	40%	25%	33%	31%	31%	38%	39%
60	38%	41%	26%	33%	33%	33%	39%	40%
90	40%	39%	23%	34%	35%	35%	39%	40%
120	38%	37%	27%	33%	31%	31%	37%	37%
180	36%	36%	27%	28%	31%	31%	37%	37%
Reaction Temperature: 250 °C								
30	40%	*45%	26%	33%	34%	34%	41%	*45%
60	-	-	23%	30%	27%	27%	-	38%
90	43%	42%	24%	35%	30%	30%	39%	38%
120	*44%	43%	24%	35%	30%	30%	39%	-
180	37%	39%	24%	34%	30%	30%	39%	40%

Table 3. Yields of lactic acid obtained after the hydrothermal reaction of NaHCO$_3$ with glucose in the presence of catalysts. The higher yields obtained after the hydrothermal reactions are marked with an asterisk (*).

Reaction Time (min)	Reaction Temperature: 200 °C							No Catalyst
	Cu	Ni	Pd/C	C	Ru/C	Fe$_3$O$_4$	Fe$_2$O$_3$	
30	34%	34%	23%	28%	33%	35%	35%	35%
60	31%	31%	22%	26%	32%	34%	34%	34%
90	36%	31%	27%	29%	30%	37%	37%	36%
120	32%	32%	27%	-	29%	34%	34%	34%
180	31%	33%	28%	34%	28%	33%	31%	31%
Reaction Time (min)	Reaction Temperature: 250 °C							No Catalyst
	Cu	Ni	Pd/C	C	Ru/C	Fe$_3$O$_4$	Fe$_2$O$_3$	
30	35%	38%	36%	39%	39%	* 43%	37%	38%
60	33%	-	33%	40%	32%	40%	40%	35%
90	36%	38%	39%	39%	34%	35%	40%	35%
120	34%	39%	34%	38%	34%	38%	37%	38%
180	31%	37%	31%	36%	29%	36%	34%	37%

Results of Hydrothermal Reactions with Ammonium Carbamate as Carbon Source

As in the previous case, the main products of the reaction with AC were FA, AA and LA. The yields achieved for each compound are shown in Tables 4–6. The experiments were repeated at least twice, the average error being around 3%.

Table 4. Yields of formic acid obtained after the hydrothermal reaction of NH$_4$[H$_2$NCO$_2$] with glucose in the presence of catalysts. The higher yields obtained after the hydrothermal reactions are marked with an asterisk (*).

Reaction Time (min)	Reaction Temperature: 200 °C							No Catalyst
	Cu	Ni	Pd/C	C	Ru/C	Fe$_3$O$_4$	Fe$_2$O$_3$	
30	16%	14%	9%	21%	9%	21%	18%	17%
60	17%	19%	7%	-	-	20%	17%	17%
90	17%	14%	3%	21%	3%	24%	18%	17%
120	14%	7%	3%	16%	3%	* 26%	19%	19%
180	16%	15%	2%	20%	3%	25%	19%	19%
Reaction Time (min)	Reaction Temperature: 250 °C							No Catalyst
	Cu	Ni	Pd/C	C	Ru/C	Fe$_3$O$_4$	Fe$_2$O$_3$	
30	10%	8%	2%	21%	2%	25%	20%	17%
60	9%	3%	3%	15%	1%	24%	17%	16%
90	8%	4%	3%	9%	1%	24%	16%	16%
120	5%	3%	3%	16%	1%	20%	12%	14%
180	5%	1%	3%	15%	1%	9%	12%	14%

After the hydrothermal reduction of AC, it was found that the highest yield of FA was 26% and was obtained by using Fe$_3$O$_4$ at 200 °C and 120 min (Table 4).

The maximum value for AA was obtained with Ni, 15% at 250 °C and 180 min. This was followed by that obtained with Cu, 14% at 200 °C and 60 min (Table 5).

Table 5. Yields of acetic acid obtained after the hydrothermal reaction of $NH_4[H_2NCO_2]$ with glucose in the presence of catalysts. The higher yields obtained after the hydrothermal reactions are marked with an asterisk (*).

Reaction Time (min)	Reaction Temperature: 200 °C							
	Cu	Ni	Pd/C	C	Ru/C	Fe_3O_4	Fe_2O_3	No Catalyst
30	9%	9%	6%	7%	9%	11%	11%	11%
60	* 14%	12%	8%	8%	9%	11%	11%	11%
90	11%	12%	8%	11%	8%	11%	11%	12%
120	11%	10%	9%	9%	9%	12%	11%	12%
180	11%	12%	9%	11%	11%	12%	12%	13%
Reaction Time (min)	Reaction Temperature: 250 °C							
	Cu	Ni	Pd/C	C	Ru/C	Fe_3O_4	Fe_2O_3	No Catalyst
30	9%	12%	8%	8%	10%	11%	10%	10%
60	9%	11%	8%	9%	10%	10%	8%	10%
90	10%	13%	9%	6%	11%	11%	11%	11%
120	10%	13%	9%	10%	11%	10%	10%	11%
180	10%	* 15%	10%	11%	13%	11%	10%	11%

Table 6. Yields of lactic acid obtained after the hydrothermal reaction of $NH_4[H_2NCO_2]$ with glucose in the presence of catalysts. The higher yields obtained after the hydrothermal reactions are marked with an asterisk (*).

Reaction Time (min)	Reaction Temperature: 200 °C							
	Cu	Ni	Pd/C	C	Ru/C	Fe_3O_4	Fe_2O_3	No Catalyst
30	8%	9%	5%	* 13%	5%	10%	9%	9%
60	7%	9%	6%	5%	5%	10%	8%	9%
90	10%	8%	4%	11%	3%	* 16%	10%	11%
120	12%	9%	6%	9%	4%	6%	9%	8%
180	9%	9%	6%	7%	5%	5%	10%	10%
Reaction Time (min)	Reaction Temperature: 250 °C							
	Cu	Ni	Pd/C	C	Ru/C	Fe_3O_4	Fe_2O_3	No Catalyst
30	5%	7%	5%	11%	4%	13%	12%	9%
60	8%	4%	4%	11%	4%	6%	9%	10%
90	7%	4%	4%	5%	3%	4%	9%	8%
120	8%	6%	4%	9%	5%	10%	8%	9%
180	6%	5%	4%	10%	3%	10%	8%	10%

For LA, the highest value was 16% and was obtained by using Fe_3O_4 at 200 °C and 90 min. This was followed by that obtained with C, 13% at 200 °C and 30 min (Table 6).

Once again, Fe_3O_4 promoted the maximum yields of FA over AA and LA in comparison to the rest of the catalysts. Only Fe_3O_4 and C improved the yield of FA compared to the sample with no catalyst.

In general, the yields of FA, AA and LA obtained by the reduction of AC are much lower (less than 25%) than those observed with SB (less than 53%). Some other works have shown that sodium bicarbonates and carbonates required high-temperature reactions to achieve higher yields of FA. SB and AC are decomposed easily into HCO^{3-}, which is the species that is going to be reduced in the reaction. In the case of AC, not only HCO^{3-} is formed. There is another step in which AC is also decomposed because the H^+ protons of the ion NH_4^+ are being donated to other compounds, and then the yield to FA is reduced because there is a competition between two reactions: the reduction of AC and the thermal decomposition of AC [13,16]. It was observed that the experiments held at 200 °C showed higher yields of FA than the reactions at 250 °C. The reduction of CO_2 is favored by the reaction in alkaline media; when the temperature rises, NH_4^+ dissociates into NH_3 and H^+, which are species that reduce the alkalinity and might reduce the solubility of CO_2 in water [10,38].

2.3. Nuclear Magnetic Resonance Spectroscopy Results

It is known that FA can be generated from sugars at lower temperatures in basic aqueous media [32–34] and can be also obtained by the reduction of SB at temperatures higher than 300 °C [29]. In order to understand the reactions, it is necessary to check whether the FA is coming from SB or from glucose and if the catalysts are favoring or disfavoring one or the other reaction. To do so, experiments with an isotope of sodium bicarbonate ($NaH^{13}CO_3$; SB-^{13}C) were performed with the different catalysts. ^{13}C-NMR analyses were carried out to identify the fraction of formic acid that possesses ^{13}C, which comes from the reduction of the carbon source, and the fraction that comes from glucose. The experiments were conducted at 250 °C and 2 h.

The fraction of formic acid coming from the SB-^{13}C when using each of the catalysts is presented in Figure 2.

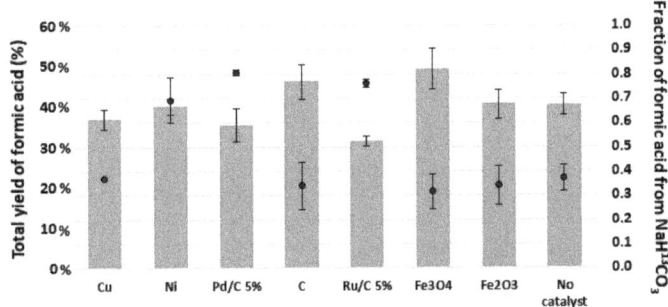

Figure 2. Fractions of formic acid coming from the reduction $NaH^{13}CO_3$ and from the oxidation of glucose for each of the catalysts at 250 °C and 2 h. Gray bars represent the total yield of formic acid of each sample (obtained by HPLC); black dots represent the fraction of formic acid coming from $NaH^{13}CO_3$ (obtained by ^{13}C-NMR). The average error in the measure of the fraction of formic acid was 5%.

It was observed that although Fe_3O_4 is the catalyst that provides the highest yield of total FA (49%, 250 °C and 2h, measured by HPLC), its proportion of reduced SB-^{13}C is lower (0.32) in comparison to the fraction of FA obtained with Pd/C 5%, Ru/C 5% and Ni (0.81, 0.76 and 0.69, respectively).

The metal supported catalysts (Pd/C 5% and Ru/C 5%) presented the highest selectivity in reducing CO_2 in comparison to the performance of the activated carbon support (C), which reached a fraction of 0.34.

There were catalysts that did not improve the reduction of SB-^{13}C; in fact, the reaction without catalyst (fraction FA-^{13}C: 0.37) showed a slightly higher capability to reduce CO_2 than Cu, Fe_3O_4 and Fe_2O_3 (0.37, 0.32 and 0.34, respectively).

The order in which catalysts were able to reduce CO_2 captured as SB-^{13}C was as follows: Pd/C 5% > Ru/C 5% > Ni > Cu > C ≈ Fe_2O_3 > Fe_3O_4.

In all the experiments, the only products that came from the direct reduction of carbon source (SB-^{13}C) were FA-^{13}C at δ = 163 ppm and an unidentified compound at δ = 173 ppm (this peak was absent in Fe_2O_3 and in the sample with no catalyst). At δ = 127 ppm, another peak was observed; according to the literature [39,40], this compound could be $^{13}CO_2$ dissolved in the sample.

An AA-^{13}C standard was injected. AA-^{13}C standard peak was observed at δ = 184 ppm, and then the possibility that the unidentified peak at δ = 173 ppm was AA-^{13}C was excluded.

Possible Mechanisms of Reaction

In the NMR spectra, it was confirmed that the reduction of the carbon source led mostly to the formation of FA, while byproducts and FA were obtained from the oxidation of glucose.

In literature were found some of the possible mechanisms of reaction of glucose at high water temperatures, subcritical and supercritical water [41–44]. Glucose can be transformed in two different ways: by following a retro-aldol condensation reaction to produce glycolaldehyde or through the isomerization of the glucose into fructose (favored by basic media) which can be dehydrated to form 5-HMF (favored by the acid media) or can produce glyceraldehyde by means of another retro-aldol condensation reaction. Finally, the glyceraldehyde can be isomerized into pyruvaldehyde, which could be a precursor of lactic acid.

Besides the retro-aldol reactions that can lead to the production of lactic acid and glycolaldehyde, 5-HMF can be transformed into formaldehyde and furfural in acid media [43], but in our case, reactions were performed in basic media, so this step may or may not be occurring.

Some other works [42] described that glucose can also dehydrate to form 1,6-anhydroglucose. This molecule can be a precursor of acids or can be transformed into D-fructose and follow a reverse aldol condensation reaction to form erythrose and glycolaldehyde that can produce acids as well. In Figure 3, the main mechanisms of oxidation of glucose are represented. According to Kabyemela et al. [42], some of the products derived from the oxidation of glucose that can be identified according to these mechanisms are fructose, erythrose, glyceraldehyde, glycolaldehyde, pyruvaldehyde, dihydroxyacetone, 1,6-anhydroglucose, 5-HMF, acetic acid and formic acid. Kabyemela et al. [42] also have identified some products of the decomposition of fructose such as pyruvaldehyde, erythrose, glyceraldehyde, dihydroxyacetone, acetic acid and formic acid. Erythrose and 1,6-anhydroglucose can also be the precursors of acetic and formic acids.

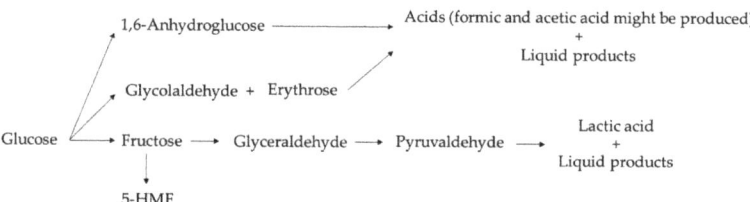

Figure 3. Mechanisms of oxidation of glucose.

Glucose has five -OH (hydroxyl) groups. In a previous work, it has been proposed that alcohol groups act as reducing agents for CO_2 [29]. According to Shen et al., the reduction of the carbon source is mainly due to the alcohol moiety [8,45]. According to other studies, compounds with primary alcohol groups presented slightly higher yields in comparison with compounds with secondary alcohol groups. Because of the steric effects, the position of the hydroxyl group in the compound could be of importance in the reduction of the carbon source [29]. Shen et al. proposed a mechanism of reduction of CO_2 through alcohol molecules in which, through a cyclic transition state, a H^- from the α-carbon of $-OH$ moiety is transferred to the ion bicarbonate and the resulting species dehydrate quickly into formate [8,29,45]. Most of the products of glucose decomposition contain alcohol groups (fructose, glyceraldehyde, glycolaldehyde, lactic acid), and Andérez et al. [29] proved that formic acid is rendered in appreciable yields.

Regarding catalysts, it can be found in the literature that in reactions that use metals as reductants, HCO_3^- is adsorbed on a Pd/C surface, promoting the formation of C_1 intermediates species to produce FA and traces of CH_4 and improving the generation of C-C bonds to form C_2 compounds [46]. Cu and Ni have showed also good performance in reducing HCO_3^- into C_1 compounds when using metals such as Fe as reducing agents [47,48]. In other studies, experiments with a mix of Fe and Fe_3O_4 were performed to reduce CO_2. In these works, Fe is reduced into Fe_3O_4, generating hydrogen. Fe_3O_4 is transformed into Fe_3O_{4-x}, and then hydrogen and C=O of HCO_3^- are adsorbed in the surface of the metal oxide and react to produce formic acid [49].

As seen before, catalysts can influence the performance of the hydrothermal reactions for CO_2 reduction and glucose oxidation.

3. Materials and Methods

3.1. Chemicals

Ammonium carbamate (AC) (99%), sodium bicarbonate (SB) (100%), sodium bicarbonate ^{13}C (SB-^{13}C) (100%) and acetic acid ^{13}C (AA-^{13}C) were used as sources of captured CO_2. D-(+)-Glucose (100%) was used as reducing agent. Fine powder of commercial Cu, Ni, Pd/C (5 wt% of metal loading), activated carbon, Ru/C (5 wt% of metal loading, 50% water wet paste), Fe_2O_3 and Fe_3O_4 were used as catalysts. Sodium bicarbonate was purchased from COFARCAS (Spain), Ru/c 5% was provided by Strem Chemicals and the rest of the chemicals were acquired from Sigma-Aldrich. Deionized water was used to prepare the dilutions.

3.2. Catalytic Experiments

Two solutions were prepared: the first one contained 0.05 M of glucose and 0.5 M of sodium bicarbonate, and the second one consisted of 0.05 M glucose and 0.5 M ammonium carbamate (ratio 1:10, glucose:carbon source in mol). Hydrothermal reactions for the conversion of CO_2 (captured as ammonium carbamate and sodium bicarbonate) were carried out in SS316 stainless steel horizontal tubular reactors of 10 mL (internal volume). The batch reactors were filled with liquid up to 45% of the total volume.

Different sets of reactions were performed at temperatures of 200 and 250 °C for 30, 60, 90, 120 and 180 min. The pressure inside the reactor corresponds to the saturation pressure of the set temperature. A fluidized bed heater was used to reach the temperature of the reactions. A temperature probe was installed inside the vessels. Reactors were placed in the heaters, achieving the working temperature within 7 min. Once the residence time was completed, each reactor was immediately extracted and placed in a bath of cold water. All samples were filtered with 0.22 µm Nylon filter and then collected in glass vials to be analyzed by HPLC or NMR spectroscopy.

The yields to formic acid, lactic acid and acetic acid were calculated as follows:

$$Y_{product} = \frac{C_{Product,f}}{C_{Glucose,i}} \quad (1)$$

where $C_{Product,f}$ is the molar concentration of formic acid at the end of the reaction and $C_{Glucose,i}$ is the initial molar concentration of glucose.

All the experiments were repeated at least 2 times, and the error was calculated with the standard deviations of the yields.

The amounts of catalyst used were as follows: For Cu and Ni, a molar relation of metal:carbon source of 5:1 was used. For Pd/C 5%, C and Ru/C 5%, 55 wt% catalyst with respect to the initial weight of carbon source was added. For Fe_3O_4 and Fe_2O_3, a molar relation of metal:carbon source of 1:1 was utilized.

3.3. Scanning Electronic Microscopy (SEM)

SEM images of the catalyst were taken in order to corroborate the particle size. An FEI QUANTA 200 FEG (ESEM) was used at high-vacuum operation ($<6 \times 10^{-4}$ Pa (4.5×10^{-2} Torr)), electron landing energy of 5 keV and spot of 3.0.

3.4. High-Pressure Liquid Chromatography (HPLC)

The liquid samples were analyzed by HPLC (Waters, Alliance separation module e2695) with RI detector (Waters, 2414 module) and a Rezex ROA-Organic Acid H+ (8%) column. As mobile phase, 25 mM H_2SO_4 was used at 0.5 mL/min of flow rate. The temperatures of the column and the detector were 40 and 30 °C.

3.5. Nuclear Magnetic Resonance Spectroscopy (NMR Spectroscopy)

Spectra of all the samples in which SB-^{13}C was used as carbon source were recorded on a 500 MHz Agilent instrument equipped with a OneNMR probe. The acquisition parameters for ^{13}C NMR spectra were as follows: 25 °C, 70 s relaxation delay between transients, 45° pulse width, spectral width of 31250 Hz, a total of 16 transients and 1.048 s acquisition time. The inverse gated decoupling technique to suppress the nuclear Overhauser effect (NOE) was used to obtain quantitative measurement. The acquisition parameters for ^{1}H NMR spectra were as follows: 25 °C, 70 s relaxation delay between transients, 90° pulse width, spectral width of 8012.8 Hz, a total of 4 transients and 2.044 s acquisition time. The sequence PRESAT was used in order to suppress the strong signal of water.

^{1}H and ^{13}C NMR chemical shifts (δ) were reported in parts per million (ppm) and referenced to tetramethylsilane (TMS).

4. Conclusions

In this work, the hydrothermal conversion of CO_2 captured as sodium bicarbonate and ammonium carbamate was studied. Glucose was used as a reducing agent, and metal and metal oxides (Cu, Ni, Pd/C 5%, Ru/C 5%, Fe_2O_3 and Fe_3O_4), as well as activated carbon (C), were used as catalysts. The main products of the reaction with ammonium carbamate were formic acid, acetic acid and lactic acid.

The yields of formic acid, acetic acid and lactic acid obtained by the reduction of ammonium carbamate were much lower (less than 25%) than those observed when sodium bicarbonate was used as the carbon source (less than 53%).

For ammonium carbamate experiments, C and Fe_3O_4 promoted higher yields of FA over AA and LA in comparison to the rest of the catalysts and improved the yield of FA in comparison to the sample without catalyst.

In the experiments with sodium bicarbonate, C and Fe_3O_4 appeared to be the most promising catalysts for improving the yield of formic acid. The origin of the carbon forming formic acid was investigated by using $NaH^{13}CO_3$. It was found that although C and Fe_3O_4 achieved the highest total formic acid yield, they seem to favor the oxidation of glucose instead of the reduction of CO_2. However, it should be noted that even though Pd/C 5%, Ni and Ru/C 5% yields of total formic acid were lower, they were shown to be more selective in producing formic acid from CO_2 than the other catalysts. This aspect is important when considering the selection of a catalyst for making a process that primarily promotes a higher conversion of the carbon source.

Author Contributions: Conceptualization, M.D.B., F.A.M., Á.M. and M.I.C.; methodology, M.D.B. and M.I.C.; formal analysis, M.I.C.; investigation, M.D.B. and M.I.C.; resources, M.D.B., F.A.M., Á.M. and M.I.C.; writing—original draft preparation, M.D.B. and M.I.C.; supervision, M.D.B., F.A.M. and Á.M.; funding acquisition, M.D.B. and Á.M. All authors have read and agreed to the published version of the manuscript.

Funding: This research was funded by Ministerio de Ciencia y Universidades by project RTI2018-097456-B-I00 and by the Regional Government of Castilla y León and the EU-FEDER program (CLU-2019-04).

Institutional Review Board Statement: Not applicable.

Informed Consent Statement: Not applicable.

Acknowledgments: M.I.C. acknowledges Universidad de Valladolid and Banco de Santander for the predoctoral grant. The authors acknowledge Laboratorio de Técnicas Instrumentales UVa for their assistance in the NMR and SEM analysis.

Conflicts of Interest: The authors declare no conflict of interest.

References

1. Intergovernmental Panel on Climate Change Global Warming of 1.5 °C. Available online: https://www.ipcc.ch/sr15/ (accessed on 21 December 2021).
2. United Nations el Acuerdo de París | CMNUCC. Available online: https://unfccc.int/es/process-and-meetings/the-paris-agreement/el-acuerdo-de-paris (accessed on 27 January 2022).
3. Khezri, M.; Heshmati, A.; Khodaei, M. Environmental Implications of Economic Complexity and Its Role in Determining How Renewable Energies Affect CO_2 Emissions. *Appl. Energy* **2022**, *306*, 117948. [CrossRef]
4. Tapia, J.F.D.; Lee, J.Y.; Ooi, R.E.H.; Foo, D.C.Y.; Tan, R.R. A Review of Optimization and Decision-Making Models for the Planning of CO_2 Capture, Utilization and Storage (CCUS) Systems. *Sustain. Prod. Consum.* **2018**, *13*, 1–15. [CrossRef]
5. Sakakura, T.; Choi, J.C.; Yasuda, H. Transformation of Carbon Dioxide. *Chem. Rev.* **2007**, *107*, 2365–2387. [CrossRef] [PubMed]
6. Dibenedetto, A.; Angelini, A.; Stufano, P. Use of Carbon Dioxide as Feedstock for Chemicals and Fuels: Homogeneous and Heterogeneous Catalysis. *J. Chem. Technol. Biotechnol.* **2014**, *89*, 334–353. [CrossRef]
7. Rafiee, A.; Rajab Khalilpour, K.; Milani, D.; Panahi, M. Trends in CO_2 Conversion and Utilization: A Review from Process Systems Perspective. *J. Environ. Chem. Eng.* **2018**, *6*, 5771–5794. [CrossRef]
8. Shen, Z.; Zhang, Y.; Jin, F. The Alcohol-Mediated Reduction of CO_2 and $NaHCO_3$ into Formate: A Hydrogen Transfer Reduction of $NaHCO_3$ with Glycerine under Alkaline Hydrothermal Conditions. *RSC Adv.* **2012**, *2*, 797–801. [CrossRef]
9. Mikkelsen, M.; Jørgensen, M.; Krebs, F.C. The Teraton Challenge. A Review of Fixation and Transformation of Carbon Dioxide. *Energy Environ. Sci.* **2010**, *3*, 43–81. [CrossRef]
10. He, C.; Tian, G.; Liu, Z.; Feng, S. A Mild Hydrothermal Route to Fix Carbon Dioxide to Simple Carboxylic Acids. *Org. Lett.* **2010**, *12*, 649–651. [CrossRef]
11. Foustoukos, D.I.; Seyfried, W.E. Hydrocarbons in Hydrothermal Vent Fluids: The Role of Chromium-Bearing Catalysts. *Science* **2004**, *304*, 1002–1005. [CrossRef]
12. del Río, J.I.; Pérez, E.; León, D.; Martín, Á.; Bermejo, M.D. Catalytic Hydrothermal Conversion of CO_2 Captured by Ammonia into Formate Using Aluminum-Sourced Hydrogen at Mild Reaction Conditions. *J. Ind. Eng. Chem.* **2021**, *97*, 539–548. [CrossRef]
13. Etiope, G.; Sherwood Lollar, B. Abiotic Methane on Earth. *Rev. Geophys.* **2013**, *51*, 276–299. [CrossRef]
14. Centi, G.; Quadrelli, E.A.; Perathoner, S. Catalysis for CO_2 Conversion: A Key Technology for Rapid Introduction of Renewable Energy in the Value Chain of Chemical Industries. *Energy Environ. Sci.* **2013**, *6*, 1711–1731. [CrossRef]
15. Ahn, C.K.; Lee, H.W.; Lee, M.W.; Chang, Y.S.; Han, K.; Rhee, C.H.; Kim, J.Y.; Chun, H.D.; Park, J.M. Determination of Ammonium Salt/Ion Speciation in the CO_2 Absorption Process Using Ammonia Solution: Modeling and Experimental Approaches. *Energy Procedia* **2011**, *4*, 541–547. [CrossRef]
16. Uhm, S.; Chung, S.T.; Lee, J. Characterization of Direct Formic Acid Fuel Cells by Impedance Studies: In Comparison of Direct Methanol Fuel Cells. *J. Power Sources* **2008**, *178*, 34–43. [CrossRef]
17. Rice, C.; Ha, S.; Masel, R.I.; Waszczuk, P.; Wieckowski, A.; Barnard, T. Direct Formic Acid Fuel Cells. *J. Power Sources* **2002**, *111*, 83–89. [CrossRef]
18. Roman-Gonzalez, D.; Moro, A.; Burgoa, F.; Pérez, E.; Nieto-Márquez, A.; Martín, Á.; Bermejo, M.D. 2Hydrothermal CO_2 Conversion Using Zinc as Reductant: Batch Reaction, Modeling and Parametric Analysisof the Process. *J. Supercrit. Fluids* **2018**, *140*, 320–328. [CrossRef]
19. Duo, J.; Jin, F.; Wang, Y.; Zhong, H.; Lyu, L.; Yao, G.; Huo, Z. $NaHCO_3$-Enhanced Hydrogen Production from Water with Fe and in Situ Highly Efficient and Autocatalytic $NaHCO_3$ Reduction into Formic Acid. *Chem. Commun.* **2016**, *52*, 3316–3319. [CrossRef]
20. Jin, F.; Gao, Y.; Jin, Y.; Zhang, Y.; Cao, J.; Wei, Z.; Smith, R.L. High-Yield Reduction of Carbon Dioxide into Formic Acid by Zero-Valent Metal/Metal Oxide Redox Cycles. *Energy Environ. Sci.* **2011**, *4*, 881–884. [CrossRef]
21. Lyu, L.; Jin, F.; Zhong, H.; Chen, H.; Yao, G. A Novel Approach to Reduction of CO_2 into Methanol by Water Splitting with Aluminum over a Copper Catalyst. *RSC Adv.* **2015**, *5*, 31450–31453. [CrossRef]
22. Takahashi, H.; Kori, T.; Onoki, T.; Tohji, K.; Yamasaki, N. Hydrothermal Processing of Metal Based Compounds and Carbon Dioxide for the Synthesis of Organic Compounds. *J. Mater. Sci.* **2008**, *43*, 2487–2491. [CrossRef]
23. Zhong, H.; Yao, G.; Cui, X.; Yan, P.; Wang, X.; Jin, F. Selective Conversion of Carbon Dioxide into Methane with a 98% Yield on an In Situ Formed Ni Nanoparticle Catalyst in Water. *Chem. Eng. J.* **2019**, *357*, 421–427. [CrossRef]
24. Chen, Y.; Jing, Z.; Miao, J.; Zhang, Y.; Fan, J. Reduction of CO_2 with Water Splitting Hydrogen under Subcritical and Supercritical Hydrothermal Conditions. *Int. J. Hydrogen Energy* **2016**, *41*, 9123–9127. [CrossRef]

25. Le, Y.; Yao, G.; Zhong, H.; Jin, B.; He, R.; Jin, F. Rapid Catalytic Reduction of $NaHCO_3$ into Formic Acid and Methane with Hydrazine over Raney Ni Catalyst. *Catal. Today* **2017**, *298*, 124–129. [CrossRef]
26. Chen, Q.; Qian, Y. Carbon Dioxide Thermal System: An Effective Method for the Reduction of Carbon Dioxide. *Chem. Commun.* **2001**, 1402–1403. [CrossRef]
27. Su, J.; Lu, M.; Lin, H. High Yield Production of Formate by Hydrogenating CO_2 Derived Ammonium Carbamate/Carbonate at Room Temperature. *Green Chem.* **2015**, *17*, 2769–2773. [CrossRef]
28. Farnetti, E.; Crotti, C. Selective Oxidation of Glycerol to Formic Acid Catalyzed by Iron Salts. *Catal. Commun.* **2016**, *84*, 1–4. [CrossRef]
29. Andérez-Fernández, M.; Pérez, E.; Martín, A.; Bermejo, M.D. Hydrothermal CO_2 Reduction Using Biomass Derivatives as Reductants. *J. Supercrit. Fluids* **2018**, *133*, 658–664. [CrossRef]
30. Kang, S.; Li, X.; Fan, J.; Chang, J. Hydrothermal Conversion of Lignin: A Review. *Renew. Sustain. Energy Rev.* **2013**, *27*, 546–558. [CrossRef]
31. Jin, F.; Zeng, X.; Jing, Z.; Enomoto, H. A Potentially Useful Technology by Mimicking Nature—Rapid Conversion of Biomass and CO_2 into Chemicals and Fuels under Hydrothermal Conditions. *Ind. Eng. Chem. Res.* **2012**, *51*, 9921–9937. [CrossRef]
32. Yun, J.; Yao, G.; Jin, F.; Zhong, H.; Kishita, A.; Tohji, K.; Enomoto, H.; Wang, L. Low-Temperature and Highly Efficient Conversion of Saccharides into Formic Acid under Hydrothermal Conditions. *AIChE J.* **2016**, *62*, 3657–3663. [CrossRef]
33. Sundqvist, B.; Karlsson, O.; Westermark, U. Determination of Formic-Acid and Acetic Acid Concentrations Formed during Hydrothermal Treatment of Birch Wood and Its Relation to Colour, Strength and Hardness. *Wood Sci. Technol.* **2006**, *40*, 549–561. [CrossRef]
34. Gao, P.; Li, G.; Yang, F.; Lv, X.N.; Fan, H.; Meng, L.; Yu, X.Q. Preparation of Lactic Acid, Formic Acid and Acetic Acid from Cotton Cellulose by the Alkaline Pre-Treatment and Hydrothermal Degradation. *Ind. Crops Prod.* **2013**, *48*, 61–67. [CrossRef]
35. Ding, K.; Le, Y.; Yao, G.; Ma, Z.; Jin, B.; Wang, J.; Jin, F. A Rapid and Efficient Hydrothermal Conversion of Coconut Husk into Formic Acid and Acetic Acid. *Process Biochem.* **2018**, *68*, 131–135. [CrossRef]
36. Cantero, D.A.; Vaquerizo, L.; Martinez, C.; Bermejo, M.D.; Cocero, M.J. Selective Transformation of Fructose and High Fructose Content Biomass into Lactic Acid in Supercritical Water. *Catal. Today* **2015**, *255*, 80–86. [CrossRef]
37. Cantero, D.A.; Bermejo, M.D.; Cocero, M.J. Kinetic Analysis of Cellulose Depolymerization Reactions in near Critical Water. *J. Supercrit. Fluids* **2013**, *75*, 48–57. [CrossRef]
38. Takahashi, H.; Liu, L.H.; Yashiro, Y.; Ioku, K.; Bignall, G.; Yamasaki, N.; Kori, T. CO_2 Reduction Using Hydrothermal Method for the Selective Formation of Organic Compounds. *J. Mater. Sci.* **2006**, *41*, 1585–1589. [CrossRef]
39. Ethier, A.L.; Switzer, J.R.; Rumple, A.C.; Medina-Ramos, W.; Li, Z.; Fisk, J.; Holden, B.; Gelbaum, L.; Pollet, P.; Eckert, C.A.; et al. The Effects of Solvent and Added Bases on the Protection of Benzylamines with Carbon Dioxide. *Processes* **2015**, *3*, 497–513. [CrossRef]
40. Merritt, M.E.; Harrison, C.; Storey, C.; Jeffrey, F.M.; Sherry, A.D.; Malloy, C.R. Hyperpolarized 13C Allows a Direct Measure of Flux through a Single Enzyme-Catalyzed Step by NMR. *Proc. Natl. Acad. Sci. USA* **2007**, *104*, 19773–19777. [CrossRef]
41. Aida, T.M.; Sato, Y.; Watanabe, M.; Tajima, K.; Nonaka, T.; Hattori, H.; Arai, K. Dehydration of D-Glucose in High Temperature Water at Pressures up to 80 MPa. *J. Supercrit. Fluids* **2007**, *40*, 381–388. [CrossRef]
42. Kabyemela, B.M.; Adschiri, T.; Malaluan, R.M.; Arai, K. Glucose and Fructose Decomposition in Subcritical and Supercritical Water: Detailed Reaction Pathway, Mechanisms, and Kinetics. *Ind. Eng. Chem. Res.* **1999**, *38*, 2888–2895. [CrossRef]
43. Sasaki, M.; Goto, K.; Tajima, K.; Adschiri, T.; Arai, K. Rapid and Selective Retro-Aldol Condensation of Glucose to Glycolaldehyde in Supercritical Water. *Green Chem.* **2002**, *4*, 285–287. [CrossRef]
44. Cantero, D.A.; Álvarez, A.; Bermejo, M.D.; Cocero, M.J. Transformation of Glucose into Added Value Compounds in a Hydrothermal Reaction Media. *J. Supercrit. Fluids* **2015**, *98*, 204–210. [CrossRef]
45. Shen, Z.; Zhang, Y.; Jin, F. From $NaHCO3$ into Formate and from Isopropanol into Acetone: Hydrogen-Transfer Reduction of $NaHCO3$ with Isopropanol in High-Temperature Water. *Green Chem.* **2011**, *13*, 820–823. [CrossRef]
46. Zhong, H.; Yao, H.; Duo, J.; Yao, G.; Jin, F. Pd/C-Catalyzed Reduction of $NaHCO_3$ into CH_3COOH with Water as a Hydrogen Source. *Catal. Today* **2016**, *274*, 28–34. [CrossRef]
47. Kudo, K.; Komatsu, K. Selective Formation of Methane in Reduction of CO_2 with Water by Raney Alloy Catalyst. *J. Mol. Catal. A Chem.* **1999**, *145*, 257–264. [CrossRef]
48. Zhong, H.; Gao, Y.; Yao, G.; Zeng, X.; Li, Q.; Huo, Z.; Jin, F. Highly Efficient Water Splitting and Carbon Dioxide Reduction into Formic Acid with Iron and Copper Powder. *Chem. Eng. J.* **2015**, *280*, 215–221. [CrossRef]
49. Liu, X.; Zhong, H.; Wang, C.; He, D.; Jin, F. CO_2 Reduction into Formic Acid under Hydrothermal Conditions: A Mini Review. *Energy Sci. Eng.* **2022**, 1–13. [CrossRef]

Article

Deep Eutectic Solvents as Phase Change Materials in Solar Thermal Power Plants: Energy and Exergy Analyses

Hamed Peyrovedin [1], Reza Haghbakhsh [2,3], Ana Rita C. Duarte [3] and Alireza Shariati [1,*]

1. School of Chemical and Petroleum Engineering, Shiraz University, Shiraz 71345-51154, Iran; hamed.p.2012@gmail.com
2. Department of Chemical Engineering, Faculty of Engineering, University of Isfahan, Isfahan 81746-73441, Iran; r.haghbakhsh@eng.ui.ac.ir
3. LAQV, REQUIMTE, Departamento de Química da Faculdade de Ciências e Tecnologia, Universidade Nova de Lisboa, 2829-516 Caparica, Portugal; ard08968@fct.unl.pt
* Correspondence: shariati@shirazu.ac.ir; Tel.: +98-713-613-3704

Abstract: Nowadays, producing energy from solar thermal power plants based on organic Rankine cycles coupled with phase change material has attracted the attention of researchers. Obviously, in such solar plants, the physical properties of the utilized phase change material (PCM) play important roles in the amounts of generated power and the efficiencies of the plant. Therefore, to choose the best PCM, various factors must be taken into account. In addition, considering the physical properties of the candidate PCM, the issue of environmental sustainability should also be considered when making the selection. Deep eutectic solvents (DESs) are novel green solvents, which, in addition to having various favorable characteristics, are environmentally sustainable. Accordingly, in this work, the feasibility of using seven different deep eutectic solvents as the PCMs of solar thermal power plants with organic Rankine cycles was investigated. By applying exergy and energy analyses, the performances of each were compared to paraffin, which is a conventional PCM. According to the achieved results, most of the investigated "DES cycles" produce more power than the conventional cycle using paraffin as its PCM. Furthermore, lower amounts of the PCM are required when paraffin is replaced by a DES at the same operational conditions.

Keywords: DES; green solvent; solar energy; Rankine cycle; PCM; exergy analysis; energy analysis

Citation: Peyrovedin, H.; Haghbakhsh, R.; Duarte, A.R.C.; Shariati, A. Deep Eutectic Solvents as Phase Change Materials in Solar Thermal Power Plants: Energy and Exergy Analyses. *Molecules* **2022**, *27*, 1427. https://doi.org/10.3390/molecules27041427

Academic Editor: Joaquín García Álvarez

Received: 28 January 2022
Accepted: 15 February 2022
Published: 20 February 2022

Publisher's Note: MDPI stays neutral with regard to jurisdictional claims in published maps and institutional affiliations.

Copyright: © 2022 by the authors. Licensee MDPI, Basel, Switzerland. This article is an open access article distributed under the terms and conditions of the Creative Commons Attribution (CC BY) license (https://creativecommons.org/licenses/by/4.0/).

1. Introduction

Power generation using fossil fuels is the most commonly used method throughout the world. One of the most significant disadvantages of using fossil fuels is the release of greenhouse gases, such as carbon dioxide, into the atmosphere [1–3]. Accordingly, various sustainable methods, such as the use of low-grade heat [4,5], geothermal energy [6,7], wind energy [8,9], and solar energy [10–14], have been applied to produce clean energy with little environmental pollution. Among these novel methods, harnessing solar energy via solar thermal power plants coupled with the Rankine cycle has gained attention [10–14]. In such plants, the collected solar energy is transformed to heat, being used by the Rankine cycle to generate power by use of a turbine [10–14]. However, the greatest disadvantage of solar energy plants is the limited availability of solar radiation on cloudy days and, also, at night. In order to overcome this issue, the utilization of thermal energy storage (TES) systems incorporating phase change materials (PCMs) was introduced to achieve uniform power generation [15,16]. Actually, PCMs consist of various groups of materials with high heat capacities, capable of storing and releasing energy using their latent and sensible heats [17]. In recent years, the potentials of different materials, such as organic and inorganic materials, were studied as PCMs in a variety of processes, including heating and cooling processes, solar energy storage, and the food industries [17–20]. In solar thermal power generation plants, different types of materials were considered as PCMs, including both organic and inorganic material and conventional eutectic mixtures [21]. However, all of the aforementioned materials have certain shortcomings. For instance, organic PCMs, such as paraffin, are flammable and their volume changes are relatively large. Inorganic PCMs, such as metallic PCMs, are mostly corrosive and have issues of high-volume change upon temperature changes. Regarding conventional eutectic mixtures, they have very high melting-point temperatures, and so, are limited to only certain high-temperature applications [21,22]. Moreover, most of the thermodynamic properties of eutectic PCMs are unknown [21].

According to the required properties for each process, various materials are available to consider as PCMs, however, nowadays, it is more vital than ever to consider only those that are environmentally friendly. One such category of sustainable material, having only recently been introduced to the research community by Abbott et al. in 2003 [23], is the Deep Eutectic Solvent (DES). These sustainable solvents also have the potential to be applied as PCMs [21]. A DES is actually a mixture of two or more components, including one hydrogen bond acceptor (HBA) and one or more hydrogen bond donors (HBD). DESs have many advantages such as low vapor pressure, biodegradability, sustainability, non-flammability, ease of preparation, and low cost. Furthermore, they are mostly nontoxic [4,5,24,25]. In addition, the most unique characteristic of DESs is the ability to tune their physical properties. Since combinations of numerous HBA and HBD components are possible, as well as various ratios of the two, countless types of DESs with different physical properties can be envisioned. Therefore, by setting the required physical properties for each specific application, the most favorable DES can be specifically engineered for the purpose. Due to the multitude of advantages, the applications of DESs in various processes are being investigated, including, for example, extraction [25,26], electrochemistry [25,27], absorption [4,5], and chemical reactions [25,28]. However, studies investigating the feasibility of using DESs as PCMs in solar thermal power plants are quite rare [26].

The only published study in open literature considering DESs as PCMs is that of Shahbaz et al., which considered the application of a calcium chloride hexahydrate-based DES as a PCM for thermal-comfort building applications. They prepared five DESs using choline chloride and $CaCl_2.6H_2O$ with different HBA to HBD molar ratios and reported their thermal properties. According to their thermal cycling tests, they claimed that the two DESs of choline chloride: $CaCl_2.6H_2O$ with the molar ratios of 1:6 and 1:8, can potentially be used for the thermal comfort processes in buildings. However, they did not consider energy and exergy analyses for their suggested process [26].

Based on the very favorable characteristics of DESs, and the benefits of replacing conventional PCMs with environmentally sustainable material in solar thermal power generation plants, the feasibility of using various DESs as PCMs in solar thermal power generation cycles was investigated in this study. For this purpose, a conventional solar thermal power generation cycle was modified, and then, by employing energy and exergy analyses, the performances of all the cycles considering seven different DESs as PCMs were studied.

2. Method

2.1. The Modified Solar Thermal Power Generation Cycle

The schematic diagram of the modified cycle under consideration is presented in Figure 1. According to this cycle, for 12 hours during the day, the heating fluid (liquid water) enters the water tank as Stream 8, which receives solar energy that is collected by collectors as heat $\dot{Q}_s$ and leaves the water tank as Stream 9. The heated liquid water in stream 9 is separated into the two streams of 10 and 6. Stream 10 enters the PCM tank, which contains a DES as the PCM for absorbing heat from entering the heated water (Stream 10) during the day. The cooled liquid water then leaves the PCM tank as Stream 7. In this mode, the PCM tank is in the "charging" state to increase its energy. The other heated water stream (Stream 6) enters the evaporator and provides the required heat for the working fluid (R134a) of the Rankine cycle during the day and leaves the evaporator with lower energy as Stream 5. This leaving stream is finally combined with Stream 7 and the resulting stream is recycled to the water tank for continuing the cycle. On the other hand, in the evaporator of the Rankine cycle, the working fluid (R134a) in Stream 4 absorbs heat from the heated water and is evaporated. Evaporated R134a, with high pressure and temperature, enters the turbine as Stream 1 and produces power, $\dot{W}_s$. Following power production, the low-temperature–low-pressure vapor of R134a enters the condenser as Stream 2 and desorbs heat, $\dot{Q}_c$, to become liquified and leave the condenser as Stream 3. The pressure of liquified R134a is increased using Pump 1 and the pressurized R134a is recycled to the evaporator as Stream 4 for receiving heat once more from the heated water and continuing the Rankine cycle. However, during the night (for 12 h), the required heat for evaporating R134a in the evaporator is provided by the PCM tank which is now in the energy discharging mode. Accordingly, during the night, the liquified R134a (Stream 4) enters the PCM tank as Stream 4′ instead of entering the evaporator as Stream 4. In the PCM tank, the liquified R134a absorbs heat, $Q_{PCM,night}$, and upon evaporation, it enters the turbine as Stream 1′. Accordingly, during the night, Streams 5–10 which are responsible for transferring solar energy to R134a in the Rankine cycle by the water tank are shut off, and so, the required energy of the Rankine cycle is provided only by the charged PCM tank. By this design, the power production process continuously operates, both day and night, at a constant rate.

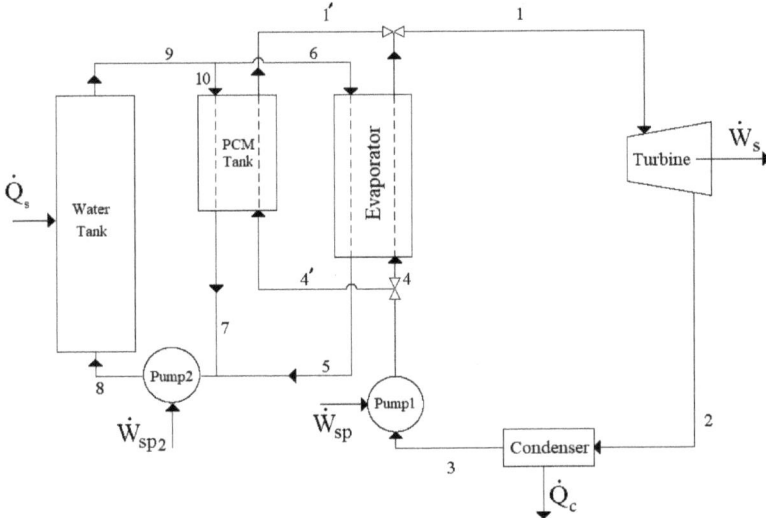

Figure 1. Schematic diagram of the modified solar thermal power generation cycle.

2.2. Energy Analysis

The energy analysis of the modified solar thermal power generation cycle is applied by considering the energy balance for all of the units of the investigated cycle. For the energy analysis, a number of usual assumptions considered in literature studies [4,5,12,29,30], are also considered here.

- The pressure drops (i.e., the required shaft work for Pump 2) in the pipes, PCM tank, condenser, and evaporator are neglected. Additionally, heat losses of the pipelines are neglected;
- Stream 3 is considered as saturated liquid R134a at the condenser pressure;
- The turbine's isentropic efficiency is considered to be equal to 0.75;
- The PCM tank is well insulated;
- The required work of Pump 1 is negligible in comparison to the produced work of the turbine;
- The outlet water from the water tank (Streams 6, 9, and 10) is saturated liquid;

- The mass flow rate of the outlet water from the water tank (Stream 9) is split equally into Streams 6 and 10. Then, mass flow rates of the water entering the PCM tank and evaporator are the same during the day;
- The general cycle properties, excluding the received solar energy, remains constant during day and night;
- Day and night hours are considered equal, as 12 h.

According to these common assumptions, by applying the first law of thermodynamics to all of the equipment of the investigated cycle, energy analysis is considered.

Equations (1)–(3) show the applied energy balances for the turbine, condenser, and evaporator, respectively.

$$\left|\dot{W}_s\right| = \dot{m}_r(h_1 - h_2) = \dot{m}_r(h_{1'} - h_2) \quad (1)$$

where h_1 and h_2 are the specific enthalpies of the inlet and outlet streams of the turbine, respectively. Additionally, h_1 and $h_{1'}$ are considered for the day and night, respectively. $\dot{m}_r$ and $\dot{W}_s$ are the mass flow rate of the working fluid and the produced power of the Rankine cycle, respectively.

For the condenser,

$$\left|\dot{Q}_c\right| = \dot{m}_r(h_2 - h_3) \quad (2)$$

where h_2 and h_3 are the specific enthalpies of the inlet and outlet streams of the condenser, respectively. $\dot{Q}_c$ is the desorbed heat from the working fluid of the Rankine cycle.

For the evaporator, which is used only during the day,

$$\dot{m}_w(h_6 - h_5) = \dot{m}_r(h_1 - h_4) \quad (3)$$

where $\dot{m}_w$ represents the mass flow rate of the heating fluid (water), and h_6 and h_5 are the specific enthalpies of the inlet and outlet heating working fluid streams (water) of the evaporator, respectively. h_4 and h_1 are the specific enthalpies of the inlet and outlet Rankine cycle working fluid streams (R134a) of the evaporator, respectively.

For the PCM tank, the energy balance is investigated separately for day and night.

A. *During the Day* The PCM tank is charged during the day by absorbing heat from the heating fluid (water). Accordingly, the energy balance of the PCM tank during the day follows Equation (4).

$$Q_{PCM,day} = m_{PCM}\Delta h_{fus,PCM} = t_{charging}\dot{m}_w(h_{10} - h_7) \quad (4)$$

where m_{PCM} is the mass of the PCM, $\Delta h_{fus,PCM}$ is the PCM enthalpy of fusion, and $t_{charging}$ is the charging time in the day, equal to 12 h. $Q_{PCM,day}$ is the heat absorbed by the PCM from the heating fluid (water) during the day. Streams 4' and 1' are shut down during the day and the only inlet and outlet streams of the PCM tank are Streams 10 and 7, whose specific enthalpies are shown as h_{10} and h_7.

B. *During the Night* The PCM tank is discharged during the night by desorbing heat to the Rankine cycle working fluid (R134a). Therefore, the energy balance of the PCM tank during the night follows Equation (5).

$$\left|Q_{PCM,night}\right| = m_{PCM}\Delta h_{fus,PCM} = t_{discharging}\dot{m}_r\left|h_{1'} - h_{4'}\right| \quad (5)$$

where $t_{discharging}$ is the discharging time during the night, equal to 12 h. $Q_{PCM,night}$ is the desorbed heat by the PCM to the Rankine cycle working fluid (R134a) during the night. Streams 10 and 7 are shut down at night, therefore, the only inlet and outlet streams of the PCM tank are Streams 4' and 1', with specific enthalpies of $h_{4'}$ and $h_{1'}$, respectively.

For the water tank, which is used only during the day, the energy balance is,

$$m_w C_{p_w} \frac{dT_w}{dt} = \dot{Q}_s + \dot{m}_{w_9}(h_8 - h_9) \tag{6}$$

where h_8 and h_9 are the specific enthalpies of the inlet and outlet streams of the water tank, respectively. $\dot{Q}_s$ is the collected solar energy. $\dot{m}_{w_9}$ is the mass flow rate of Stream 9 and based on the proposed assumptions, it is twice the mass flow rate of Streams 10 (or 6). Therefore,

$$\dot{m}_{w_9} = 2\dot{m}_w \tag{7}$$

In Equations (6) and (7), $\dot{m}_w$ and T_w are the total mass and the temperature of water in the water tank, respectively, and C_{p_w} is the heat capacity of water. In this study, it is assumed that the collected solar energy is controlled carefully using controlling collectors, therefore, the water tank during the day is at a thermal steady state. In this way, the unsteady state term of Equation (6) can be neglected. This assumption is, in fact, easily obtainable because during the day, the amount of collected solar energy which is transferred to the water tank is controlled in a way to keep the water at its boiling point, and since a pure component boils at a constant temperature, the temperature of water in the water tank remains constant. In this way, there is no temperature change in the water tank. So, during the day, Equation (6) can be simplified as follows.

$$\dot{Q}_s = \dot{m}_{w_9}(h_9 - h_8) \tag{8}$$

2.3. Exergy Analysis

Exergy analysis is a way to define how far a system operates from ideal conditions. Exergy indicates the maximum amount of work that a system can generate under the second law of thermodynamics. Consequently, since all real systems are far from their ideal state, they cannot produce the maximum theoretical amount of work, and some of the theoretical maximum is wasted as exergy destruction [31].

For a steady-state process, the destruction of exergy for equipment i ($\dot{E}_{d,i}$) is generally determined based on Equations (9) and (10) [32,33].

$$\dot{E}_{d,i} = \sum_j (\dot{m}_j e_j)_{in} - \sum_k (\dot{m}_k e_k)_{out} + \sum \dot{W}_{in} - \sum \dot{W}_{out} + \sum [\dot{Q}(1 - \frac{T_0}{T})]_{in} - \sum [\dot{Q}(1 - \frac{T_0}{T})]_{out} \tag{9}$$

$$e_i = (h_i - h_0) - T_0(s_i - s_0) \tag{10}$$

In Equation (9), the first and second terms of the right-hand side show the input and output exergies by the streams for equipment i. The third and fourth terms show the exergy changes of equipment i owing to the work transferred, and, finally, the last two terms of the right-hand side of Equation (9), represent the exergy changes due to the heat transferred [32,33]. In this equation, T_0 is the surrounding temperature, considered as 273.15 K, which is also the selected reference temperature. T is the temperature of the equipment. In Equation (10), h_0 and s_0 are the enthalpy and entropy of the environment, considered at the reference conditions (i.e., at the reference temperature of T_0 and reference pressure of P_0), and h_i and s_i are the enthalpy and entropy, respectively, of stream i at temperature T and pressure P.

In addition to the exergy destruction of equipment i, in the cycle, the contribution of exergy destruction, $E_{cont,i}$, in the total exergy loss of the cycle, $\dot{E}_{d,tot}$ can be determined based on Equation (11).

$$E_{cont,i} = \frac{\dot{E}_{d,i}}{\dot{E}_{d,tot}} = \frac{\dot{E}_{d,i}}{\sum \dot{E}_{d,i}} \tag{11}$$

In this manner, for each equipment of the investigated cycle, the exergy analysis is applied according to Equations (9) and (10) [31–33].

For the turbine, because it was considered to follow an isentropic process, there is no heat transfer. Then, Equations (9) and (10) are simplified to Equation (12) for the exergy destruction by the turbine, $\dot{E}_{d,turb}$,

$$\dot{E}_{d,turb} = \dot{m}_r(h_1 - h_2) - \dot{m}_r T_0(s_1 - s_2) - \dot{W}_s \tag{12}$$

For the condenser, the exergy destruction, $\dot{E}_{d,C}$, is derived by,

$$\dot{E}_{d,C} = \dot{m}_r(h_2 - h_3) - \dot{m}_r T_0(s_2 - s_3) - \dot{Q}_C\left(1 - \frac{T_0}{T_{LS}}\right) \tag{13}$$

where T_{LS} is the heat sink temperature which absorbs $\dot{Q}_C$, and is considered as 298.15 K.

For the evaporator, the exergy destruction, $\dot{E}_{d,e}$ is calculated by Equation (14) (during the day).

$$\dot{E}_{d,e} = \dot{m}_r[(h_4 - h_1) - T_0(s_4 - s_1)] + \dot{m}_w[(h_6 - h_5) - T_0(s_6 - s_5)] \tag{14}$$

For the PCM tank, it is important to consider the assumption of insulation of the tank. Therefore, the exergy destructions of the PCM tank are determined based on Equations (15) and (16) for day and night, respectively.

$$E_{d,PCM,day} = \dot{m}_w((h_{10} - h_7) - T_0(s_{10} - s_7)) \tag{15}$$

$$E_{d,PCM,night} = \dot{m}_r((h_{4'} - h_{1'}) - T_0(s_{4'} - s_{1'})) \tag{16}$$

In Equations (15) and (16), $E_{d,PCM,day}$ and $E_{d,PCM,night}$ are the exergy destructions of the PCM tank during the day and night, respectively. As a result, for a 24-h period, the exergy destruction of the PCM, $E_{d,PCM}$, can be calculated based on Equation (17) [34,35].

$$E_{d,PCM} = E_{d,PCM,day} + E_{d,PCM,night} \tag{17}$$

Finally, according to Equations (9) and (10), the exergy destruction of the water tank, $\dot{E}_{d,wt}$, is calculated based on Equation (18).

$$\dot{E}_{d,wt} = \dot{m}_{w_9}(h_8 - h_9) - \dot{m}_{w_9} T_0(s_8 - s_9) + \dot{Q}_S\left(1 - \frac{T_0}{T_{HW}}\right) \tag{18}$$

where T_{HW} is the heat source temperature and equal to $0.75 T_{sun}$ [29]. Moreover, for calculating the enthalpy and entropy changes of liquid water at constant pressure in the water tank, Equations (19) and (20) are used.

$$\Delta h = \int_{T_8}^{T_9} C_{pw} dT \tag{19}$$

$$\Delta s = \int_{T_8}^{T_9} \frac{C_{pw}}{T} dT \tag{20}$$

In these equations, C_{pw} is the heat capacity of water, and T_8 and T_9 are the inlet and outlet temperatures of the water streams of the water tank.

After determining the exergy destruction of all of the equipment, the total exergy destruction of the cycle, which includes the non-idealities of the system, can be determined according to Equation (21).

$$\dot{E}_{d,tot} = \sum \dot{E}_{d,i} \quad (21)$$

According to this equation, the total exergy destruction of the cycle is actually the sum of the exergy destruction of each equipment in the cycle.

2.4. Investigated DESs

In this study, seven DESs, as well as paraffin, were considered as PCMs to study a solar thermal power generation cycle. The information of the studied DESs, including the HBA and HBD components, and their molar ratios and molecular weights are presented in Table 1.

Table 1. The HBA, HBD, and molar ratios of the investigated DESs in this study.

DES Code	HBA	HBD	HBA:HBD Molar Ratio	DES Molecular Weight (g/mol)
DES1	Choline chloride	Suberic acid	1:1 [1]	156.92
DES2	Choline chloride	Urea	1:0.9 [2]	102.22
DES3	Choline chloride	Gallic acid	1:0.5 [1]	149.79
DES4	Choline chloride	4-Hydroxybenzoic acid	1:0.5 [1]	139.13
DES5	Choline chloride	Oxalic acid	1:0.8 [2]	117.81
DES6	Choline chloride	Itaconic acid	1:1 [1]	134.87
DES7	Choline chloride	p-Coumaric acid	1:0.5 [1]	147.81

[1] Reference [36] [2] Reference [37].

3. Results and Discussion

The first step for performing the calculations in the presented modified cycle, is determining the physical properties of the DESs. The enthalpy of fusion and melting point are required for each DES. Table 2 presents the values of enthalpies of fusion for the HBA and HBD components, as well as the melting points of the investigated DESs. In order to calculate the enthalpies of fusion of the DESs, a simple thermodynamic mixing rule was used for the HBA and HBD components, as given by Equation (22) [38].

$$\Delta h_{fus,PCM} = y_{HBA}\Delta h_{fus,HBA} + y_{HBD}\Delta h_{fus,HBD} \quad (22)$$

where y_{HBA} and y_{HBD} are the mole fractions of the HBA and HBD components, respectively, and $\Delta h_{fus,HBA}$ and $\Delta h_{fus,HBD}$ are their corresponding enthalpies of fusion, respectively. The calculated values of enthalpies of fusion for the investigated DESs are also reported in Table 2.

Table 2. Enthalpies of fusion and melting points of the investigated DESs in this study.

DES	HBA to HBD molar ratio	$\Delta h_{fus,HBA}(\frac{kJ}{mol})$	$\Delta h_{fus,HBD}(\frac{kJ}{mol})$	$\Delta h_{fus,DES}(\frac{kJ}{mol})$	$\Delta h_{fus,DES}(\frac{J}{g})$	$T_{m,DES}(°C)$
DES1	1:1	29.76 [1]	30.70 [2]	30.23	192.65	93 [4]
DES2	1:0.9	29.76 [1]	13.61 [2]	22.17	216.89	80 [5]
DES3	1:0.5	29.76 [1]	30.96 [3]	30.17	201.42	77 [4]
DES4	1:0.5	29.76 [1]	32.00 [2]	30.50	219.22	87 [4]
DES5	1:0.8	29.76 [1]	12.31 [3]	22.08	187.42	73 [5]
DES6	1:1	29.76 [1]	17.49 [3]	23.62	175.13	57 [4]
DES7	1:0.5	29.76 [1]	24.78 [3]	28.10	190.11	67 [4]
Paraffin	-	-	-	-	189.00 [6]	68 [6]

[1] Reference [39]; [2] Reference [40]; [3] Calculated using the Joback–Reid method [41]; [4] Reference [36]; [5] Reference [37]; [6] Reference [42].

In addition to the studied DESs, paraffin, with a carbon number range of 21 to 50 and a melting point of 68 °C, with an enthalpy of fusion equal to 189 J/g, was considered as a conventional PCM [42].

All of the required properties of R134a (the working fluid of Rankine cycle) and water (the working fluid of the heating cycle), including enthalpies, entropies, vapor pressures at different temperatures and pressures, and heat capacities were obtained from the NIST database [40].

In order to have a fair investigation of all the DESs, the operational conditions of the studied cycles for each DES were considered the same. Table 3 reports the operational conditions of the investigated cycles.

Table 3. The operational conditions for the investigated cycles.

Water Tank Outlet Temperature, T_9	Condenser Temperature Range (°C)	R134a Outlet Temperature of PCM tank, $T_{1'}$	Evaporator Pressure (kPa)	Mass Flow Rate of Water, $\dot{m}_w$ (kg/s)	Mass Flow Rate of R134a $\dot{m}_r$ (kg/s)
$T_{m,PCM} + 5$	30–55	$T_{m,PCM} - 5$	1000–2000	1.5	0.1

According to the presented operational conditions, the outlet water-temperature from the water tank, T_9, for all the investigated cycles was assumed to be higher than the melting-point temperature of the investigated DESs (PCMs), to ascertain the transfer of heat from hot water to the DES. Moreover, the outlet temperature of R134a from the PCM tank, $T_{1'}$ was considered to be lower than the PCM melting point temperature, in order to be sure of heat transfer from the PCM to R134a. Moreover, the condenser temperature, the evaporator pressure, and the mass flow rates of water and R134a were selected according to the thermodynamic properties of the working fluid and the selected PCMs, as well as taking into account the values given in previously published studies [15,33].

After obtaining all of the required information for the investigated cycles, the performances of the cycles using the investigated DESs as PCMs were investigated by energy and exergy analyses.

The most important equipment in the investigated cycles, which are flexible in changing the operational conditions, are the condenser and evaporator. Therefore, by changing the condenser temperature and evaporator pressure (according to Table 3), the performances of the investigated cycles were studied, with a focus on the produced power, the required mass of DES, and the total exergy loss of the cycle.

3.1. Method of Calculation

To calculate the cycle's characteristics, such as power production, required mass of PCM, and exergy losses, the following calculation steps were followed:

Step 1. Based on the selected condenser temperature, evaporator pressure, and the provided assumptions, the enthalpies and entropies of Streams 1 (1′), 3, and 4 (4′) were determined;

Step 2. The entropy and enthalpy of Stream 2 were calculated based on the turbine's isentropic efficiency, which was considered as 0.75 in this work;

Step 3. Using the calculated enthalpies, the produced power and the required mass of PCM were calculated based on Equations (1) and (5);

Step 4. According to the given exergy analysis method, the exergy losses were determined.

3.2. Effect of the Condenser Temperature

The effects of changing the condenser temperature on the produced power, the required mass of PCM, and the total exergy loss of each cycle are shown in Figures 2–5 for all of the studied cycles. However, it is important to keep in mind that the inlet R134a to the turbine should be at a superheated vapor state, therefore, the evaporator pressures of each cycle will be different. The values of the evaporator pressure in each cycle are also shown in Figures 2–5.

Based on the achieved results, it can be seen that by increasing the condenser temperature, the produced power and the required mass of PCM both decrease. In fact, by increasing the condenser temperature while all the other operational conditions of the cycle are constant, the enthalpy of Stream 1 (or 1'), which is a function of the evaporator pressure and temperature, T_1 (or $T_{1'}$), remains constant for each cycle. Moreover, increasing the condenser temperature increases the condenser pressure as well. Accordingly, Stream 2 leaves the turbine at a higher pressure and temperature. Therefore, the enthalpy of Stream 2 will increase when the condenser temperature is increased. Based on Equation (1), for a constant mass flow rate of the working fluid, $\dot{m}_r$, and a constant enthalpy, h_1 (or $h_{1'}$), the produced power decreases by increasing h_2. This can be seen in all of the studied cycles in Figure 2. By comparing the different DESs investigated, it is shown that except for DES6 and DES7, the other DESs produce higher amounts of power than the conventional paraffin PCM at the same operational conditions.

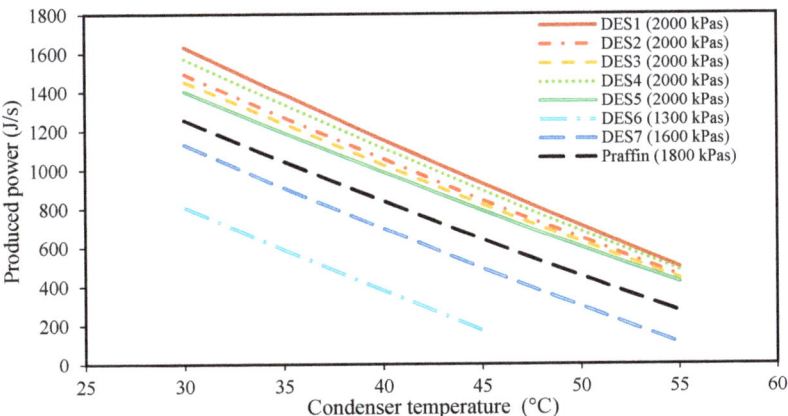

Figure 2. The effect of condenser temperature on the produced power. (The evaporator pressure for each system is shown in the legend for each PCM).

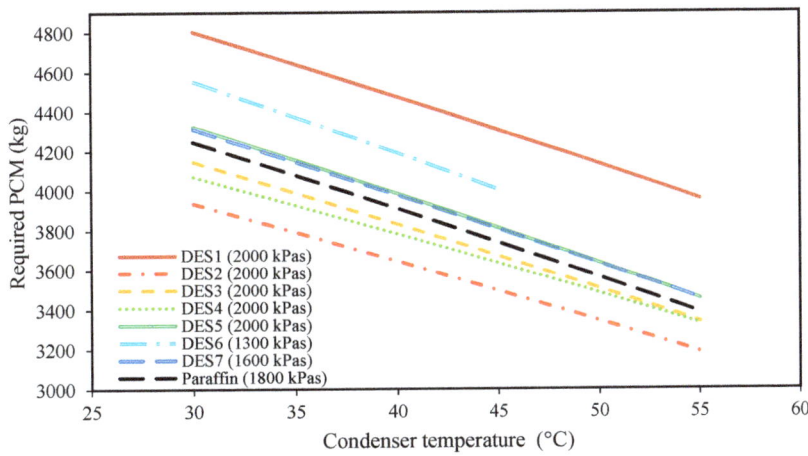

Figure 3. The effect of condenser temperature on the required amount of DES. (The evaporator pressure of each system is shown in the legend of each PCM).

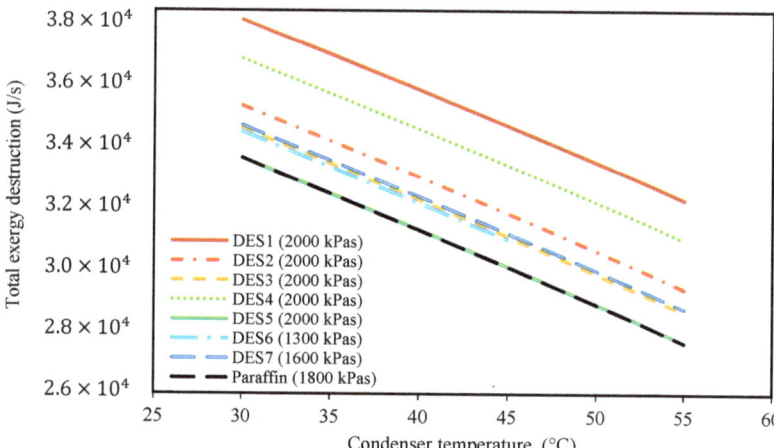

Figure 4. The effect of condenser temperature on the total exergy destruction. (The evaporator pressure of each system is shown in the legend of each PCM).

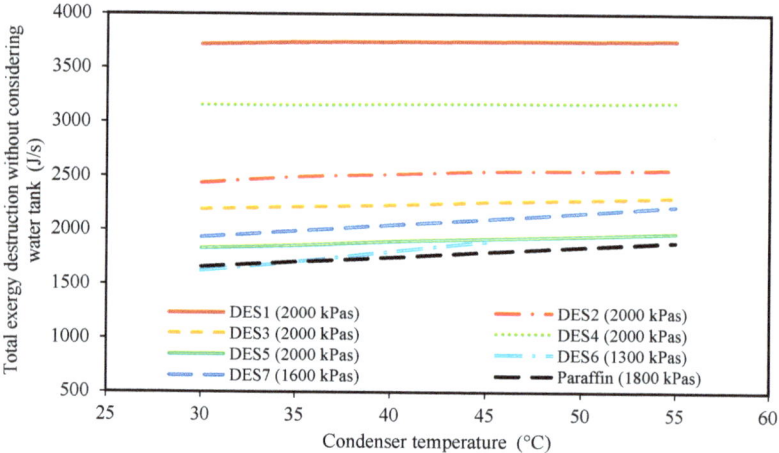

Figure 5. The effect of condenser temperature on the total exergy destruction without considering the water tank exergy loss. (The evaporator pressure of each system is shown in the legend of each PCM).

Additionally, as discussed earlier, increasing the condenser temperature does not have any effect on the properties of Streams 1 and 1'. However, increasing the condenser temperature increases T_3, and then, $T_{4'}$ as well, which means that Stream 4 reaches higher enthalpy values. Accordingly, based on Equation (5), a lower amount of PCM is required at higher condenser temperatures, which is also evidenced by Figure 3.

Another important finding from this figure is the lower required amounts of DES4, DES3, and DES2 as the PCMs with respect to paraffin. The other investigated cycles require greater amounts of DES than paraffin. In fact, one of the most important properties that play a vital role in the performance of a cycle is the enthalpy of fusion of the PCM. By comparing the enthalpies of fusion of the studied DESs, it is seen that DES2, DES3, and DES4, have the highest enthalpies of fusion among the PCMs. Accordingly, in the cycles with either DES2, DES3, or DES4 as the PCM, a lower mass of PCM is required to provide a desired amount of power, in comparison to other cycles.

Additionally, Figure 4, demonstrates the effect of condenser temperature on the total exergy destruction of the investigated cycles. Based on the results of this figure, by increasing the condenser temperature, the total exergy destruction of each of the studied cycles decreases. Indeed, by increasing the condenser temperature, the produced power and the required amount of PCM both decrease. Accordingly, the required amount of input heat to the water tank decreases as well. Based on Equation (18), by increasing the condenser temperature, the exergy destruction of the water tank also decreases. By comparing the investigated cycles, it is shown that only the cycle of DES5 has a similar total exergy destruction to the paraffin cycle.

Moreover, it is common practice to study the total exergy destruction of only the Rankine cycle instead of the whole cycle. For this purpose, Figure 5 is presented. This figure demonstrates the effect of condenser temperature on the total exergy destruction of the cycle without considering the exergy loss of the water tank. Based on the achieved results, at higher condenser temperatures, the total exergy destruction is higher.

In fact, when the difference between the condenser and the surrounding temperatures increases, the process of discarding heat $\dot{Q}_c$ to the surrounding moves further away from a "reversible" process. Accordingly, the total exergy destruction increases at higher condenser temperatures. By comparing the results of Figures 4 and 5, it can be seen that the effect of condenser temperature on the total exergy destruction of the whole cycle is the exact opposite of the results of Figure 4. In fact, it can be concluded that the exergy loss of the water tank is much greater than the other parts of the cycle, and, thus, controls the behavior of total exergy destruction of the cycle. Therefore, as discussed earlier, when the condenser temperature increases, lower amounts of heat are necessary for increasing the enthalpy of Stream 4, so the temperature change of water in the evaporator decreases, leading to lower exergy destruction of the water tank, which has the highest effect on the total exergy losses.

In general, based on the achieved results of Figures 2–5, it can be concluded that lower condenser temperatures of the investigated cycles are more favorable from the point of view of produced power. However, the condenser temperature cannot be lower than a specific value. In fact, to ensure that the discarding of heat, $\dot{Q}_C$, to the surrounding does indeed occur, the condenser temperature should not be lower than the surrounding temperature. However, it should be noted that decreasing the evaporator temperature leads to higher exergy destructions, and also, larger amounts of required DES. Therefore, based on these findings, the temperature of 30 °C can be suggested as a suitable condenser temperature for all of the studied cycles to achieve high power production.

3.3. Effect of the Evaporator Pressure

To study the effect of evaporator pressure (according to Table 3) on the performances of the investigated cycles, the produced power, the required amount of PCM, and the total exergy destruction upon evaporator pressure changes were studied and the results are presented in Figures 6–9, respectively.

These investigations were carried out at a condenser temperature of 30 °C, which was proposed above as a possible optimum condenser temperature.

Based on Figure 6, by increasing the evaporator pressure, the produced power increases for all of the studied cycles. Indeed, increasing the evaporator pressure does not have any effect on the pressure of Stream 2, while it does increase the pressure of Stream 1 during the day. Therefore, the inlet pressure of the turbine increases while the outlet pressure remains constant, so the produced power increases while considering a constant working fluid mass flow rate. Additionally, since we assumed that the cycle's operational conditions are the same during night and day, the same scenario can be assumed for the pressures of Streams 1 and 2 during the night, which leads to the production of more power during the night as well. Moreover, it can be seen that all of the investigated DESs, except for DES6 and DES7, produce greater, or at least the same amount of power as paraffin. The reason that DES6 and DES7 produce lower power in comparison to the other DESs and the studied paraffin, is their smaller enthalpies of fusion.

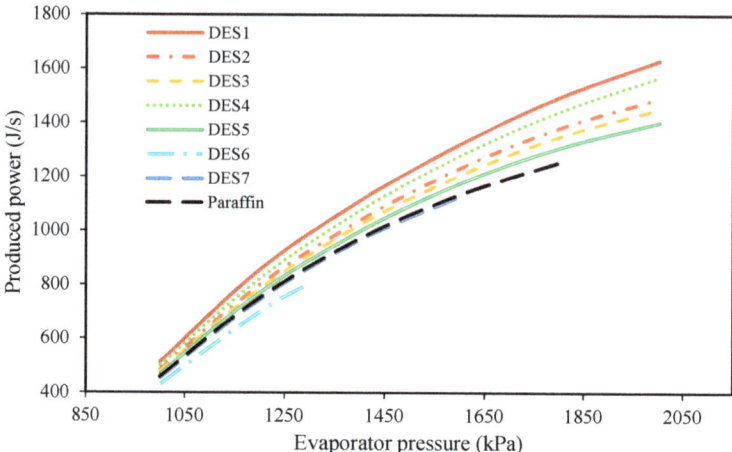

Figure 6. The effect of evaporator pressure on the produced power.

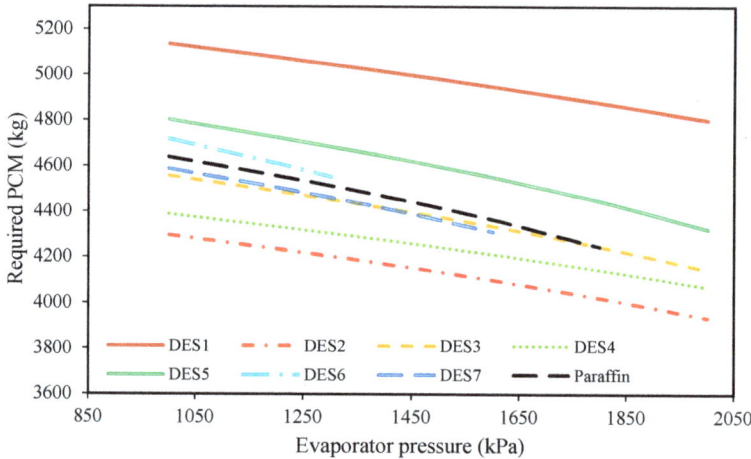

Figure 7. The effect of evaporator pressure on the required amount of DES.

Indeed, as mentioned earlier, the enthalpy of fusion of a PCM is an important factor whose value affects the behavior of the cycle. In fact, by the increased enthalpy of fusion of a PCM, a higher amount of energy can be stored within a fixed period of time. Therefore, a PCM with a high enthalpy of fusion can provide greater energy to the refrigerant of the Rankine cycle. Subsequently, and based on the performance of the Rankine cycle, a greater amount of power can be achieved when a larger amount of energy is added to its refrigerant.

Additionally, according to Figure 7, it can be seen that by increasing the evaporator pressure, the required mass of PCM decreases for all of the investigated cycles. Because the pressures of Streams 1 and 1' are the same during day and night, increasing the evaporator pressure at a constant evaporator temperature leads to reduced enthalpies of Streams 1 and 1'. Accordingly, based on Equation (6), for a constant mass flow rate of the working fluid, smaller amounts of the PCM are required. Additionally, based on the results of Figure 7, it can be seen that except for DES1, DES5, and DES6, the required amount of PCM for the investigated cycle is either lower or the same as the cycle which uses paraffin, due to the differences between the enthalpies of fusion.

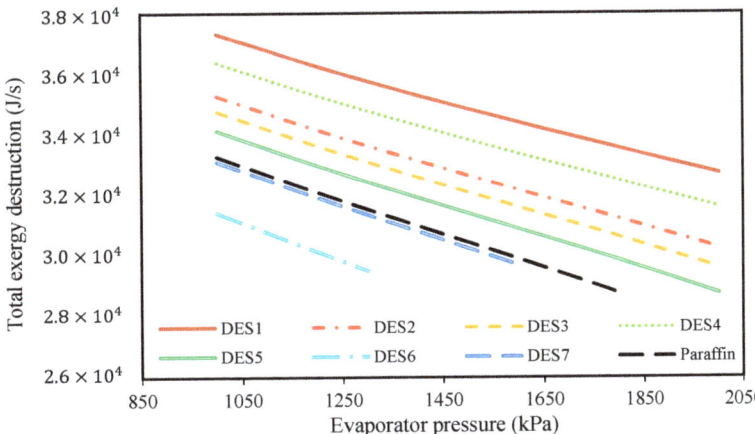

Figure 8. The effect of evaporator pressure on the total exergy destructions of the investigated cycles.

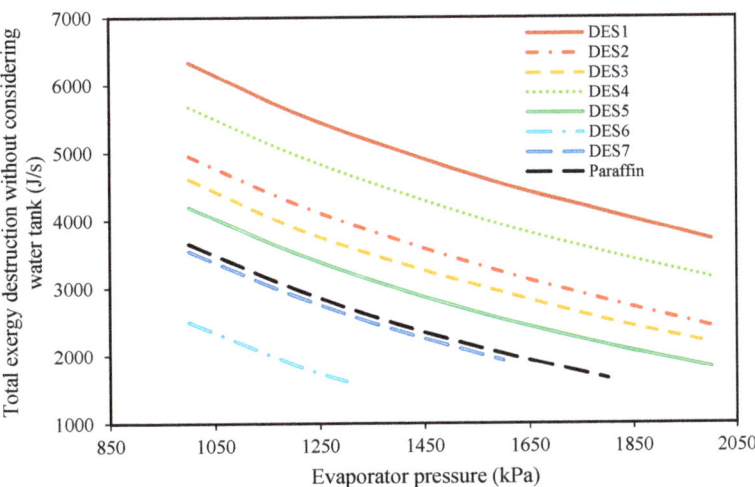

Figure 9. The effect of evaporator pressure on the total exergy loss of the cycle without considering the exergy destructions of the water tank.

In addition to the required PCM and the produced power, the effect of changing of evaporator pressure on the total exergy destructions of the investigated cycles was studied and shown in Figure 8.

According to the results, increasing the evaporator pressure decreases the total exergy destruction of all the studied cycles. Additionally, in Figure 9, the effect of changing evaporator pressure on the total exergy destruction of the investigated cycles, without considering the exergy destruction of the water tank, is presented. Based on the results, increasing the evaporator pressure leads to decreases in the total exergy destruction (without the water tank) for all of the studied cycles. Actually, it was shown that by increasing the evaporator pressure, the required mass of the PCMs consequently decreases, which means that the working fluid (R134a) requires lower amounts of heat for vaporization. In other words, since it was assumed that the investigated cycle's operational conditions are the same during night and day, by increasing the evaporator pressure, the required amount of heat which is required for vaporizing R134a decreases during the day. Actually, in the daytime, water is responsible for providing the required amount of heat for the evaporation of R134a, and by increasing the evaporator pressure, the temperature-change of water decreases. From a thermodynamics point of view, by decreasing the water temperature, the evaporator tends toward a reversible process, so its exergy destruction decrease.

Based on the achieved results, it can be concluded that increasing the evaporator pressure is favorable for the cycle's performance and the highest possible evaporator pressure should be chosen, however, since the inlet fluid to the turbine should be superheated vapor, there is a limit on evaporator pressure increase. Additionally, evaporator pressure is restricted by safety protocols and operational limitations.

In Figure 9, the effect of the melting point temperature of the PCM is shown on the cycle performance. DES6, DES7, and paraffin have a lower melting-point temperatures in comparison to the other studied PCMs, and since the temperature of the Rankine cycle's refrigerant is equal to $T_{m,PCM} - 5$, the outlet refrigerant from the evaporator cannot be superheated vapor at high evaporator pressures when a PCM with a low melting-point is used. Based on this limitation, it is suggested to consider the evaporator pressure as 2000 kPa. By comparing the results of the investigated cycles in Table 4, it can be seen that the cycle which uses DES2 (1 Choline chloride: 0.9 urea) as its PCM requires the lowest amount of DES. Additionally, the cycle which uses DES5 (1 Choline chloride:0.8 oxalic acid) as its PCM has the lowest total exergy destruction.

Table 4. The results of exergy and energy analyses for all of the investigated cycles at the condenser temperature of 30 °C and their evaporator pressure.

Cycle	Evaporator Pressure (kPa)	Produced Power (J/s)	Required Mass of PCM (kg)	Total Exergy Destruction (J/s)	Total Exergy Destruction Without the Water Tank (J/s)
DES1	2000	1630.5	4803.57	32,717.11	3710.68
DES2	2000	1491.75	3936.70	30,249.98	2427.12
DES3	2000	1452.75	4146.14	29,600.00	2182.53
DES4	2000	1569.75	4071.55	31,599.43	3148.88
DES5	2000	1402.01	4324.36	28,758.28	1826.06
DES6	1300	807.75	4553.63	29,515.55	1619.44
DES7	1600	1128.75	4311.56	29,697.07	1928.57
Paraffin	1800	1254.14	4246.63	28,763.60	1658.43

However, the cycle which uses DES1 (1 Choline chloride:1 suberic acid) as the PCM has the highest produced power. Nevertheless, according to the results of Table 4, choosing DES4 (1 Choline chloride:0.5 4-hydroxybenzoic acid) as the PCM is the most rational because following DES1, it has the highest power production while the required mass of DES is much lower than DES1. Furthermore, from the exergy destruction point of view, its total exergy destruction is in the same order as the other cycles. For a detailed examination, the contribution of all of the equipment of the cycle using DES4 as the PCM is shown in Figure 10. Moreover, the exergy destruction contribution of the other investigated cycles is also given in Figures S1–S6 of the Supplementary Materials.

Based on Figure 10, it is obvious that the water tank has the highest contribution in comparison to the other equipment. One of the most important sources of such high exergy destruction in the water tank is the temperature of the heat source of the water tank. Accordingly, for improving the performance of the cycle, the operation of the water tank should be optimized.

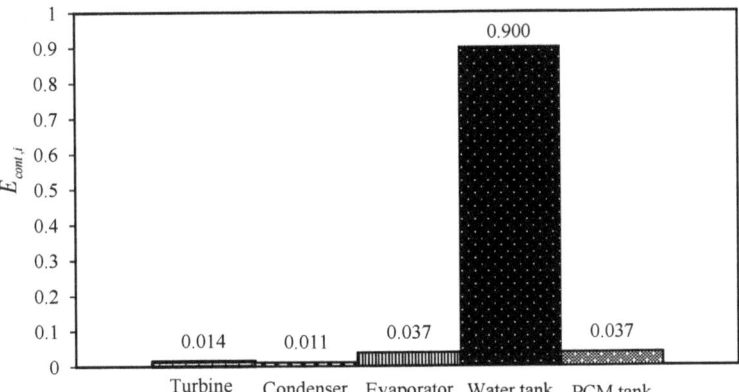

Figure 10. The contribution of each unit of the cycle in the exergy destruction.

3.4. Effects of the Melting Point Temperature and the Enthalpy of Fusion

In the previous sections, the performances of the investigated DESs were compared to one another at various operational conditions. In this section, we discuss the effects of the melting point temperature and the enthalpy of fusion of a DES on the cycle performance. According to the achieved results, a DES with a higher melting point temperature and higher enthalpy of fusion, such as DES1, DES3, and DES4, is more favorable and leads to a better cycle performance. A higher melting point temperature of a DES leads to a higher enthalpy of the fluid entering the turbine. However, melting point temperature is not the only criteria for selecting a suitable DES. In this work, it was shown that DES4 can potentially be the best DES among the investigated DESs according to performance, however, its melting-point temperature is lower than that of DES1. Actually, the heat of fusion of a DES is also an important factor that should be considered for selecting the best DES. In fact, increasing the enthalpy of fusion of a DES leads to lower required amounts of DES for the same amount of power generation. In general, when choosing an appropriate DES for power generation in the given cycle, the melting-point temperature and enthalpy of fusion of the DES should be high enough, while some other operational conditions, such as viscosity, should be considered as well.

4. Conclusions

In this work, a modified cycle was introduced for a solar thermal power plant that uses a PCM tank for storing solar energy during the day and releases the energy during the night. Based on the modified cycle, power generation based on solar energy can occur continuously not only during the day, but also, throughout the night. Additionally, in order to investigate the feasibility of replacing conventional PCMs with green and sustainable materials, various DESs were considered as novel PCMs for use in solar thermal power plants. The feasibility study was carried out by applying exergy and energy analyses to the modified cycles. For this purpose, seven different DESs were suggested as potential PCMs, to be compared with paraffin as a conventional PCM. Based on the considered PCMs, the optimum operating conditions of the modified solar thermal power plant cycles were investigated by studying the effects of changing the condenser temperature and evaporator pressure on the produced power, the required amount of DES, and the total exergy destruction of the cycles. Based on the achieved results, it was suggested that the highest of cycle performances can potentially be achieved at a condenser temperature of 30 °C and an evaporator pressure of 2000 kPa. At these suggested operational conditions, the cycle which uses DES4 (Choline chloride:4-hydroxybenzoic acid 1:0.5) as its PCM shows the best performance. By comparing the achieved results, it was found that some of the selected DESs have better performance than paraffin from the points of view of energy and exergy analyses. Due to the larger enthalpy of fusion of DES4 in comparison to paraffin, the cycle which operates with DES4 produces 25% more power in comparison to the cycle which uses paraffin as the PCM, together with a lower required amount of DES (175 kg lower), and their total exergy losses are in the same order.

Additionally, by comparing the contributions of each equipment of the solar thermal power plant cycle in the aspect of total exergy destruction, it was concluded that the water tank which absorbs the solar energy, has the highest contribution to the total exergy destruction of the cycle.

Based on the results of this work, it can be concluded that DES4 has the potential to be used as a PCM in solar power plants due to its suitable performance in comparison to paraffin, in addition to its environmental benefits.

Supplementary Materials: The following supporting information can be downloaded online: Figure S1 to Figure S7.

Author Contributions: Methodology, conceptualization, software, validation, formal analysis, Writing—Original draft preparation, H.P.; Conceptualization, formal analysis, methodology, software, Writing—Review and editing, validation, R.H.; funding acquisition, supervision, Writing—Review and editing, A.R.C.D.; supervision, validation, Writing—Review and editing, A.S. All authors have read and agreed to the published version of the manuscript.

Funding: This research was funded by European Union Horizon 2020, grant number ERC-2016-CoG 725034 (ERC Consolidator Grant Des.solve). This work was also supported by the Associate Laboratory for Green Chemistry- LAQV which is financed by national funds from FCT/MCTES (UID/QUI/50006/2019).

Institutional Review Board Statement: Not applicable.

Informed Consent Statement: Not applicable.

Data Availability Statement: Not applicable.

Acknowledgments: The authors are grateful to Shiraz University, University of Isfahan and Universidade Nova de Lisboa for providing facilities.

Conflicts of Interest: The authors declare no conflict of interest.

References

1. Vural, G. How do output, trade, renewable energy and non-renewable energy impact carbon emissions in selected Sub-Saharan African Countries? *Resour. Policy* **2020**, *69*, 101840. [CrossRef]
2. Fathi Assi, D.A.; Isiksal, A.Z.; Tursoy, T. Renewable energy consumption, financial development, environmental pollution, and innovations in the ASEAN +3 group: Evidence from (P-ARDL) model. *Renew. Energy* **2020**, *165*, 689–700. [CrossRef]
3. Levenda, A.M.; Behrsin, I.; Disano, F. Renewable energy for whom? A global systematic review of the environmental justice implications of renewable energy technologies. *Energy Res. Soc. Sci.* **2021**, *71*, 101837. [CrossRef]
4. Haghbakhsh, R.; Peyrovedin, H.; Raeissi, S.; Duarte, A.R.C.; Shariati, A. Investigating the performance of novel green solvents in absorption refrigeration cycles: Energy and exergy analyses. *Int. J. Refrig.* **2020**, *113*, 174–186. [CrossRef]
5. Haghbakhsh, R.; Peyrovedin, H.; Raeissi, S.; Duarte, A.R.C.; Shariati, A. Energy conservation in absorption refrigeration cycles using DES as a new generation of green absorbents. *Entropy* **2020**, *22*, 409. [CrossRef] [PubMed]
6. Walch, A.; Mohajeri, N.; Gudmundsson, A.; Scartezzini, J.-L. Quantifying the technical geothermal potential from shallow borehole heat exchangers at regional scale. *Renew. Energy* **2021**, *165*, 369–380. [CrossRef]
7. Barbier, E. Geothermal energy technology and current status: An overview. *Renew. Sustain. Energy Rev.* **2002**, *6*, 3–65. [CrossRef]
8. Daut, I.; Razliana, A.R.N.; Irwan, Y.M.; Farhana, Z. A Study on the wind as renewable energy in Perlis, Northern Malaysia. *Energy Procedia* **2012**, *18*, 1428–1433. [CrossRef]
9. Joselin Herbert, G.M.; Iniyan, S.; Sreevalsan, E.; Rajapandian, S. A review of wind energy technologies. *Renew. Sustain. Energy Rev.* **2007**, *11*, 1117–1145. [CrossRef]
10. Behar, O. Solar thermal power plants—A review of configurations and performance comparison. *Renew. Sustain. Energy Rev.* **2018**, *92*, 608–627. [CrossRef]
11. Javed, M.S.; Ma, T.; Jurasz, J.; Amin, M.Y. Solar and wind power generation systems with pumped hydro storage: Review and future perspectives. *Renew. Energy* **2020**, *148*, 176–192. [CrossRef]
12. Shankar Ganesh, N.; Srinivas, T. Design and modeling of low temperature solar thermal power station. *Appl. Energy* **2012**, *91*, 180–186. [CrossRef]

13. Chowdhury, M.T.; Mokheimer, E.M.A. Recent developments in solar and low-temperature heat sources assisted power and cooling systems: A design perspective. *J. Energy Resour. Technol.* **2019**, *142*, 040801. [CrossRef]
14. Singh, N.; Kaushik, S.C.; Misra, R.D. Exergetic analysis of a solar thermal power system. *Renew. Energy* **2000**, *19*, 135–143. [CrossRef]
15. Dragomir-Stanciu, D.; Luca, C. Solar power generation system with low temperature heat storage. *Procedia Technol.* **2016**, *22*, 848–853. [CrossRef]
16. Kargar, M.R.; Baniasadi, E.; Mosharaf-Dehkordi, M. Numerical analysis of a new thermal energy storage system using phase change materials for direct steam parabolic trough solar power plants. *Sol. Energy* **2018**, *170*, 594–605. [CrossRef]
17. Pielichowska, K.; Pielichowski, K. Phase change materials for thermal energy storage. *Prog. Mater. Sci.* **2014**, *65*, 67–123. [CrossRef]
18. Zalba, B.; Marín, J.M.; Cabeza, L.F.; Mehling, H. Review on thermal energy storage with phase change: Materials, heat transfer analysis and applications. *Appl. Therm. Eng.* **2003**, *23*, 251–283. [CrossRef]
19. Koca, A.; Oztop, H.F.; Koyun, T.; Varol, Y. Energy and exergy analysis of a latent heat storage system with phase change material for a solar collector. *Renew. Energy* **2008**, *33*, 567–574. [CrossRef]
20. Gürtürk, M.; Koca, A.; Öztop, H.F.; Varol, Y.; Şekerci, M. Energy and exergy analysis of a heat storage tank with a novel eutectic phase change material layer of a solar heater system. *Int. J. Green Energy* **2017**, *14*, 1073–1080. [CrossRef]
21. Mofijur, M.; Mahlia, T.M.I.; Silitonga, A.S.; Ong, H.C.; Silakhori, M.; Hasan, M.H.; Putra, N.; Rahman, S.M.A. phase change materials (PCM) for solar energy usages and storage: An overview. *Energies* **2019**, *12*, 3167. [CrossRef]
22. Pirasaci, T.; Goswami, D.Y. Influence of design on performance of a latent heat storage system for a direct steam generation power plant. *Appl. Energy* **2016**, *162*, 644–652. [CrossRef]
23. Abbott, A.P.; Boothby, D.; Capper, G.; Davies, D.L.; Rasheed, R.K. Deep Eutectic solvents formed between choline chloride and carboxylic acids: Versatile alternatives to ionic liquids. *J. Am. Chem. Soc.* **2004**, *126*, 9142–9147. [CrossRef] [PubMed]
24. Zhao, B.-Y.; Xu, P.; Yang, F.-X.; Wu, H.; Zong, M.-H.; Lou, W.-Y. Biocompatible deep eutectic solvents based on choline chloride: Characterization and application to the extraction of rutin from *Sophora japonica*. *ACS Sustain. Chem. Eng.* **2015**, *3*, 2746–2755. [CrossRef]
25. Zhang, Q.; de Oliveira Vigier, K.; Royer, S.; Jérôme, F. Deep eutectic solvents: Syntheses, properties and applications. *Chem. Soc. Rev.* **2012**, *41*, 7108–7146. [CrossRef] [PubMed]
26. Shahbaz, K.; AlNashef, I.M.; Lin, R.J.T.; Hashim, M.A.; Mjalli, F.S.; Farid, M.M. A novel calcium chloride hexahydrate-based deep eutectic solvent as a phase change materials. *Sol. Energy Mater. Sol. Cells* **2016**, *155*, 147–154. [CrossRef]
27. Brett, C.M.A. Deep eutectic solvents and applications in electrochemical sensing. *Curr. Opin. Electrochem.* **2018**, *10*, 143–148. [CrossRef]
28. Khandelwal, S.; Tailor, Y.K.; Kumar, M. Deep eutectic solvents (DESs) as eco-friendly and sustainable solvent/catalyst systems in organic transformations. *J. Mol. Liq.* **2016**, *215*, 345–386. [CrossRef]
29. Freeman, J.; Hellgardt, K.; Markides, C.N. An assessment of solar-powered organic Rankine cycle systems for combined heating and power in UK domestic applications. *Appl. Energy* **2015**, *138*, 605–620. [CrossRef]
30. Freeman, J.; Guarracino, I.; Kalogirou, S.A.; Markides, C.N. A small-scale solar organic Rankine cycle combined heat and power system with integrated thermal energy storage. *Appl. Therm. Eng.* **2017**, *127*, 1543–1554. [CrossRef]
31. Dincer, I. *Refrigeration Systems and Applications*, 3rd ed.; John Wiley & Sons: New York, NY, USA, 2017.
32. Aman, J.; Ting, D.S.K.; Henshaw, P. Residential solar air conditioning: Energy and exergy analyses of an ammonia–water absorption cooling system. *Appl. Therm. Eng.* **2014**, *62*, 424–432. [CrossRef]
33. Yataganbaba, A.; Kilicarslan, A.; Kurtbaş, İ. Exergy analysis of R1234yf and R1234ze as R134a replacements in a two evaporator vapour compression refrigeration system. *Int. J. Refrig.* **2015**, *60*, 26–37. [CrossRef]
34. Hoseini Rahdar, M.; Emamzadeh, A.; Ataei, A. A comparative study on PCM and ice thermal energy storage tank for air-conditioning systems in office buildings. *Appl. Therm. Eng.* **2016**, *96*, 391–399. [CrossRef]
35. Navidbakhsh, M.; Shirazi, A.; Sanaye, S. Four E analysis and multi-objective optimization of an ice storage system incorporating PCM as the partial cold storage for air-conditioning applications. *Appl. Therm. Eng.* **2013**, *58*, 30–41. [CrossRef]
36. Maugeri, Z.; Domínguez de María, P. Novel choline-chloride-based deep-eutectic-solvents with renewable hydrogen bond donors: Levulinic acid and sugar-based polyols. *RSC Adv.* **2012**, *2*, 421–425. [CrossRef]
37. Pyykkö, P. Simple estimates for eutectic behavior. *Chemphyschem A Eur. J. Chem. Phys. Phys. Chem.* **2019**, *20*, 123–127. [CrossRef]
38. Raud, R.; Bell, S.; Adams, K.; Lima, R.; Will, G.; Steinberg, T.A. Experimental verification of theoretically estimated composition and enthalpy of fusion of eutectic salt mixtures. *Sol. Energy Mater. Sol. Cells* **2018**, *174*, 515–522. [CrossRef]
39. López-Porfiri, P.; Brennecke, J.F.; Gonzalez-Miquel, M. Excess molar enthalpies of deep eutectic solvents (DESs) composed of quaternary ammonium salts and glycerol or ethylene glycol. *J. Chem. Eng. Data* **2016**, *61*, 4245–4251. [CrossRef]
40. Linstorm, P.J.; Mallard, W.G. *NIST Chemistry WebBook, NIST Standard Reference Database Number 69*; National Institute of Standards and Technology: Gaithersburg, MD, USA. Available online: https://webbook.nist.gov/ (accessed on 1 January 2021).
41. Joback, K.G.; Reid, R.C. Estimation of pure-component properties from group-contributions. *Chem. Eng. Commun.* **1987**, *57*, 233–243. [CrossRef]
42. Ukrainczyk, N.; Kurajica, S.; Šipušić, J. Thermophysical comparison of five commercial paraffin waxes as latent heat storage materials. *Chem. Biochem. Eng. Q.* **2010**, *24*, 129–137.

Article

The Response Surface Optimization of Supercritical CO_2 Modified with Ethanol Extraction of *p*-Anisic Acid from *Acacia mearnsii* Flowers and Mathematical Modeling of the Mass Transfer

Graciane Fabiela da Silva, Edgar Teixeira de Souza Júnior, Rafael Nolibos Almeida, Ana Luisa Butelli Fianco, Alexandre Timm do Espirito Santo, Aline Machado Lucas, Rubem Mário Figueiró Vargas and Eduardo Cassel *

Unit Operations Laboratory (LOPE), School of Technology, Pontifical Catholic University of Rio Grande do Sul, Av Ipiranga 6681, Building 30, Block F, Room 208, Porto Alegre ZC 90619-900, RS, Brazil; gracianenh@gmail.com (G.F.d.S.); edgar.souza@edu.pucrs.br (E.T.d.S.J.); rnolibos@gmail.com (R.N.A.); lu_fianco@hotmail.com (A.L.B.F.); alexandre.santo@acad.pucrs.br (A.T.d.E.S.); aline.lucas@pucrs.br (A.M.L.); rvargas@pucrs.br (R.M.F.V.)
* Correspondence: cassel@pucrs.br; Tel.: +55-51-3353-4585

Abstract: A widely disseminated native species from Australia, *Acacia mearnsii*, which is mainly cultivated in Brazil and South Africa, represents a rich source of natural tannins used in the tanning process. Many flowers of the *Acacia* species are used as sources of compounds of interest for the cosmetic industry, such as phenolic compounds. In this study, supercritical fluid extraction was used to obtain non-volatile compounds from *A. mearnsii* flowers for the first time. The extract showed antimicrobial activity and the presence of *p*-anisic acid, a substance with industrial and pharmaceutical applications. The fractionation of the extract was performed using a chromatographic column and the fraction containing *p*-anisic acid presented better minimum inhibitory concentration (MIC) results than the crude extract. Thus, the extraction process was optimized to maximize the *p*-anisic acid extraction. The response surface methodology and the Box–Behnken design was used to evaluate the pressure, temperature, the cosolvent, and the influence of the particle size on the extraction process. After the optimization process, the *p*-anisic acid yield was 2.51% w/w and the extraction curve was plotted as a function of time. The simulation of the extraction process was performed using the three models available in the literature.

Keywords: *Acacia mearnsii*; supercritical fluid extraction; *p*-anisic acid; response surface methodology; mathematical modeling

1. Introduction

Plants are a source of natural products that have several applications and are an important raw material for obtaining compounds of interest to the pharmaceutical and food industries [1,2]. Several species of *Acacia* have been studied and their secondary metabolites have anti-inflammatory [3–5], antifungal [6–8], antibacterial [9–12], antioxidant [13–15], anticancer [16,17], antidepressant [18], and antifeedant properties [19], among others. The extracts of different *Acacia* species, such as *A. catechu*, *A. concinna*, *A. dealbata*, *A. decurrens*, *A. farnesiana*, and *A. senegal* are used in cosmetics, according to the International Cosmetic Ingredient Dictionary and Handbook [20].

The *Acacia mearnsii* De Wild, a member of the Leguminosae family (subfamily Mimosoideae), is widely grown in Brazil and South Africa, with an estimated 540,000 hectares cultivated worldwide [21], representing the main source of vegetable tannins. Although the biological activities of other *Acacia* genus species have already been studied and the use of its bark and wood is already common, the *A. mearnsii* flowers have not yet been industrially explored.

Supercritical fluid extraction (SFE) is an important process used to obtain bioactive compounds [22] and it stands out in the food, pharmaceutical, and cosmetic industries due to its capacity to extract compounds with a high purity without thermal degradation, as well as its use of toxic solvents [23,24]. SFE is considered a clean technology when carbon dioxide, which is considered a green solvent, is used as a solvent, offering advantages over traditional methods, such as the extraction by steam distillation and hydrodistillation methods [25,26].

In SFE, operational conditions such as temperature, pressure, and particle size influence the process efficiency [22]. The design of experiments and response surface methodologies are commonly employed for the identification and optimization of variables [27–35]. The methodology allows evaluating the influence of several variables regarding one or more responses with a reduced number of experiments, therefore, reducing the time and cost [36]. After the extraction process has been optimized, the mathematical modeling of the extraction process dynamic is an important step for the prediction and process scale-up [37,38]. The mathematical modeling of the extraction is substantial, to evaluate the influence of the operational parameters in the technical and economic viability of an industrial process, with a reduced number of laboratory experiments [38–40].

In this work, supercritical carbon dioxide extraction was applied to the *Acacia mearnsii* flowers, a part of the plant with potential that is yet to be explored. The antibacterial activity of the extract was evaluated in the crude extract and the fractions obtained by column chromatography; *p*-anisic acid was identified in the fraction with the best antibacterial activity. *p*-anisic acid, also known as draconic acid or 4-methoxybenzoic acid.

(IUPAC), is an important substance that has digestive, diuretic, and expectorant properties and it is used as an aroma component in the food and cosmetic industries as a flavor, a preservative, and an antiseptic agent [41,42]. It also has importance in medical science for the treatment of Parkinson's disease, hepatitis B and C viruses, liver diseases, the post-radiation treatment of breast cancer, and skin desquamation [43]. Finally, it is an important substance in the production of pharmaceutical intermediates and pharmaceutical products, agrochemicals, and dyes [44]. The solubility of *p*-anisic acid in water is low, but it is highly soluble in alcohols and is soluble in ethers, as well as ethyl acetate [45].

The extraction evaluation can be carried out in terms of either the overall yield or the selectivity of a target component. The selectivity aspect is important since a high purity product for the desired analyte does not require subsequent purification operations. These purification steps make the process more expensive, in addition to exposing the extract to solvents that may not be compatible with the outcome of the product [46–48]. Thus, the response surface methodology was used for the optimization of *p*-anisic acid selectivity in the extract obtained by supercritical carbon dioxide extraction. The effects of pressure, temperature, and particle size were evaluated using a Box–Behnken design. For the optimized conditions within the framework investigated, the mass transfer parameters of three mathematical models were estimated to support the extraction process simulation.

2. Results

2.1. Step I—Design of the Experiments and the Chemical and Biological Evaluations

2.1.1. Factorial Design

First of all, a factorial design 2^2 was performed evaluating the pressure and modifier effects in the global yield of the extraction process. The experiments resulted in 7 extracts with a global yield between 1.25 and 2.49% w/w (extract/plant). These results are presented in Table 1 together with the experimental design matrix, the levels of each factor, and their combinations determined by the factorial design.

Table 1. Factorial 2^2 design matrix and observed responses.

Standard Run Order	Codified Variables		Uncodified Variables		Global Extract Yield	S/F [b]
	Pressure	Modifier	Pressure (bar)	Modifier	(% w/w) [a]	($g_{solv.}/g_{plant}$)
1	−1	−1	120	Water	1.25	24.2
2	−1	1	120	Ethanol	1.70	24.0
3	1	−1	240	Water	1.71	24.2
4	1	1	240	Ethanol	2.49	24.0
5	0	0	180	Water:Ethanol (1:1 v/v)	2.27	24.1
6	0	0	180	Water:Ethanol (1:1 v/v)	2.33	24.1
7	0	0	180	Water:Ethanol (1:1 v/v)	2.30	24.1

[a] gram of crude extract from 100 g of dried flowers; [b] S/F is the solvent-to-feed ratio.

The lowest yield value was obtained at the lowest pressure (120 bar), using water as the cosolvent, while the largest yield was obtained at the extraction process with the highest pressure (240 bar), using ethanol as the cosolvent. The factorial design data were processed using the Minitab® statistical software and the analysis of variance (ANOVA) proved that both factors, pressure, and the cosolvent were statistically significant in the global yield of the supercritical fluid extraction. The effects were evaluated using a linear regression and the contour plot can be viewed in Figure 1.

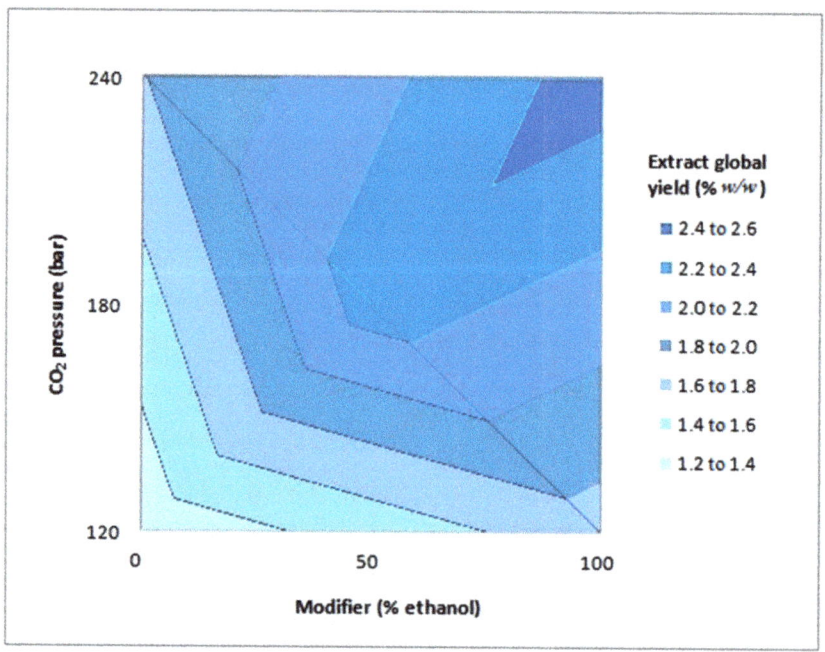

Figure 1. Contour plot for global extraction yield as a function of CO_2 pressure and modifier.

The model fitted to the experimental data was presented as a coefficient of determination, R^2, which was equal to 0.9981. The adjusted coefficient of determination had a value of 0.9944, which means that only 0.56% of the variations were not explained by the model used for the contour plot, which is presented in Equation (1):

$$global\ yield\ \left(\%\frac{w}{w}\right) = 1.81281 + 0.32906\ P + 0.29094\ M + 0.06094\ P\ M \qquad (1)$$

where P is the pressure and M is the modifier (values for coded variables).

2.1.2. Extract Purification and Chemical Analyses

The 7 extracts were analyzed by HPLC and in all of them, a compound with a retention time of about 10.6 min was detected (Supplementary Material—Figures S1–S7). In the first step of this study, the compound was only analyzed qualitatively. After a comparative analysis with different standards, the most common compound was identified as *p*-anisic acid. Once the compound was identified in all extracts, the following steps involved its purification by silica gel column chromatography, thin-layer chromatography (TLC), and HPLC for each fraction.

Most of the identified compounds in the raw extracts, including the *p*-anisic acid, were only identified in the ethyl acetate fraction. Therefore, this fraction was submitted once again to column chromatography with a gradient of solvents in increasing order of polarity, and 10 fractions were collected. According to the results of the HPLC analysis, *p*-anisic acid was detected in subfractions 4 and 5, obtained with a ratio of hexane-to-ethyl acetate of 40:60 and 20:80 v/v as solvents, respectively (Supplementary Material—Figures S8–S9). In subfraction 2, a compound that presented as orange in color in the TLC analysis (Supplementary Material—Figure S10) after staining with sulfuric vanillin was not identified in the HPLC analysis under the studied conditions (Supplementary Material). Nevertheless, this subfraction was chosen for the antimicrobial activity tests, along with subfraction 4, the ethyl acetate fraction, and all 7 crude extracts. We did not use subfraction 5 for the antimicrobial assays due to the lower yield compared to subfraction 4, since both have a similar chromatographic profile.

2.1.3. The Antimicrobial Activity of *A. mearnsii* Supercritical Extracts

Bioautography

The antimicrobial activity of the extracts was evaluated using the bioautography method against *Staphylococcus aureus* (ATCC 25923) and *Escherichia coli* (ATCC 25922). All the crude extracts inhibited the growth of the Gram-positive microorganism *S. aureus*, while none of the crude extracts inhibited the growth of the Gram-negative microorganism *E. coli*. These results are in agreement with those reported for the extracts of *Acacia podalyriifolia*, which also showed an inhibition for *S. aureus* and presented no activity against *E. coli* [49].

Minimum Inhibitory Concentration

The minimum inhibitory concentration (MIC) was then determined only against *S. aureus*. MIC tests were performed with the 7 crude extracts, obtained according to factorial design, and with some fractions separated by column chromatography. The best result for subfraction 4 (Table 2) was attributed to the higher *p*-anisic acid concentration verified by HPLC analyses. The activity of phenolic acids against *S. aureus* has been verified in several studies [50–52]. According to Basri et al. [53], the minimum inhibitory concentration (MIC) of *p*-anisic acid against *S. aureus* is MIC = 15.0.

Table 2. Minimal inhibitory concentration (MIC) of *A. mearnsii* crude extracts and its purified fractions against *S. aureus*.

Sample	MIC (mg·mL^{-1})
Extract 2 (P = 120 bar; cosolvent: ethanol)	24
Extract 7 (P = 180 bar; cosolvent: ethanol:water 1:1 v/v)	24
Extracts 1, 3, 4, 5, and 6	>24
Ethyl acetate fraction	59.2
Subfraction 2 (solvent: hexane:ethyl acetate 80:20)	35.9
Subfraction 4 (solvent: hexane:ethyl acetate 40:60)	11.8

Although the MIC of the *A. mearnsii* flower extracts obtained with supercritical fluid was high, its combination with synthetic and classic antibiotics may enhance the extract of *A. mearnsii* activity. According to studies, the synergistic effect of the association of antibiotics with plant extracts against resistant bacteria leads to new options for the treatment of infectious diseases. [54]. For example, the extracts of *Solanum paludosum Moric* obtained by the supercritical fluid process, tested by Siqueira [55], have not presented antibacterial activity but have shown modulating activity that reduced the antibiotic MIC up to eight times. Olajuyigbe & Afolayan [56] tested the effect of the methanolic extract of *A. mearnsii* bark and its synergistic effect when combined with antibiotics against 8 bacteria of clinical relevance. Concerning *S. aureus* (ATCC 6538), the minimum inhibitory concentration of the extract was 0.313 mg·mL^{-1} and it has presented synergism with antibiotics such as erythromycin, metronidazole, amoxicillin, chloramphenicol, and kanamycin. Thus, despite the low effect of *A. mearnsii* flower extracts, there is the potential for exploring its combined use with traditional drugs. These results also suggest future tests against other Gram-positive bacteria such as *Staphylococcus epidermidis*, one of the main causative agents of hospital infection [57].

2.2. Step II—The Optimization of SFE of p-Anisic Acid and the Mathematical Modeling of Mass Transfer

Once the *p*-anisic acid was identified in the extract, the next step was to maximize the compound selectivity due to its wide applicability. Until this point, there was no data available in the literature on the supercritical extraction of flowers from *A. mearnsii*, which justified the study.

From the Step I results, a new study regarding the extraction process conditions was evaluated. The use of ethanol as the cosolvent was maintained, considering the results obtained from the factorial design. As larger extract yields were observed for higher CO_2 pressures set in the factorial design, the pressure range was increased from 200 to 300 bar. The solvability and diffusivity of the supercritical fluid are directly related to its density, which is a function of temperature and pressure. Thus, another factor selected for the optimization process was the temperature range, from 40 to 60 °C. The final parameter evaluated was the particle size. The flowers were ground and passed through a series of six sieves (24, 32, 42, 60, 150, and 325 mesh). Based on the amount retained in each sieve, the fractions retained in the 42, 60, and 150 mesh sieves (0.423, 0.303, and 0.125 mm, respectively) were used. From these variables, a Box–Behnken design was established and the results for the global extract yield and the *p*-anisic acid selectivity are presented in Table 3, where extractions have the solvent-to-feed ratio (S/F) equal to 61.8 $g_{solvent}/g_{plant}$.

Table 3. Design matrix in the Box–Behnken model and observed responses.

Run Order	Uncodified Variables			Responses	
	Pressure (bar)	Temperature (°C)	Medium Particle Size (mesh)	*p*-Anisic Acid Yield (% w/w) [a]	Global Extract Yield (% w/w) [b]
1	250	50	60	1.83	1.76
2	200	60	60	0.57	7.64
3	300	40	60	2.48	2.98
4	200	40	60	2.19	5.11
5	250	50	60	2.11	6.65
6	250	60	150	1.86	4.45
7	300	50	150	0.82	2.51
8	200	50	150	1.96	2.49
9	200	50	42	1.21	4.85
10	300	60	60	0.84	3.26
11	300	50	42	2.35	0.86
12	250	40	42	2.29	1.88
13	250	40	150	1.85	2.92
14	250	50	60	2.17	3.54
15	250	60	42	1.12	2.44

[a] grams of *p*-anisic acid in 100 g of crude extract; [b] grams of crude extract from 100 g of dried flowers.

The p-anisic acid selectivity ranged from 0.57 to 2.48% w/w (g p-anisic acid/g extract) in the crude extracts and the crude extract yield ranged from 0.86 to 7.84% w/w (g extract/g plant). The global extract yield using the Box–Behnken design had a significant increase since the highest yield was 2.49% w/w in the factorial design.

To evaluate the response surface model, an analysis of variance (ANOVA) was performed with the statistical software Minitab® using the results of Table 3. The statistical significance and the influence of the extraction parameters were estimated by the analysis of variance regarding the p-anisic acid selectivity, which are presented in Table 4.

Table 4. Analysis of variance of the p-anisic acid selectivity from the crude extract.

Source	DF	Seq SS	Adj SS	Adj MS	F	p
Regression	9	0.000501	0.000501	0.000056	5.79	0.034
Linear	3	0.000252	0.000059	0.00002	2.05	0.226
T	1	0.000244	0.000001	0.000001	0.09	0.773
P	1	0.000004	0.000057	0.000057	5.92	0.059
G	1	0.000004	0.000003	0.000003	0.34	0.587
Square	3	0.000054	0.000054	0.000018	1.87	0.252
T∗T	1	0.000007	0.00001	0.00001	1.02	0.359
P∗P	1	0.000046	0.000047	0.000047	4.9	0.078
G∗G	1	0.000001	0.000001	0.000001	0.1	0.76
Interaction	3	0.000195	0.000195	0.000065	6.75	0.033
T∗P	1	0	0	0	0	0.975
T∗G	1	0.000072	0.000072	0.000072	7.47	0.041
P∗G	1	0.000123	0.000123	0.000123	12.79	0.016
Residual error	5	0.000048	0.000048	0.00001		
Lack-of-fit	3	0.000041	0.000041	0.000014	4.18	0.199
Pure error	2	0.000007	0.000007	0.000003		
Total	14	0.000549				

DF: Degrees of freedom; Seq SS: sequential sum of squares; Adj SS: adjusted sum of squares; F: F-statistics; p: p-value. T, P, and G correspond to the variables: temperature, pressure, and medium particle size, respectively. ∗: the interaction between factors.

According to the ANOVA, only the interaction between pressure (P) and the average particle size (G) and the interaction between the temperature (T) and particle size (G) were significant ($p < 0.05$), considering a significance level of 95% ($\alpha = 0.05$). The analysis also indicated that the regression was statistically significant and could be applied to describe the variation in the p-anisic acid amounts in the extract. However, p-values larger than 0.05 were obtained for the quadratic and linear regressions, indicating that part of the data behaved linearly and in a partly quadratic fashion. Thus, the regression coefficients from the response surface were estimated for the full quadratic Box–Behnken model [58], resulting in Equation (2) (for variables not coded):

$$\text{selectivity}\left(\%\frac{W_{p-\text{anisic acid}}}{W_{\text{extract}}}\right) = -9.742 + 8.782 \times 10^{-2} P + 5.495 \times 10^{-2} T + 1.660 \times 10^{-2} G - 1.429 \times 10^{-4} P^2 - 1.631 \times 10^{-3} T^2 - 3.523 \times 10^{-5} G^2 - 1.003 \times 10^{-5} PT - 1.858 \times 10^{-4} PG - 7.097 \times 10^{-4} TG \quad (2)$$

The model given by Equation (2) fits the experimental data with a coefficient of determination equal to 0.9125. The validity of the model was further confirmed by the non-significant value ($p = 0.199 > 0.05$), which indicated the quadratic model as a statistically significant model for the response. Through this equation, the response surfaces shown in Figure 2 were generated. The higher selectivity of p-anisic acid was obtained at lower temperatures, smaller mean particle sizes, and higher-pressures, as can be observe in the Figure 2. The combination of high pressure with smaller particles led to bed compaction and preferential flow paths, so a better result was observed with high pressure and larger particle size (smaller mesh).

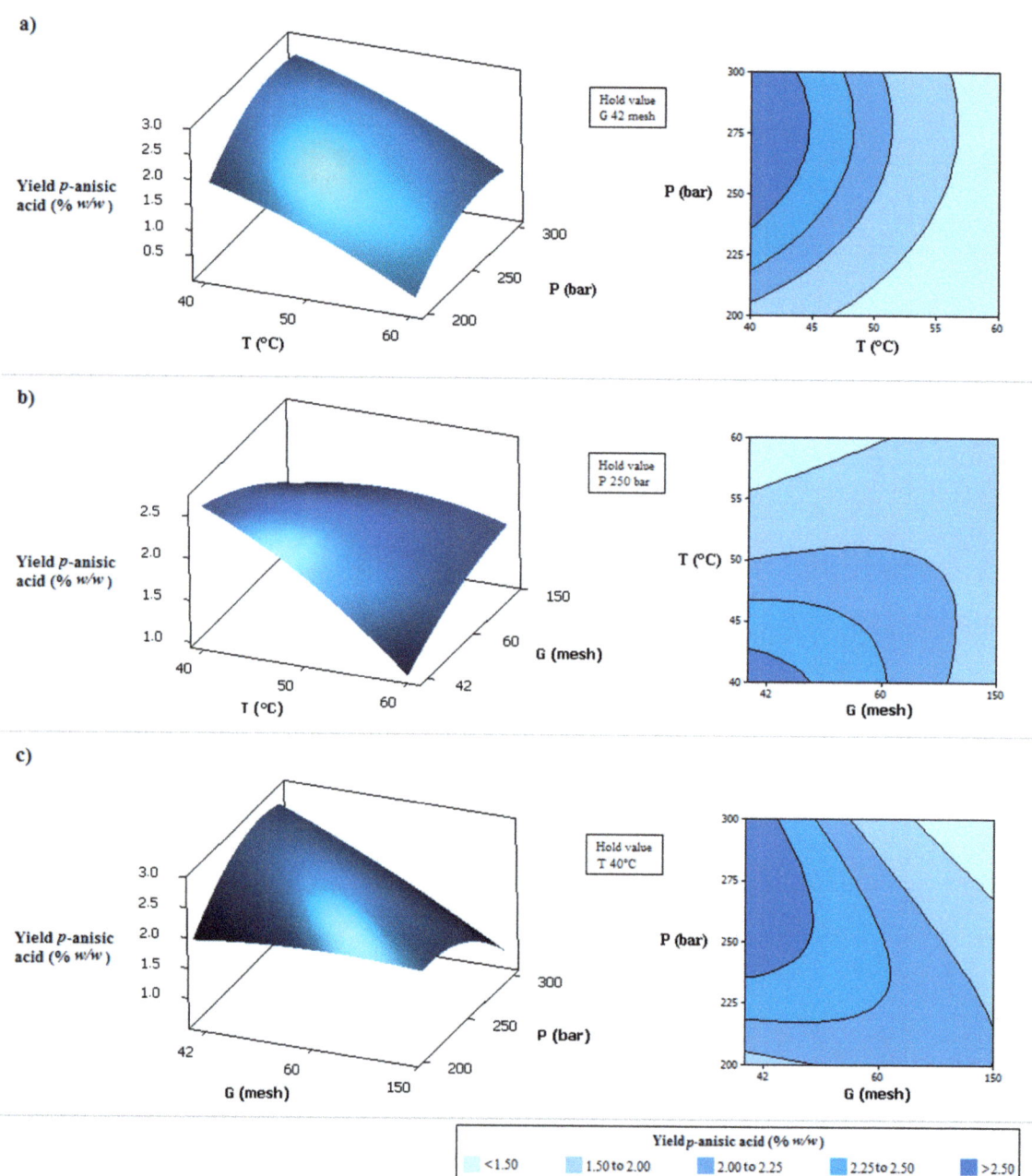

Figure 2. Response surfaces and contour plots for effects of two independent variables on the yield of *p*-anisic acid in the extract obtained by supercritical extraction: (**a**) CO_2 pressure (P) and CO_2 temperature (T); (**b**) CO_2 temperature (T) and medium particle size of milled flowers (G); (**c**) CO_2 pressure (P) and medium particle size of milled flowers (G).

From the response optimizer of the Minitab® software, the optimal parameters to maximize the *p*-anisic acid selectivity were defined as 278.8 bar, 40 °C, and 42 mesh and, thus, the yield would be 2.76% (Figure 3). Under these operating conditions, adjusting the pressure to 279 bar, the extraction was carried out in triplicate and the yield curves versus time were determined. The average selectivity of *p*-anisic acid was 2.51%, indicating an error of 9.06% to the value estimated by the model, confirming again the good fit. This result is also very close to the value found for extraction in the conditions of 300 bar, 40 °C, and 60 mesh (*p*-anisic acid selectivity of 2.48%) due to the slight variation in the extraction conditions. However, when working with lower pressure, there is an increase in the energy efficiency involved in the process, which is another important factor regarding process optimization.

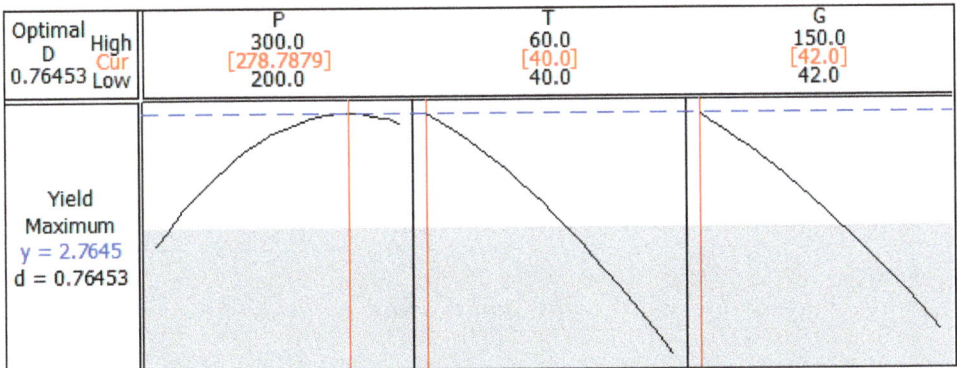

Figure 3. Process parameters optimized to maximum yield of *p*-anisic acid.

The mass transfer mathematical modeling was performed using the three selected models and the global yield versus time curve. The experimental data and the fitted models are shown in Figure 4. The modeling was performed considering the plant particle as a sphere and its average diameter as the average particle size (mesh).

The three models presented a good fit for the experimental data. However, as shown in Figure 4, some differences between the models were noticed. The Sovová model notes that the grinding process breaks the cell walls, making the solute easily accessible, while the Reverchon model notes that there is little solute that is easily accessible. Comparing the models with the experimental data, the Sovová model fits better in the initial extraction stage, which suggests that the grinding of *A. mearnsii* flowers increases the solute availability. The linear behavior at the beginning of extraction is observed by several authors [40,59–63] and is associated with the saturation of the particle surface, caused by grinding and its subsequent exposure to the extract. Despite being a simplified model, the model proposed by Crank presented a better fit with the data. Considering the good fit of the models, it is possible to say that internal diffusion controls the supercritical fluid extraction process of *A. mearnsii* flowers.

The MATLAB® optimization tool was used for the Crank model fitting. The order of magnitude for the Crank model internal diffusion coefficient was the same as was found by Goto et al. [64] and Hornovar et al. [65]. The same software was also used to estimate the four parameters of the Sovová model: Z, W, x_k, and y_r, which were obtained by the least-squares method and were minimized by the Nelder–Mead simplex method [66]. Once these parameters were estimated, the mass transfer coefficients for solid and liquid phases were calculated. The solid phase mass transfer coefficient presented an order of magnitude of 10^{-9} (m/s) which agrees with the values obtained by Scopel et al. [67] and Nagy et al. [68]. The mass transfer coefficient for the solvent was found to equal 9.6×10^{-10} m/s, whose order of magnitude is the same as found by Gallo et al. [69] when studying the supercritical extraction of *pyrethrum* flowers. The Reverchon model was implemented in the simulation software EMSO [70] and the system of equations was solved by an integrator of multiple steps, optimized by a flexible polyhedron. Thus, the values of 5.8×10^{-4} (s^{-1}) for the internal mass transfer coefficient and 5.3×10^{-3} for the equilibrium constant were estimated by the least-squares method. The order of magnitude found for the parameters coincides with the values determined by Silva et al. [32], Garcez et al. [71], Almeida et al. [60], Scopel et al. [67], and Campos et al. [72]. All the parameters and the coefficients of determination for each model are presented in Table 5.

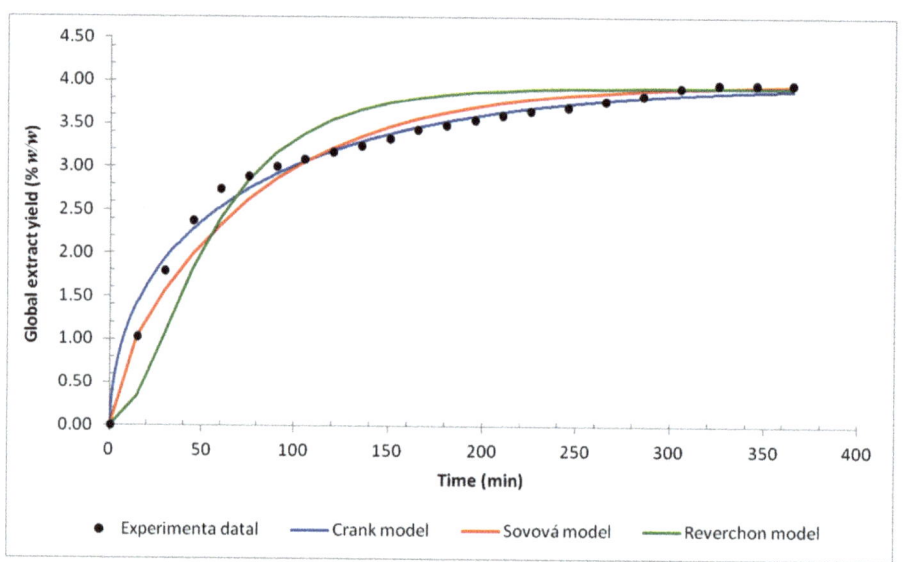

Figure 4. The yield curve for supercritical fluid extraction at 40 °C, 279 bar, and milled flower (42 mesh): mathematical models and experimental data.

Table 5. Adjusted and calculated parameters for mathematical models of mass transfer.

Model	Adjusted Parameters				Calculated Parameters	
Crank (R^2 = 0.9865)	D (m$^2 \cdot$s^{-1}) 8.424×10^{-10}					
Sovová (R^2 = 0.9772)	Z 4.721×10^{-2}	W 7.753×10^{-2}	x_k 3.532×10^{-2}	y_r 5.368×10^{-1}	k_s (m·s^{-1}) 1.206×10^{-9}	k_f (m·s^{-1}) 9.684×10^{-10}
Reverchon (R^2 = 0.9420)	t_i (s) 1710	K (m$^3 \cdot$kg^{-1}) 5.294×10^{-3}			D_i (m$^2 \cdot$s^{-1}) 1.228×10^{-12}	k_{TM} (m·s^{-1}) 5.848×10^{-4}

3. Discussion

In this work, the application of supercritical fluid extraction to obtain extracts of *A. mearnsii* flowers was studied. *A. mearnsii* flowers are a widely available feedstock since the tree is extensively cultivated due to the industrial interest in the production of tannins and wood chips. The investigation of antimicrobial activity revealed that the extracts have activity against *S. aureus*. The best result was observed for the fraction containing a higher concentration of *p*-anisic acid, suggesting a relationship between the activity and this compound. Even if the minimal inhibitory concentration values are larger than for other extracts reported in the literature, there is the potential to explore its combined use with traditional drugs. Furthermore, *p*-anisic acid is a raw material used in the cosmetic, food, and pharmaceutical industries. Thus, the extraction process was optimized to maximize the extraction of the compound of interest.

The use of factorial and Box–Behnken designs was an efficient tool to study the influences of the process parameters in the extraction yield. The maximum global yield in the factorial design was 2.49%, while in the Box–Behnken design it was 7.84%, showing an increment of more than three times. On the other hand, this study shows that a higher global yield is not directly linked to a higher p-anisic acid yield.

Furthermore, the three mass transfer models fitted well with the experimental data and demonstrated that diffusion is the main mechanism associated with this process. The mass transfer parameters related to the description of the extraction process obtained in this work may be useful in future work in the scale-up and optimization of processes for obtaining the supercritical carbon dioxide extract of *A. mearnsii* flowers. Thus, this work contributes to the increasing interest in the potential use of extracts of *Acacia mearnsii* flowers.

4. Materials and Methods

4.1. Plants

The flowers of *A. mearnsii* were supplied by Tanac S/A (RS, Brazil) and were harvested in Piratini-RS (31°17′51″ S, 53°13′29″ W) during the spring (October). The Tanac S/A is a world leader in the production of tanning plant extracts; the company has approximately 23,000 hectares of planted forests and plants around 2000 trees per hectare. All the flowers were oven-dried at 40 °C for 48 h. In the first step of the extractions, the plant material was not milled. In the process optimization with Box–Behnken design, the plant was ground in a Wiley mill. The particle size analysis was performed using a set of standard Tyler Series sieves. One hundred grams of the milled plant was added to a set of 6 sieves with 24, 32, 42, 60, 150, and 325 mesh and they were shaken for 15 min using a vibrating stirrer. The fractions with particle sizes of 42, 60, and 150 mesh were selected for the experiments [73].

4.2. Supercritical Fluid Extraction (SFE)

The extraction process was carried out in a pilot plant described in detail in previous works [74,75]. All experiments were conducted in the 500 mL vessels loaded with 80 g of *A. mearnsii* dried flowers. The carbon dioxide mass flow was 800 g·h^{-1} with a cosolvent flow rate of 0.5 mL·min^{-1}.

4.3. Experimental Designs

4.3.1. Factorial Design

Initially, the supercritical fluid pressure and cosolvent effects in the global extraction yield were investigated; for that, a 2^2 factorial design was used. The pressure was evaluated at 120, 180, and 240 bar and ethanol, water, and a mixture of ethanol and water (1:1 v/v) were used as cosolvents, aiming to promote the extraction of polar compounds. The choice of these variables was supported by reports on the optimization of supercritical extraction [71,76,77]. In this step, the carbon dioxide temperature was 40 °C for all extractions. The experiments were performed in triplicate at the factorial design's central point. The response surface was then associated with the regression equation. The evaluation of the effects of the variables and their interactions was achieved through an analysis of variance (ANOVA) [78].

4.3.2. Box–Behnken Design

The second step of this study was comprised of a Box–Behnken design, which was used for the extraction process optimization to obtain the compound, p-anisic acid. This experimental design consisted of a fractional factorial design method in three levels [55]. Three variables were evaluated, resulting in a total of 15 experiments. The pressure was evaluated at 200, 250, and 300 bar; the temperature was evaluated at 40, 50, and 60 °C; and an average particle size of 42, 60, and 150 mesh (0.423, 0.303, and 0.125 mm, respectively) were used. All 15 experiments used 4.7% (w/w) ethanol as the cosolvent since p-anisic acid is highly soluble in this solvent. These process variables were chosen based on previous studies that used RSM to analyze the SFE of plant material [32,79–81].

A polynomial equation was fitted into the experimental data. The optimal extraction condition was achieved in terms of the highest selectivity of the interest compound (*p*-anisic acid). The matrix design was developed in Minitab® software.

4.4. Chemical Analysis

To investigate the chemical composition of the extracts, each was separated by column chromatography. Both the extracts and the fractions were analyzed by thin-layer chromatography (TLC) and high-performance liquid chromatography (HPLC).

4.4.1. Silica Gel Column Chromatographic Separation

Forty-two grams of the silica gel 60 (Merck) was placed into a glass column with a silica height of 15 cm. The extract was eluted with 250 mL of the following organic solvents: hexane, dichloromethane, ethyl acetate, and methanol, in increasing orders of polarity. The separation occurred in a vacuum, with 0.5 g of the dry extract [82]. In the second step, the ethyl acetate fraction was eluted in a vacuum using, as a mobile phase, the gradients shown in Table 6. The ethyl acetate fraction was chosen based on the TLC and HPLC results.

Table 6. Gradient solvent system used in column chromatography.

Solvent	Ratio (% v/v)	Subfraction Collected
Hexane	100	1
Hexane:Ethyl acetate	80:20	2
Hexane:Ethyl acetate	60:40	3
Hexane:Ethyl acetate	40:60	4
Hexane:Ethyl acetate	20:80	5
Ethyl acetate	100	6
Ethyl acetate:Dichloromethane	50:50	7
Dichloromethane	100	8
Dichloromethane:Methanol	50:50	9
Methanol	100	10

4.4.2. High-Performance Liquid Chromatography (HPLC)

The extracts were analyzed in Agilent 1200 Series liquid chromatography equipped with a U.V. detector. The separation was carried out in a C18 column (4.6 mm × 250 mm × 5 µm). The mobile phase used a binary system consisting of water (A) and acetonitrile (B), both with 2% acetic acid, in gradient mode of 80–20% of B in 30 min, with a 1.0 mL·min^{-1} flow. The injected sample volume was 5 µL and the detector was set at 258 nm. The content of *p*-anisic acid in the extracts was evaluated with a calibration curve ($R^2 = 0.9986$) containing 0.6, 0.45, 0.3, 0.15, and 0.06 mg·mL^{-1} of standard *p*-anisic acid (99% purity, Sigma Aldrich). The extracts were diluted in acetonitrile at 10 mg·mL^{-1} for their analysis by HPLC.

4.5. Antimicrobial Activity

The indirect bioautography method [83] was used to indicate the antimicrobial activity of the supercritical fluid extracts from *A. mearnsii* flowers. The analysis was performed to evaluate the activity against the microorganisms *Staphylococcus aureus* (ATCC 25923) and *Escherichia coli* (ATCC 25922). In this analysis, the extracts were applied to a thin layer chromatography (TLC) plate and were submitted to a run with dichloromethane as the mobile phase. After the solvent evaporation, the TLC plates were plunged into the culture medium inoculated with the microorganisms and were incubated for 24 h at 37 °C. The inoculum was prepared with a suspension of the microorganisms of 1.0×10^4 CFU·mL^{-1} and was incorporated in the Mueller–Hinton agar. After the growth phase, a solution of

INT (p-iodonitrotetrazolium violet) was added for better visualization of the inhibition halos [84]. Amoxicillin 0.1 mg·mL^{-1} was used as qualitative positive control while the culture medium inoculated with the microorganism was used as a negative control, without the presence of the extracts.

The antimicrobial activity was determined by the minimum inhibitory concentration (MIC) using a dilution method on microplates [85]. The microorganism inoculum was prepared with the colonies in a saline solution and the Mueller–Hinton broth, resulting in a final concentration of 1.0×10^4 CFU·mL^{-1}. In each microplate well, 100 µL of the Mueller–Hinton broth containing inoculums, followed by 100 µL of the extract solubilized in Tween 20, as well as water (at final concentrations of 0.75, 1.5, 3.0, 6.0, 12.0, and 24 mg·mL^{-1}), were introduced. Other concentrations were used for the fractions obtained after the column chromatographic separation, defined according to each fraction weight. For the ethyl acetate, the final concentrations were 1.8, 3.6, 7.3, 14.7, 29.5, and 59.2 mg·mL^{-1}; for the fraction obtained with hexane:ethyl acetate (80:20), named "subfraction 2", the final concentrations were 2.2, 4.4, 8.9, 17.9, 35.9, and 71.3 mg·mL^{-1}; and the fraction obtained with hexane:ethyl acetate (60:40), named "subfraction 4", was tested at concentrations of 0.73, 1.4, 2.9, 5.9, 11.8, and 23.6 mg·mL^{-1}.

4.6. Mass Transfer Mathematical Modeling

The description of the mass transfer phenomena involved in the supercritical fluid extraction was performed and evaluated using three models, as described below.

4.6.1. Crank (1975) Model

The Crank Model [86] was developed from Fick's Second Law, which considers the diffusion of a single particle in the form of a sphere. This model describes the behavior of the bed, as a whole, from the simplified mass transfer process in a single particle. The model shows that the sphere is initially at a uniform concentration and that the surface concentration is maintained as a constant. The total amount of the diffusing substance entering or leaving the sphere can be written as [86]:

$$\frac{M_t}{M_\infty} = 1 - \frac{6}{\pi^2} \sum_{n=1}^{\infty} \frac{1}{n^2} \exp\left(-\frac{Dn^2\pi^2 t}{r^2}\right) \quad (3)$$

where M_t and M_∞ are the mass in a determined time and an infinite time (maximum mass obtained in the extraction), respectively, D is the diffusivity of the solute inside the particle (m^2·s^{-1}), t is the extraction time (s), r is the particle radius (m) and n is the number of the series expansion.

4.6.2. Sovová (1994) Model

The model proposed by Sovová [63] considers that the extraction of the solute by supercritical CO$_2$ can be divided into three periods. The first period of extraction considers that only the easily accessible solute can be extracted, which has direct contact with the solvent; the second period considers that the easily accessible solute is gradually depleted from the inlet to the outlet of the bed, and the third period includes solutes that are difficult to access, contained within the particles. Therefore, the extract mass initially present in the solid phase (O) is the sum of the easily accessible mass of solute (P) and the inaccessible mass of solute contained within the solid particles (K). The solid phase, free from the solute (N), is the constant during the extraction and relates to the initial concentrations of the solute:

$$x(t=0) = x_0 = \frac{O}{N} = x_p + x_k = \frac{P}{N} + \frac{K}{N} \quad (4)$$

The mass balance in the fluid phase and solid phase for a bed element is described by two differential equations that were analytically solved by Sovová [63], who simplified a few hypotheses. The final expression is given by Equation (5) in terms of the extract mass relative to the extract-free solid mass:

$$e = \begin{cases} qy_r[1 - \exp(-Z)] \\ y_r[q - q_m \exp(z_w - Z)] \\ x_0 - \frac{y_r}{W} \ln\left\{1 + \left[\exp\left(W\frac{x_0}{y_r} - 1\right)\exp[W(q_m - q)]\frac{x_k}{x_0}\right]\right\} \end{cases} \quad (5)$$

where:

$$q = \frac{Q\,t}{N} \quad (6)$$

$$q_m = \frac{(x_0 - x_k)}{y_r\,Z} \quad (7)$$

$$q_n = q_m + \frac{1}{W} \ln \frac{x_k + (x_0 - x_k)\exp(Wx_0/y_r)}{x_0} \quad (8)$$

$$\frac{z_w}{Z} = \frac{y_r}{w\,x_0} \ln \frac{x_0 \exp[W(q - q_m)] - x_k}{x_0 - x_k} \quad (9)$$

$$Z = \frac{k_f a_0 \rho}{\dot{q}(1-\varepsilon)\rho_s} \quad (10)$$

$$W = \frac{k_s a_0}{\dot{q}(1-\varepsilon)} \quad (11)$$

In the above equations, y_r is the solubility; Z and W are the adjustable parameters for fast and slow periods, respectively, and are directly proportional to the mass transfer coefficients of each phase; the term z_w corresponds to the boundary coordinate between fast and slow extraction; and k_f and k_s are the mass transfer coefficients of fluid and solid phases, respectively. The unknown quantities x_k, y_r, k_s, and k_f were estimated by the least-squares method.

4.6.3. Reverchon (1996) Model

The model proposed by Reverchon [25] was developed from a mass balance for the solid and fluid phases, according to Equations (12) and (13). In the study, the extract is considered a pseudo component, which is not readily available on the surface of the particles after the milling process. As a result, the mass transfer is controlled by internal resistance. The axial dispersion is considered negligible. The density and solvent flow rate are said to be constant along the bed:

$$uV\frac{\partial c}{\partial h} + \varepsilon V\frac{\partial c}{\partial t} + (1-\varepsilon)V\frac{\partial q}{\partial t} = 0 \quad (12)$$

$$(1-\varepsilon)V\frac{\partial q}{\partial t} = -A_p k_{TM}(q - q^*) \quad (13)$$

where u is the interstitial velocity of the fluid, ε is the bed porosity, and ρ_s is the plant density. The previous differential equations satisfy the initial conditions described by $c(h,0) = 0$ and $q(h,0) = q_0$ for all h, and the following boundary conditions $C(0,t) = 0$ for all t. A linear relationship describes the equilibrium behavior between the phases during the supercritical fluid extraction process [25]:

$$q^* = K\,C \quad (14)$$

where K is the volumetric partition coefficient of the extract between fluid and solid phases at the equilibrium condition. Reverchon (1996) sets the internal diffusion time as:

$$t_i = \frac{(1-\varepsilon)V}{A_p k_{TM}}. \quad (15)$$

and Equation (13) can be rewritten as:

$$\frac{\partial q}{\partial t} = -\frac{1}{t_i}(q - q^*); \qquad (16)$$

and the internal diffusion time (Equation (17)) is related to the internal diffusion coefficient (D_i):

$$t_i = \frac{\mu \, l^2}{D_i} \qquad (17)$$

where μ is a constant related to the particle geometry (equal to 3/5 for spherical particles) and l is the characteristic dimension given by the ratio between the particle volume and the particle superficial area.

5. Conclusions

The Factorial 2^2 design indicated that the crude extract yield was higher when ethanol was used as the cosolvent, and the maximum tested pressure (240 bar) was applied in the extraction. The crude extract obtained showed antimicrobial action against *S. aureus*. The purification by silica gel column chromatography generated a fraction rich in a compound identified as *p*-anisic acid and this fraction improved the antimicrobial performance against *S. aureus*. The use of the selectivity criterion as an objective function in the response surface method demonstrated that the condition that includes 278.8 bar, 40 °C, and 42 mesh is the condition that produces the highest amount of *p*-anisic acid per unit of extract mass. This result is important, as this optimal condition is obtained with the use of eco-friendly solvents. The three mathematical models used to simulate the extraction kinetics were adequate for the supercritical CO_2 extraction, with aqueous ethanol as the cosolvent, from *A. meanrsii* flowers. Thus, the supercritical extraction is an adequate and clean method to obtain *p*-anisic acid from *Acacia mearnsii* flowers and this work demonstrates that the potential usage of this plant material is abundant but not exploitative. As a new source of bioactive compounds, the use of this flower extract also contributes to reducing the volume of solid waste generated by the cultivation of this plant in forests.

Supplementary Materials: The following supporting information is available online, Figure S1: chromatogram of extract 1 (P = 120 bar, cosolvent: water), Figure S2: chromatogram of extract 2 (P = 120 bar, cosolvent: ethanol), Figure S3: chromatogram of extract 3 (P = 240 bar, cosolvent: water), Figure S4: chromatogram of extract 4 (P = 240 bar, cosolvent: ethanol), Figure S5: chromatogram of extract 5 (P = 180 bar, cosolvent: ethanol:water), Figure S6: chromatogram of extract 6 (P = 180 bar, cosolvent: ethanol:water), Figure S7: chromatogram of extract 7 (P = 180 bar, cosolvent: ethanol:water), Figure S8: chromatograms of (a) subfraction 4 and (b) subfraction 5 obtained by silica gel column chromatography separation, Figure S9: TLC for the ethyl acetate fraction and subfractions 1–10 after application of sulfuric vanillin as color reagent (the marks drawn represent the spots visualized under U.V. light).

Author Contributions: Conceptualization, G.F.d.S., E.C. and R.M.F.V.; data curation and formal analysis, G.F.d.S., E.T.d.S.J., R.N.A., A.L.B.F., A.T.d.E.S. and A.M.L.; investigation, G.F.d.S., E.T.d.S.J., A.L.B.F. and A.M.L.; writing of the first draft, G.F.d.S.; writing—review and editing, G.F.d.S., R.N.A., E.C. and R.M.F.V.; supervision, A.M.L., E.C. and R.M.F.V. All authors have read and agreed to the published version of the manuscript.

Funding: This work was supported by the National Council for Scientific and Technological Development (CNPq-425933/2018-0) and the National Council for the Improvement of Higher Education (CAPES-Finance Code 001).

Institutional Review Board Statement: Not applicable.

Informed Consent Statement: Not applicable.

Data Availability Statement: Not applicable.

Acknowledgments: The authors are grateful to the CNPq and CAPES for financial support and to TANAC S/A for flower supply.

Conflicts of Interest: The authors declare no conflict of interest.

Sample Availability: Samples of the extracts are not available from the authors.

Nomenclature

a_0	Superficial area	m^{-1}
A_p	Total area of particles	m^2
c	Extract concentration on fluid phase	$kg \cdot m^{-3}$
D	Diffusivity of the solute inside the particle	$m^2 \cdot s^{-1}$
D_i	Internal diffusion coefficient	$m^2 \cdot s^{-1}$
e	Extract mass in regard to N	-
G	Particle size	mesh
h	Bed height	m
h	Axial coordinate	m
J	Mass transfer rate	$kg \cdot s^{-1} \cdot m^{-3}$
K	Inaccessible solute mass inside the solid particles	kg
K	Equilibrium coefficient	$m^3 \cdot kg^{-1}$
k_f	Mass transfer coefficient for fluid phase	$m \cdot s^{-1}$
k_s	Mass transfer coefficient for solid phase	$m \cdot s^{-1}$
k_{TM}	Internal mass transfer coefficient	$m \cdot s^{-1}$
l	Characteristic dimension	m
M_t	Mass extracted at a given time	g
M_∞	Mass extracted in an infinite time	g
N	Solid phase mass free from solute	kg
n	Number of the series expansion	-
O	Initial solute mass in solid phase	kg
P	Easily accessible solute mass	kg
P	Pressure	bar
p	Descriptive p-value level	-
Q	Solvent flow rate	$g \cdot s^{-1}$
q	Specific quantity of solvent	-
q	Extract fraction in solid phase	$kg \cdot kg^{-1}$
q_m	Specific amount of solvent at the beginning of extraction in the interior of particles	-
q_n	Specific amount of solvent at the end of the extraction of easily accessible solute	-
q_0	Initial concentration of extract in solid phase	$kg \cdot kg^{-1}$
q^*	Concentration in the solid–fluid interface	$kg \cdot kg^{-1}$
$\dot{q}$	Solvent mass flow in regard to N	s^{-1}
r	Particle radius	m
T	Temperature	°C
t	Extraction time	s
t_i	Time for internal diffusion	s
u	Solvent superficial velocity	$m \cdot s^{-1}$
V	Extractor volume	m^3
x	Fraction of solute in the solid phase (solute free basis)	
x_0	Initial concentration of free solute in the solid phase (mass fraction)	
x_k	Initial extract concentration inside the solid particles (mass fraction)	
x_p	Easily accessible solute concentration (mass fraction)	
y	Solute fraction in fluid phase (free basis)	
y_r	Extract solubility on solvent	
W	Sovová model parameter for the slow extraction period	
Z	Sovová model parameter for the rapid extraction period	
z_w	Boundary coordinate between fast and slow extraction	

Greek Letters

ε	Porosity	-
μ	Constant related to the particle geometry	-
ρ	Solvent density	kg·m^{-3}
ρ_s	Solid phase density	kg·m^{-3}

References

1. Gurib-Fakim, A. Medicinal plants: Traditions of yesterday and drugs of tomorrow. *Mol. Asp. Med.* **2006**, *27*, 1–93. [CrossRef] [PubMed]
2. Muhammad, A.; Feng, X.; Rasool, A.; Sun, W.; Li, C. Production of plant natural products through engineered Yarrowia lipolytica. *Biotechnol. Adv.* **2020**, *43*, 107555. [CrossRef] [PubMed]
3. García-Lafuente, A.; Guillamón, E.; Villares, A.; Rostagno, M.A.; Martínez, J.A. Flavonoids as anti-inflammatory agents: Implications in cancer and cardiovascular disease. *Inflamm. Res.* **2009**, *58*, 537–552. [CrossRef] [PubMed]
4. Li, R.W.; Myers, S.P.; Leach, D.N.; Lin, G.D.; Leach, G. A cross-cultural study: Anti-inflammatory activity of Australian and Chinese plants. *J. Ethnopharmacol.* **2003**, *85*, 25–32. [CrossRef]
5. Chaubal, R.; Mujumdar, A.M.; Misar, A.; Deshpande, V.H.; Deshpande, N.R. Structure-activity relationship study of androstene steroids with respect to local anti-inflammatory activity. *Arzneim. Forsch./Drug Res.* **2006**, *56*, 394–398. [CrossRef]
6. Barry, K.M.; Mihara, R.; Davies, N.W.; Mitsunaga, T.; Mohammed, C.L. Polyphenols in *Acacia mangium* and *Acacia auriculiformis* heartwood with reference to heart rot susceptibility. *J. Wood Sci.* **2005**, *51*, 615–621. [CrossRef]
7. Meena, P.D.; Meena, R.L.; Chattopadhyay, C.; Kumar, A. Identification of critical stage for disease development and biocontrol of alternaria blight of Indian mustard (*Brassica juncea*). *J. Phytopathol.* **2004**, *152*, 204–209. [CrossRef]
8. Mihara, R.; Barry, K.M.; Mohammed, C.L.; Mitsunaga, T. Comparison of antifungal and antioxidant activities of *Acacia mangium* and *A. auriculiformis* heartwood extracts. *J. Chem. Ecol.* **2005**, *31*, 789–804. [CrossRef]
9. García, S.; Alarcón, G.; Rodríguez, C.; Heredia, N. Extracts of *Acacia farnesiana* and *Artemisia ludoviciana* inhibit growth, enterotoxin production and adhesion of *Vibrio cholerae*. *World J. Microbiol. Biotechnol.* **2006**, *22*, 669–674. [CrossRef]
10. Mandal, P.; Sinha Babu, S.P.; Mandal, N.C. Antimicrobial activity of saponins from *Acacia auriculiformis*. *Fitoterapia* **2005**, *76*, 462–465. [CrossRef]
11. Mutai, C.; Abatis, D.; Vagias, C.; Moreau, D.; Roussakis, C.; Roussis, V. Cytotoxic lupane-type triterpenoids from *Acacia mellifera*. *Phytochemistry* **2004**, *65*, 1159–1164. [CrossRef]
12. Hoffmann, J.J.; Timmermann, B.N.; Mclaughlin, S.P.; Punnapayak, H. Potential antimicrobial activity of plants from the southwestern United States. *Pharm. Biol.* **1993**, *31*, 101–115. [CrossRef]
13. Sulaiman, C.T.; Gopalakrishnan, V.K.; Balachandran, I. Phenolic Compounds and Antioxidant Properties of Selected *Acacia* species. *J. Biol. Act. Prod. Nat.* **2014**, *4*, 316–324. [CrossRef]
14. Maldini, M.; Montoro, P.; Hamed, A.I.; Mahalel, U.A.; Oleszek, W.; Stochmal, A.; Piacente, S. Strong antioxidant phenolics from *Acacia nilotica*: Profiling by ESI-MS and qualitative-quantitative determination by LC-ESI-MS. *J. Pharm. Biomed. Anal.* **2011**, *56*, 228–239. [CrossRef] [PubMed]
15. Chang, S.T.; Wu, J.H.; Wang, S.Y.; Kang, P.L.; Yang, N.S.; Shyur, L.F. Antioxidant activity of extracts from *Acacia confusa* Bark and Heartwood. *J. Agric. Food Chem.* **2001**, *49*, 3420–3424. [CrossRef] [PubMed]
16. Zhang, J.; Yang, J.; Duan, J.; Liang, Z.; Zhang, L.; Huo, Y.; Zhang, Y. Quantitative and qualitative analysis of flavonoids in leaves of *Adinandra nitida* by high-performance liquid chromatography with UV and electrospray ionization tandem mass spectrometry detection. *Anal. Chim. Acta* **2005**, *532*, 97–104. [CrossRef]
17. Singh, B.N.; Singh, B.R.; Sarma, B.K.; Singh, H.B. Potential chemoprevention of N-nitrosodiethylamine-induced hepatocarcinogenesis by polyphenolics from *Acacia nilotica* bark. *Chem. Biol. Interact.* **2009**, *181*, 20–28. [CrossRef]
18. Nisar, M.; Khan, S.; Dar, A.; Rehman, W.; Khan, R.; Jan, I. Antidepressant screening and flavonoids isolation from Eremostachys laciniata (L) Bunge. *Afr. J. Biotechnol.* **2011**, *10*, 9. [CrossRef]
19. Drijfhout, F.P.; Morgan, E.D. Terrestrial natural products as antifeedants. In *Comprehensive Natural Products II*; Elsevier Science: Oxford, UK, 2010; Volume 4, pp. 457–501. ISBN 9780080453828. [CrossRef]
20. Johnson, W. Final Report of the Safety Assessment of Acacia Catechu Gum, Acacia Concinna Fruit Extract, Acacia Dealbata Leaf Extract, Acacia Dealbata Leaf Wax, Acacia Decurrens Extract, Acacia Farnesiana Extract, Acacia Farnesiana Flower Wax, Acacia Farnesiana Gum, Acacia Senegal Extract, Acacia Senegal Gum, and Acacia Senegal Gum Extract1. *Int. J. Toxicol.* **2005**, *24*, 75–118. [CrossRef]
21. Griffin, A.R.; Midgley, S.J.; Bush, D.; Cunningham, P.J.; Rinaudo, A.T. Global uses of Australian acacias-recent trends and future prospects. *Divers. Distrib.* **2011**, *17*, 837–847. [CrossRef]
22. da Silva, R.P.F.F.; Rocha-Santos, T.A.P.; Duarte, A.C. Supercritical fluid extraction of bioactive compounds. *TrAC-Trends Anal. Chem.* **2016**, *76*, 40–51. [CrossRef]
23. Pereira, C.G.; Meireles, M.A.A. Supercritical fluid extraction of bioactive compounds: Fundamentals, applications and economic perspectives. *Food Bioprocess. Technol.* **2010**, *3*, 340–372. [CrossRef]
24. Herrero, M.; Cifuentes, A.; Ibañez, E. Sub- and supercritical fluid extraction of functional ingredients from different natural sources: Plants, food by-products, algae and microalgae-A review. *Food Chem.* **2006**, *98*, 136–148. [CrossRef]

25. Reverchon, E. Mathematical Modeling of Supercritical Extraction of Sage Oil. *AIChE J.* **1996**, *42*, 1765–1771. [CrossRef]
26. Anklam, E.; Berg, H.; Mathiasson, L.; Sharman, M.; Ulberth, F. Supercritical fluid extraction (SFE) in food analysis: A review. *Food Addit. Contam.* **1998**, *15*, 729–750. [CrossRef]
27. Sharif, K.M.; Rahman, M.M.; Azmir, J.; Mohamed, A.; Jahurul, M.H.A.; Sahena, F.; Zaidul, I.S.M. Experimental design of supercritical fluid extraction-A review. *J. Food Eng.* **2014**, *124*, 105–116. [CrossRef]
28. Herrero, M.; Mendiola, J.A.; Cifuentes, A.; Ibáñez, E. Supercritical fluid extraction: Recent advances and applications. *J. Chromatogr. A* **2010**, *1217*, 2495–2511. [CrossRef]
29. Aydar, A.Y. Utilization of Response Surface Methodology in Optimization of Extraction of Plant Materials. *Stat. Approaches Emphas. Des. Exp. Appl. Chem. Processes* **2018**, 157–169. [CrossRef]
30. Li, B.; Xu, Y.; Jin, Y.X.; Wu, Y.Y.; Tu, Y.Y. Response surface optimization of supercritical fluid extraction of kaempferol glycosides from tea seed cake. *Ind. Crops Prod.* **2010**, *32*, 123–128. [CrossRef]
31. do Espirito Santo, A.T.; Siqueira, L.M.; Almeida, R.N.; Vargas, R.M.F.; Franceschini, G.d.N.; Kunde, M.A.; Cappellari, A.R.; Morrone, F.B.; Cassel, E. Decaffeination of yerba mate by supercritical fluid extraction: Improvement, mathematical modeling and infusion analysis. *J. Supercrit. Fluids* **2021**, *168*, 105096. [CrossRef]
32. da Silva, G.F.; Gandolfi, P.H.K.; Almeida, R.N.; Lucas, A.M.; Cassel, E.; Vargas, R.M.F. Analysis of supercritical fluid extraction of lycopodine using response surface methodology and process mathematical modeling. *Chem. Eng. Res. Des.* **2015**, *100*, 353–361. [CrossRef]
33. Mokhtari, L.; Ghoreishi, S.M. Supercritical carbon dioxide extraction of trans-anethole from *Foeniculum vulgare* (fennel) seeds: Optimization of operating conditions through response surface methodology and genetic algorithm. *J. CO_2 Util.* **2019**, *30*, 1–10. [CrossRef]
34. Morcelli, A.; Cassel, E.; Vargas, R.; Rech, R.; Marcílio, N. Supercritical fluid (CO_2 + ethanol) extraction of chlorophylls and carotenoids from Chlorella sorokiniana: COSMO-SAC assisted prediction of properties and experimental approach. *J. CO_2 Util.* **2021**, *51*, 101649. [CrossRef]
35. Pinto, D.; De La Luz Cádiz-Gurrea, M.; Sut, S.; Ferreira, A.S.; Leyva-Jimenez, F.J.; Dall'acqua, S.; Segura-Carretero, A.; Delerue-Matos, C.; Rodrigues, F. Valorisation of underexploited *Castanea sativa* shells bioactive compounds recovered by supercritical fluid extraction with CO_2: A response surface methodology approach. *J. CO_2 Util.* **2020**, *40*, 101194. [CrossRef]
36. Anderson-Cook, C.M.; Borror, C.M.; Montgomery, D.C. Response surface design evaluation and comparison. *J. Stat. Plan. Inference* **2009**, *139*, 629–641. [CrossRef]
37. Vargas, R.M.F.; Barroso, M.S.T.; Neto, R.G.; Scopel, R.; Falcão, M.A.; da Silva, C.F.; Cassel, E. Natural products obtained by subcritical and supercritical fluid extraction from Achyrocline satureioides (Lam) D.C. using CO_2. *Ind. Crops Prod.* **2013**, *50*, 430–435. [CrossRef]
38. Melreles, M.A.A.; Zahedi, G.; Hatami, T. Mathematical modeling of supercritical fluid extraction for obtaining extracts from vetiver root. *J. Supercrit. Fluids* **2009**, *49*, 23–31. [CrossRef]
39. Cassel, E.; Vargas, R.M.F.; Martinez, N.; Lorenzo, D.; Dellacassa, E. Steam distillation modeling for essential oil extraction process. *Ind. Crops Prod.* **2009**, *29*, 171–176. [CrossRef]
40. Mouahid, A.; Bombarda, I.; Claeys-Bruno, M.; Amat, S.; Myotte, E.; Nisteron, J.P.; Crampon, C.; Badens, E. Supercritical CO_2 extraction of *Moroccan argan* (Argania Spinosa L.) oil: Extraction kinetics and solubility determination. *J. CO_2 Util.* **2021**, *46*, 101458. [CrossRef]
41. Saha, R.; Ghosh, A.; Saha, B. Kinetics of micellar catalysis on oxidation of p-anisaldehyde to p-anisic acid in aqueous medium at room temperature. *Chem. Eng. Sci.* **2013**, *99*, 23–27. [CrossRef]
42. Jänichen, J.; Petersen, W.; Jenny, R.; Nobis, M. Process to Manufacture 4-Methoxybenzoic Acid from Herbal Anethole and the Use of 4-Methoxybenzoic Acid in Cosmetic and Dermatologic Products as Well as Foodstuffs. U.S. Patent 7728168, 1 June 2010.
43. Gandhi, P.J.; Murthy, Z.V.P. Transmission of p-anisic acid through nanofiltration and goat membranes. *Desalination* **2013**, *315*, 46–60. [CrossRef]
44. Gandhi, P.J.; Talia, Y.H.; Murthy, Z.V.P. Production of p-Anisic acid by modified williamson etherification reaction using design of experiments. *Chem. Prod. Process. Modeling* **2010**, *5*. [CrossRef]
45. Joint FAO/WHO Expert Committee Evaluation of Certain Food Additives and Contaminants. World Health Organ. Tech. Rep. Ser. 2002. Available online: http://apps.who.int/iris/handle/10665/42578 (accessed on 6 January 2022).
46. Hawthorne, S.B.; Grabanski, C.B.; Martin, E.; Miller, D.J. Comparisons of Soxhlet extraction, pressurized liquid extraction, supercritical fluid extraction and subcritical water extraction for environmental solids: Recovery, selectivity and effects on sample matrix. *J. Chromatogr. A* **2000**, *892*, 421–433. [CrossRef]
47. Pourmortazavi, S.M.; Hajimirsadeghi, S.S. Supercritical fluid extraction in plant essential and volatile oil analysis. *J. Chromatogr. A* **2007**, *1163*, 2–24. [CrossRef] [PubMed]
48. Gamlieli-Bonshtein, I.; Korin, E.; Cohen, S. Selective separation of cis-trans geometrical isomers of β-carotene via CO_2 supercritical fluid extraction. *Biotechnol. Bioeng.* **2002**, *80*, 169–174. [CrossRef] [PubMed]
49. Andrade, C.A. de Estudo Químico e Biológico das Flores e das Folhas de *Acacia podalyriifolia* A. Cunn. Ex, G. Don, Leguminosae-Mimosoideae [Chemical and Biologic Study of Flowers and Leaves of *Acacia podalyriifolia* A. Cunn. Ex, G. Don, Leguminosae-Mimosoideae]. Ph.D. Thesis, Universidade Federal do Paraná, Curitiba, Brazil, 2010.

50. Kwon, Y.I.; Apostolidis, E.; Labbe, R.G.; Shetty, K. Inhibition of *Staphylococcus aureus* by phenolic phytochemicals of selected clonal herbs species of *Lamiaceae* family and likely mode of action through proline oxidation. *Food Biotechnol.* **2007**, *21*, 71–89. [CrossRef]
51. Fernández, M.A.; García, M.D.; Sáenz, M.T. Antibacterial activity of the phenolic acids fractions of *Scrophularia frutescens* and *Scrophularia sambucifolia*. *J. Ethnopharmacol.* **1996**, *53*, 11–14. [CrossRef]
52. Cueva, C.; Moreno-Arribas, M.V.; Martín-Álvarez, P.J.; Bills, G.; Vicente, M.F.; Basilio, A.; Rivas, C.L.; Requena, T.; Rodríguez, J.M.; Bartolomé, B. Antimicrobial activity of phenolic acids against commensal, probiotic and pathogenic bacteria. *Res. Microbiol.* **2010**, *161*, 372–382. [CrossRef] [PubMed]
53. Basri, D.F.; Zin, N.M.; Bakar, N.S.; Rahmat, F.; Mohtar, M. Synergistic effects of phytochemicals and oxacillin on laboratory passage-derived vancomycin-intermediate *Staphylococcus aureus* strain. *J. Med. Sci.* **2008**, *8*, 131–136. [CrossRef]
54. Nascimento, G.G.F.; Locatelli, J.; Freitas, P.C.; Silva, G.L. Antibacterial activity of plant extracts and phytochemicals on antibiotic-resistant bacteria. *Braz. J. Microbiol.* **2000**, *31*, 247–256. [CrossRef]
55. Siqueira, S.; Falcão-Silva, V.D.S.; Agra, M.D.F.; Dariva, C.; De Siqueira-Júnior, J.P.; Fonseca, M.J.V. Biological activities of *Solanum paludosum* Moric. Extracts obtained by maceration and supercritical fluid extraction. *J. Supercrit. Fluids* **2011**, *58*, 391–397. [CrossRef]
56. Olajuyigbe, O.O.; Afolayan, A.J. Synergistic interactions of methanolic extract of *Acacia mearnsii* de wild. with antibiotics against bacteria of clinical relevance. *Int. J. Mol. Sci.* **2012**, *13*, 8915–8932. [CrossRef] [PubMed]
57. Uckay, I.; Pittet, D.; Vaudaux, P.; Sax, H.; Lew, D.; Waldvogel, F. Foreign body infections due to *Staphylococcus epidermidis*. *Ann. Med.* **2009**, *41*, 109–119. [CrossRef] [PubMed]
58. Box, G.E.P.; Behnken, D.W. Some new three level designs for study of quantitative variables Box Behnken. *Technometrics* **1960**, *2*, 455–475. [CrossRef]
59. Barros, F.M.C.; Silva, F.C.; Nunes, J.M.; Vargas, R.M.F.; Cassel, E.; Von Poser, G.L. Supercritical extraction of phloroglucinol and benzophenone derivatives from *Hypericum carinatum*: Quantification and mathematical modeling. *J. Sep. Sci.* **2011**, *34*, 3107–3113. [CrossRef] [PubMed]
60. Almeida, R.N.; Neto, R.G.; Barros, F.M.C.; Cassel, E.; von Poser, G.L.; Vargas, R.M.F. Supercritical extraction of *Hypericum caprifoliatum* using carbon dioxide and ethanol+water as co-solvent. *Chem. Eng. Processing Process Intensif.* **2013**, *70*, 95–102. [CrossRef]
61. Da Silva, C.G.F.; Lucas, A.M.; Do, E.; Santo, A.T.; Almeida, R.N.; Cassel, E.; Vargas, R.M.F. Sequential processing of *Psidium guajava* L. Leaves: Steam distillation and supercritical fluid extraction. *Braz. J. Chem. Eng.* **2019**, *36*, 487–496. [CrossRef]
62. Scopel, R.; da Silva, C.F.; Lucas, A.M.; Garcez, J.J.; do Espirito Santo, A.T.; Almeida, R.N.; Cassel, E.; Vargas, R.M.F. Fluid phase equilibria and mass transfer studies applied to supercritical fluid extraction of *Illicium verum* volatile oil. *Fluid Phase Equilibria* **2016**, *417*, 203–211. [CrossRef]
63. Sovová, H. Rate of the vegetable oil extraction with supercritical CO_2—I. Modeling of extraction curves. *Chem. Eng. Sci.* **1994**, *49*, 409–414. [CrossRef]
64. Goto, M.; Roy, B.C.; Kodama, A.; Hirose, T. Modeling supercritical fluid extraction process involving solute-solid interaction. *J. Chem. Eng. Jpn.* **1998**, *31*, 171–177. [CrossRef]
65. Honarvar, B.; Sajadian, S.A.; Khorram, M.; Samimi, A. Mathematical modeling of supercritical fluid extraction of oil from canola and sesame seeds. *Braz. J. Chem. Eng.* **2013**, *30*, 159–166. [CrossRef]
66. Lagarias, J.C.; Reeds, J.A.; Wright, M.H.; Wright, P.E. Convergence properties of the Nelder–mead simplex method in low dimensions. *SIAM J. Optim.* **1998**, *9*, 112–147. [CrossRef]
67. Scopel, R.; Falcão, M.A.; Lucas, A.M.; Almeida, R.N.; Gandolfi, P.H.K.; Cassel, E.; Vargas, R.M.F. Supercritical fluid extraction from *Syzygium aromaticum* buds: Phase equilibrium, mathematical modeling, and antimicrobial activity. *J. Supercrit. Fluids* **2014**, *92*, 223–230. [CrossRef]
68. Nagy, B.; Simándi, B.; Dezso András, C. Characterization of packed beds of plant materials processed by supercritical fluid extraction. *J. Food Eng.* **2008**, *88*, 104–113. [CrossRef]
69. Gallo, M.; Formato, A.; Ianniello, D.; Andolfi, A.; Conte, E.; Ciaravolo, M.; Varchetta, V.; Naviglio, D. Supercritical fluid extraction of pyrethrins from pyrethrum flowers (*Chrysanthemum cinerariifolium*) compared to traditional maceration and cyclic pressurization extraction. *J. Supercrit. Fluids* **2017**, *119*, 104–112. [CrossRef]
70. Soares, R.D.P.; Secchi, A.R. EMSO: A new environment for modeling, simulation, and optimization. *Comput. Aided Chem. Eng.* **2003**, *14*, 947–952.
71. Garcez, J.J.; da Silva, C.G.F.; Lucas, A.M.; Fianco, A.L.; Almeida, R.N.; Cassel, E.; Vargas, R.M.F. Evaluation of different extraction techniques in the processing of *Anethum graveolens* L. seeds for phytochemicals recovery. *J. Appl. Res. Med. Aromat. Plants* **2020**, *18*, 100263. [CrossRef]
72. Campos, L.M.A.S.; Michielin, E.M.Z.; Danielski, L.; Ferreira, S.R.S. Experimental data and modeling the supercritical fluid extraction of marigold (*Calendula officinalis*) oleoresin. *J. Supercrit. Fluids* **2005**, *34*, 163–170. [CrossRef]
73. Carvalho, R.N.; Moura, L.S.; Rosa, P.T.V.; Meireles, M.A.A. Supercritical fluid extraction from rosemary (*Rosmarinus officinalis*): Kinetic data, extract's global yield, composition, and antioxidant activity. *J. Supercrit. Fluids* **2005**, *35*, 197–204. [CrossRef]
74. Cassel, E.; Vargas, R.M.F.; Brun, G.W.; Almeida, D.E.; Cogoi, L.; Ferraro, G.; Filip, R. Supercritical fluid extraction of alkaloids from *Ilex paraguariensis* St. Hil. *J. Food Eng.* **2010**, *100*, 656–661. [CrossRef]

75. Scopel, R.; Neto, R.G.; Falcão, M.A.; Cassel, E.; Vargas, R.M.F. Supercritical CO_2 extraction of *Schinus molle* L with co-solvents: Mathematical modeling and antimicrobial applications. *Braz. Arch. Biol. Technol.* **2013**, *56*, 513–519. [CrossRef]
76. Liu, J.; Liu, J.; Lin, S.; Wang, Z.; Wang, C.; Wang, E.; Zhang, Y. Supercritical fluid extraction of flavonoids from *Maydis stigma* and its nitrite-scavenging ability. *Food Bioprod. Processing* **2011**, *89*, 333–339. [CrossRef]
77. Cobb, B.F.; Kallenbach, J.; Hall, C.A.; Pryor, S.W. Optimizing the Supercritical Fluid Extraction of Lutein from Corn Gluten Meal. *Food Bioprocess Technol.* **2018**, *11*, 757–764. [CrossRef]
78. Montgomery, D.C.; Runger, G.C. *Applied Statistics and Probability for Engineers*, 3th ed.; John Wiley & Sons Inc.: New York, NY, USA, 2003.
79. Sodeifian, G.; Sajadian, S.A.; Saadati Ardestani, N. Supercritical fluid extraction of omega-3 from *Dracocephalum kotschyi* seed oil: Process optimization and oil properties. *J. Supercrit. Fluids* **2017**, *119*, 139–149. [CrossRef]
80. Garcez, J.J.; Barros, F.; Lucas, A.M.; Xavier, V.B.; Fianco, A.L.; Cassel, E.; Vargas, R.M.F. Evaluation and mathematical modeling of processing variables for a supercritical fluid extraction of aromatic compounds from *Anethum graveolens*. *Ind. Crops Prod.* **2017**, *95*, 733–741. [CrossRef]
81. Chai, Y.H.; Yusup, S.; Ruslan, M.S.H.; Chin, B.L.F. Supercritical fluid extraction and solubilization of *Carica papaya* Linn. leaves in ternary system with CO_2 + ethanol solvents. *Chem. Eng. Res. Des.* **2020**, *156*, 31–42. [CrossRef]
82. Lucas, A.M.H.; Bento, A.F.M.L.; Vargas, R.M.F.; Scheffel, T.B.; Rockenbach, L.; Diz, F.M.; Capellari, A.R.; Morrone, F.B.; Cassel, E. Use of supercritical CO_2 to obtain *Baccharis uncinella* extracts with antioxidant and antitumor activity. *J. CO_2 Util.* **2021**, *49*, 101563. [CrossRef]
83. Valgas, C.; de Souza, S.M.; Smânia, E.F.A.; Smânia Jr., A. Screening methods to determine antibacterial activity of natural products. *Braz. J. Microbiol.* **2007**, *38*, 369–380. [CrossRef]
84. Nostro, A.; Germanò, M.P.; D'Angelo, V.; Marino, A.; Cannatelli, M.A. Extraction methods and bioautography for evaluation of medicinal plant antimicrobial activity. *Lett. Appl. Microbiol.* **2000**, *30*, 379–384. [CrossRef]
85. Falcão, M.A.; Fianco, A.L.B.; Lucas, A.M.; Pereira, M.A.A.; Torres, F.C.; Vargas, R.M.F.; Cassel, E. Determination of antibacterial activity of vacuum distillation fractions of lemongrass essential oil. *Phytochem. Rev.* **2012**, *11*, 405–412. [CrossRef]
86. Crank, J. *The Mathematics of Diffusion*, 2nd ed.; Oxford University Press: New York, NY, USA, 1975.

Article

Experimental Investigation on Thermophysical Properties of Ammonium-Based Protic Ionic Liquids and Their Potential Ability towards CO_2 Capture

Nur Hidayah Zulaikha Othman Zailani [1], Normawati M. Yunus [1,*], Asyraf Hanim Ab Rahim [1] and Mohamad Azmi Bustam [2]

[1] Department of Fundamental and Applied Sciences, Centre of Research in Ionic Liquids (CORIL), Institute of Contaminant Management for Oil and Gas, Universiti Teknologi PETRONAS, Seri Iskandar 32610, Malaysia; zulaikha95.nhz@gmail.com (N.H.Z.O.Z.); asyrafhanim92@gmail.com (A.H.A.R.)

[2] Department of Chemical Engineering, Centre of Research in Ionic Liquids (CORIL), Institute of Contaminant Management for Oil and Gas, Universiti Teknologi PETRONAS, Seri Iskandar 32610, Malaysia; azmibustam@utp.edu.my

* Correspondence: normaw@utp.edu.my; Tel.: +60-5368-7689

Citation: Zailani, N.H.Z.O.; Yunus, N.M.; Ab Rahim, A.H.; Bustam, M.A. Experimental Investigation on Thermophysical Properties of Ammonium-Based Protic Ionic Liquids and Their Potential Ability towards CO_2 Capture. *Molecules* 2022, 27, 851. https://doi.org/10.3390/molecules27030851

Academic Editors: Reza Haghbakhsh, Sona Raeissi and Rita Craveiro

Received: 31 December 2021
Accepted: 24 January 2022
Published: 27 January 2022

Publisher's Note: MDPI stays neutral with regard to jurisdictional claims in published maps and institutional affiliations.

Copyright: © 2022 by the authors. Licensee MDPI, Basel, Switzerland. This article is an open access article distributed under the terms and conditions of the Creative Commons Attribution (CC BY) license (https://creativecommons.org/licenses/by/4.0/).

Abstract: Ionic liquids, which are extensively known as low-melting-point salts, have received significant attention as the promising solvent for CO_2 capture. This work presents the synthesis, thermophysical properties and the CO_2 absorption of a series of ammonium cations coupled with carboxylate anions producing ammonium-based protic ionic liquids (PILs), namely 2-ethylhexylammonium pentanoate ([EHA][C5]), 2-ethylhexylammonium hexanoate ([EHA][C6]), 2-ethylhexylammonium heptanoate ([EHA][C7]), bis-(2-ethylhexyl)ammonium pentanoate ([BEHA][C5]), bis-(2-ethylhexyl)ammonium hexanoate ([BEHA][C6]) and bis-(2-ethylhexyl)ammonium heptanoate ([BEHA][C7]). The chemical structures of the PILs were confirmed by using Nuclear Magnetic Resonance (NMR) spectroscopy while the density (ρ) and the dynamic viscosity (η) of the PILs were determined and analyzed in a range from 293.15K up to 363.15K. The refractive index (n_D) was also measured at T = (293.15 to 333.15) K. Thermal analyses conducted via a thermogravimetric analyzer (TGA) and differential scanning calorimeter (DSC) indicated that all PILs have the thermal decomposition temperature, T_d of greater than 416K and the presence of glass transition, T_g was detected in each PIL. The CO_2 absorption of the PILs was studied up to 29 bar at 298.15 K and the experimental results showed that [BEHA][C7] had the highest CO_2 absorption with 0.78 mol at 29 bar. The CO_2 absorption values increase in the order of [C5] < [C6] < [C7] anion regardless of the nature of the cation.

Keywords: ammonium-based protic ionic liquids; density; viscosity; refractive index; phase transition; thermal expansion coefficient; standard entropy; lattice potential energy; CO_2 absorption

1. Introduction

Natural gas is a naturally occurring hydrocarbon that consists of methane gas primarily followed by other mixtures of higher alkanes such as ethane, propane and butane. Generally, natural gas is widely used as a fuel and a raw material in the petrochemical industry [1,2]. Despite its mixture of combustible hydrocarbons content, trace quantities of argon (Ar), hydrogen (H), helium (He), nitrogen (N_2) as well as carbon dioxide (CO_2) and hydrogen sulfide (H_2S) are also present in natural gas [3]. Sour gas, such as CO_2, is undesirable due to its acidic property that causes corrosion in the gas pipeline [4]. Apart from that, the existence of CO_2 also reduces the fuel value of natural gas due to its non-combustible nature. Therefore, CO_2 removal in the refining process is crucial to improving the value of natural gas and the utilization of amine-based solvents, namely monoethanolamine (MEA), which had been widely practiced on industrial scales to capture CO_2 in natural gas. This chemical absorption of CO_2 by MEA is considered to be the most reliable and efficient technology for capturing CO_2 [5–8]. Gómez-Díaz and his team had compared the ability of their blended amine solvent, which is diamine (N,N-dimethylethylenediamine [DMEDA]) with MEA, towards CO_2 capture in which changes in the amine ratios did not lead to important changes in the absorption curve [6]. Despite the outstanding performance of amine-based solvents, it is also known to have a high vapor pressure and high energy input for regeneration. Therefore, studies related to the utilization of solid adsorbents such as zeolites, activated carbon, amine-functionalized adsorbents and metal organic frameworks (MOFs) had been conducted for CO_2 adsorption due to their uniqueness as they can be personalized to capture CO_2 from either post- or pre-combustion gas streams, depending upon several factors [9]. Current examples of adsorbents for CO_2 adsorption are zeolites, activated carbon, amine-functionalized adsorbents and metal organic frameworks (MOFs) [10–12]. Nonetheless, further analysis using adsorbents showed poor adsorption characteristics at low CO_2 partial pressures [13,14]. Furthermore, membrane separation processes are also used commercially for CO_2 removal from natural gas. However, a single-stage membrane system is not capable of capturing CO_2 with high efficiency [15–20]. Due to the given issue, this had encouraged researchers to find alternative solvents that can capture CO_2.

Recently, ionic liquids have been recognized as promising solvents for CO_2 capture from natural gas. The uniqueness of their properties, specifically their non-detectable vapor pressure, high thermal stability, and high affinity for CO_2, enables ILs to be used as solvents for CO_2 capture at elevated temperatures and pressures [13,21–25]. Moreover, the chemical and physical properties of ILs can be altered due to the availability of countless cation and anion combinations. Several studies involving mainly binary systems of imidazolium-based ILs-CO_2 or imidazolium-based ILs-other gas have revealed the significant solubility of CO_2 in ionic liquids when compared to other gases. For example, comparison studies of CO_2 absorption in individual solvents of 30 wt% of 1-(3-aminopropyl)-3-(2-aminoethyl)imidazolium hydroxide [Apaeim][OH], 30 wt% of 1-(3-aminopropyl)-3-(2-aminoethyl)imidazolium alaninate ([Apaeim][ala]) and 30 wt% monoethanolamine (MEA), have shown that both ILs displayed higher CO_2 absorption capacities than that of MEA solvent by the value of 2.2-fold. This has further proven that ILs are promising solvents for CO_2 capture [26]. In addition, a different class of ionic liquids namely fluorine-based protic ILs (FPILs) have displayed competitive properties for selective removal of CO_2 from flue gas and natural gas [27]. Regardless of the promising performance of CO_2 capture demonstrated by these types of ionic liquids, they are relatively expensive, they require several steps in the synthesis process, and the utilization of volatile organic solvents is inevitable during the purification process. Recently, protic ionic liquids (PILs) have attracted great interest because of their low cost and simple synthesis pathway. Generally, PILs can be conveniently prepared from stoichiometric neutralization between Brönsted acids and bases. Besides this, PILs display similar CO_2 absorptivity with other classes of ionic liquids [28–33]. In addition, Zhu and his team have synthesized a new PIL from superbase 1,8-diazabicyclo [5.4.0]- undec-7-ene (DBU) with imidazole, and they found that the PIL could reversibly capture about 1 mole of CO_2 per mole ionic liquid [34]. However, prior to utilization of ionic liquids for any applications, their precise and reliable basic thermophysical properties such as density, viscosity, thermal stability and thermal expansion data are vital for the design and scale up of process equipment. For example, density and thermal expansion data are essential for equipment sizing while thermal stability is required to ensure the practicality of the operating temperature range [35]. In addition, data on solvents' viscosity is important for the designing of industrial processes related to heat and mass transfer as well as dissolution of compounds in solvents [36]. Several research groups have also investigated and provided discussion on the temperature-dependent properties of protic ionic liquids prior to the utilization of ionic liquids in various applications [19,22,31].

Despite promising results of CO_2 absorption by protic ionic liquids published in the literature [34], our current work is focusing on the utilization of much cheaper starting reagents, namely amine solutions, for the production of new ammonium-based protic ionic liquids. This work serves as a continuation from our previous work on CO_2 absorption utilizing ammonium-based protic ionic liquids (PILs) [37]. Previously, the CO_2 absorption of ammonium-based PILs utilizing bis (2-ethylhexyl) ammonium, tributylammonium and ethanolammonium cations coupled with acetate and butyrate anions have been reported. The motivation to further investigate this type of ionic liquid for CO_2 capture has risen after we discovered that the PILs could be prepared via a simple synthesis procedure and their capability to absorb CO_2 under experimental conditions. To further study the binary system of PILs–CO_2, the synthesis of six new ammonium-based PILs, namely 2-ethylhexylammonium pentanoate ([EHA][C5]), 2-ethylhexylammonium hexanoate ([EHA][C6]), 2-ethylhexylammonium heptanoate ([EHA][C7]), bis-(2-ethylhexyl) ammonium pentanoate ([BEHA][C5]), bis-(2-ethylhexyl)ammonium hexanoate ([BEHA][C6]) and bis-(2-ethylhexyl)ammonium heptanoate ([BEHA][C7]) and their performance towards CO_2 absorption in a pressure range from 1 bar to 29 bar at 298.15K, are reported in this work.

2. Results and Discussion

2.1. Characterization of Synthesized PILs

All six ammonium-based PILs in this work exist as liquids at room temperature. The ammonium cation, [EHA] and [BEHA] were combined with anions from organic acids, [C5], [C6] and [C7] through acid-base neutralization reactions. The NMR results and the water contents for all six ammonium-based PILs, [EHA][C5], [EHA][C6], [EHA][C7], [BEHA][C5], [BEHA][C6] and [BEHA][C7], are presented in this sub section while all NMR spectra of PILs (Figures S1–S12) are available in the Supplementary Materials. The reported water content for all PILs is between 1.04% and 8.70%. Based on reported data by Chen et al., PILs are highly hygroscopic and own higher hydrophilicity in comparison to aprotic ionic liquids [38]. Meanwhile, the presence of water molecules lowers the electrostatic attractions between the ions and consequently reduces the viscosity of ILs [39]. Nonetheless, the thermophysical properties of our PILs are solely reported by using these water content values.

[EHA][C5]: ^{1}H NMR (500 MHz, $CDCl_3$): δ 0.902 [t, 9H (R-CH_3)], δ 1.518 [m, 13H (R-CH, R-CH_2)], δ 2.120 [t, 2H (COOH-CH_2)], δ 2.710 [m, 2H (NH_2-CH_2)]. ^{13}C NMR (125 MHz, $CDCl_3$): δ 181.05, 42.48, 37.97, 37.82, 30.17, 28.66, 28.43, 23.22, 22.88, 22.76, 13.98, 13.91, 10.14. Water content: 6.37%.

[EHA][C6]: ^{1}H NMR (500 MHz, $CDCl_3$): δ 0.880 [t, 9H (R-CH_3)], δ 1.516 [m, 15H (R-CH, R-CH_2)], δ 2.118 [t, 2H (COOH-CH_2)], δ 2.690 [m, 2H (NH_2-CH_2)]. ^{13}C NMR (125 MHz, $CDCl_3$): δ 181.22, 42.61, 38.09, 37.97, 31.92, 30.20, 28.14, 23.22, 22.89, 22.54, 13.97, 13.88, 10.16. Water content: 8.70%.

[EHA][C7]: ^{1}H NMR (500 MHz, $CDCl_3$ δ 0.900 [t, 9H (R-CH_3)], δ 1.508 [m, 17H (R-CH, R-CH_2)], δ 2.086 [t, 2H (COOH-CH_2)], δ 2.688 [m, 2H (NH_2-CH_2)]. ^{13}C NMR (125 MHz, $CDCl_3$): δ 181.11, 42.51, 38.22, 38.02, 31.79, 30.19, 29.45, 28.44, 26.55, 23.21, 22.90, 22.58, 14.00, 13.97, 10.14. Water content: 8.43%.

[BEHA][C5]: ^{1}H NMR (500 MHz, $CDCl_3$): δ 0.836 [m, 15H (R-CH_3)], 1.265 [m, 16H (R-CH_2, -CH)], δ 1.523 [m, 4H (R-CH_2)], δ 2.149 [t, 2H (CH_2-COO-)], δ 2.576 [d, 4H (CH_2-NH)]. ^{13}C NMR (125 MHz, $CDCl_3$): δ 179.31, 51.75, 37.15, 36.37, 30.80, 28.53, 28.02, 23.95, 22.96, 22.57, 14.02, 13.86, 10.38. Water content: 1.52%.

[BEHA][C6]: ^{1}H NMR (500 MHz, $CDCl_3$): δ 0.811 [m, 15H (R-CH_3)], δ 1.282 [m, 18H (R-CH_2, -CH)], δ 1.534 [m, 4H (R-CH_2)], δ 2.124 [t, 2H (CH_2-COO-)], δ 2.585 [d, 4H (CH_2-NH)]. ^{13}C NMR (125 MHz, $CDCl_3$): δ 179.25, 51.74, 37.15, 36.71, 31.74, 30.79, 28.51, 25.63, 23.94, 22.96, 22.50, 14.01, 13.96, 10.35. Water content: 1.04%.

[BEHA][C7]: ^{1}H NMR (500 MHz, $CDCl_3$): δ 0.865 [m, 15H (R-CH_3)], δ 1.280 [m, 20H (R-CH_2, -CH)], δ 1.540 [m, 4H (R-CH_2)], δ 2.126 [t, 2H (CH_2-COO-)], δ 2.582 [d, 4H (CH_2-NH)]. ^{13}C NMR (125 MHz, $CDCl_3$): δ 179.29, 51.65, 37.00, 36.85, 31.67, 30.74, 29.20, 28.48, 28.47, 25.93, 23.90, 22.92, 22.53, 13.96, 10.27. Water content: 1.37%.

2.2. TGA and DSC Analysis

Thermogravimetric analyzer (TGA) was used to study the thermal stability of the PILs. Table 1 shows the thermal stability data while Figure 1 displays the TGA profiles of the synthesized PILs. It could be observed from the data that the thermal stability of PILs is in the range of 416 to 437 K. For a common cation, lengthening the alkyl chain branch in the anion caused an increment in the thermal decomposition (T_d) of the PIL. This could be evidenced by the relatively high T_d of [EHA][C7] and [BEHA][C7] as compared to the others. A similar observation was reported by Bhattacharyya et al. in which the thermal stability of an amino acid ionic liquid with longer alkyl chain attached to the nitrogen, [$N_{1,1,14,2O12}$][Lys], has a higher thermal stability than another amino acid ionic liquid, [$N_{1,1,6,2O12}$][Lys], with a relatively shorter alkyl chain branch [40]. According to Keshapolla et al. and Bandres et al., the relationship between high thermal stability and long alkyl chain attached to the ionic liquid could be attributed to the presence of strong intermolecular and intramolecular forces in the alkyl chain [41,42]. In addition, it was observed that the T_d values of PILs with a common cation are relatively close to one another and a similar observation was recorded by Cai et al., involving a series of ionic liquids namely triethanolamine methanesulfonate [TEA][mesy], triethanolamine trifluoromethanesulfonate [TEA][OTf] and triethanolamine benzenesulfonate [TEA][Bsa] [43]. On the other hand, the thermal stability of the ammonium-based PILs synthesized in this work is relatively low when compared to other types of ionic liquids. For instances, the thermal stability of an imidazolium-based ionic liquid (1-butylimidazolium dicyanamide, [BMIM][DCA]) and a phosphonium-based ionic liquid, (phosphonium bis-dicarbollylcobalt (III) [PC6C6C6C14][CoCB]) are greater than 300 °C [35,44,45]. Xu and Cheng have summarized that the thermal stability of imidazolium ionic liquids was improved by increasing the degree of substitution of hydrogen by alkyl groups on the imidazolium ring [46].

Table 1. Thermal decomposition, T_d; glass transition, T_g; melting point, T_m.

Ionic Liquids	T_d	T_g	T_m
	K	°C	°C
[EHA][C5]	416.46	−97.00	-
[EHA][C6]	421.85	−98.37	-
[EHA][C7]	424.28	−96.91	-
[BEHA][C5]	428.47	−96.95	−68.34
[BEHA][C6]	431.55	−91.43	−66.82
[BEHA][C7]	437.47	−90.89	−66.69

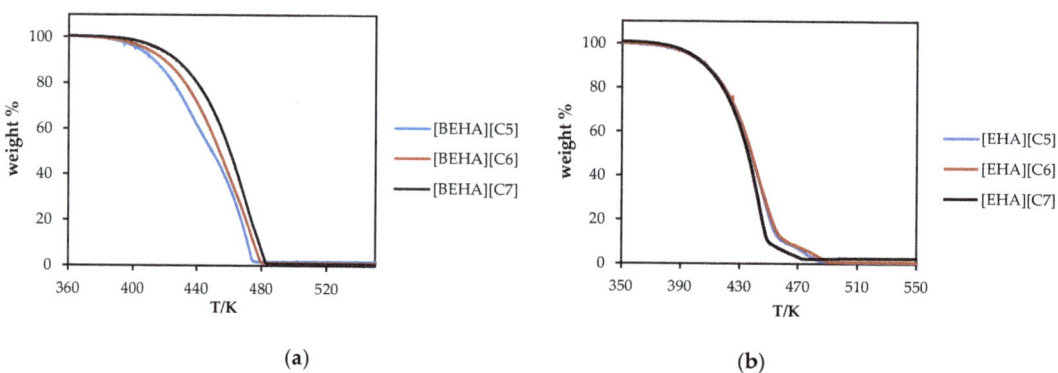

Figure 1. Thermal decomposition curves of (**a**) [EHA][C5], [EHA][C6], [EHA][C7], and (**b**) [BEHA][C5], [BEHA][C6], [BEHA][C7] at heating rate of 10 °C.min^{-1}.

The phase transitions which are glass transition temperature (T_g), and melting point (T_m) of the ammonium-based PILs were investigated by using a Differential Scanning Calorimeter (DSC) from $-150\ °C$ to $50\ °C$ and the results are tabulated in Table 1. This temperature range was chosen based on the fact that many ILs exhibit glass transition at low temperatures even beyond $-100\ °C$ [47]. Apart from providing the fundamental information, the study of phase transition of PILs at this condition is crucial due to demand in other technological areas with extreme environments. For example, in space-related applications, ILs is potentially being used as hypergolic fluids in orbiting satellites, manned spacecraft and deep-space probes [48]. Figure 2 shows the examples of DSC curves for the ammonium-based PILs synthesized in this study. Data show that all PILs possess a glass transition temperature (T_g) ranging from $-98.37\ °C$ to $-90.89\ °C$, which indicates that all PILs experience the flow of heat from amorphous glass to liquid state [19]. As T_g represents the cohesive energy of the sample, PILs that exhibit T_g values have low cohesive energy that could contribute to advantageous physiochemical properties such as low viscosity and high ionic conductivity [47]. A similar trend of marginal difference in the T_g values for the ammonium-based PILs was also observed and discussed by other researchers employing ammonium-based ionic liquids as well [47]. In contrast, only ammonium-based PILs with [BEHA] cation exhibited a melting temperature (T_m) in which all T_m are in the range of $-68.34\ °C$ to $-66.69\ °C$. Only a minimal increment in the T_m values was observed when the alkyl chain of anion increases [C5] to [C7]. Primarily, the T_m of PIL is dependent on the crystal lattice strength in the PIL. The low T_m of the PIL could be related to the low crystal lattice energy due to poor packing efficiency in the crystal lattice of PIL itself [43,49]. The data obtained in this work suggests that [BEHA][C7] has a better packing of the counterions in its structure than that of [BEHA][C5] and [BEHA][C6].

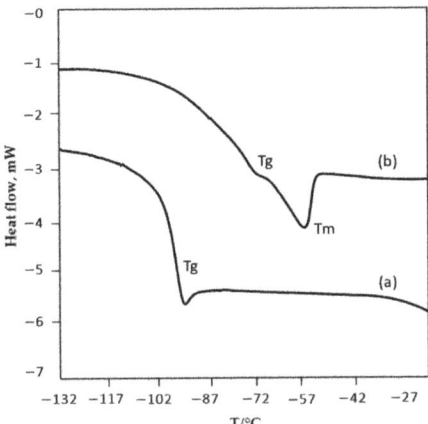

Figure 2. Differential scanning calorimetry (DSC) curves of (a) [EHA][C7], and (b) [BEHA][C6] at a heating rate of $10\ °C\ min^{-1}$.

2.3. Density (ρ), Thermal Expansion Coefficient (α_p), Standard Entropy ($S°$) and Lattice Potential Energy (U_{pot}) Measurement

The density of ammonium-based PILs was studied at temperatures ranging from 293.15 to 363.15 K. The plots of the experimental density of the PILs are shown in Figure 3 while the experimental data and the plots with standard errors are available in Table S1 and Figure S13, respectively, in the Supplementary Material. As illustrated in Figure 3, the densities for all six ammonium-based PILs decreased linearly with temperature. Experimental data also indicates that the density of the PILs deceases as the alkyl chain of the anion increases for both [EHA] and [BEHA] PILs. The results are in accordance with published results in literature for PILs with diethylammonium and dibutylammonium cations with the density values ranging approximately from 0.82 g.cm^{-3} to 0.94 g.cm^{-3} [50]. A similar observation was also found by other researchers when the densities of their tetrabutylammonium ionic liquids were analyzed over a temperature range of 283.4 to 333.4 K [51]. As temperature increases, the volume of ionic liquids increases, and the density of the ionic liquids decreases accordingly. At higher temperatures, the intermolecular forces between the constituent ions weaken, and this increases the mobility of the ions which in turn increases the volume of these ions [37,52,53]. Further analysis also revealed that [EHA][C5] has the highest density values compared to the rest of the ammonium-based PILs. The small size of [EHA] cation compared to [BEHA] cation affects local packing of the PIL structure and thus contributes to the increase in the density values of [EHA][C5] [37,54]. Comparable observations using PIL with ethylammonium cation were also found by several researchers, in which an increasing trending packing efficiency was proportional with the decreasing of molecular weight [51,55]. Notably, the increased alkyl chain length in both cation and anion of the PIL has promoted the steric hindrance and asymmetric nature in the PIL structure as bigger and bulkier PILs result in a lower density value for the PILs [40,41]. This trend can be observed in [BEHA][C6] and [BEHA][C7] as they exhibit the lowest density values.

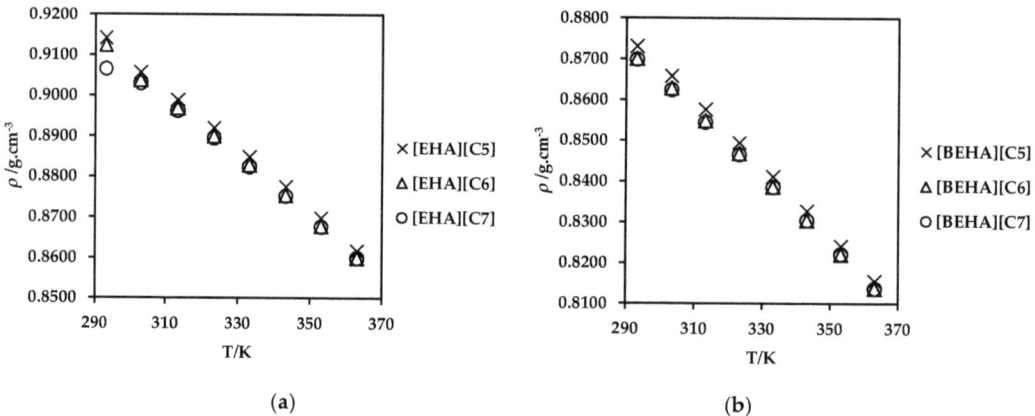

Figure 3. Density (ρ) values of (**a**) [EHA][C5], [EHA][C6], [EHA][C7], and (**b**) [BEHA][C5], [BEHA][C6], [BEHA][C7] as a function of temperature.

The thermal expansion coefficient can provide information about the intermolecular interaction in the PILs, and it can be calculated from the experimental values of density, ρ by using Equation (1). The calculated data is tabulated in Table 2. Thermal expansion coefficients, α_p for the ammonium-based PILs can be defined as [37,53,56]:

$$\alpha_p = -1/\rho \cdot (\delta\rho/\delta T) = -(A_2)/(A_1 + A_2 T) \tag{1}$$

The calculated values in Table 2 show that the thermal expansion coefficients vary only slightly with the increase of C-numbers in the structure of the PILs. PILs with [BEHA] cation has higher α_p than that of PILs with [EHA] cation. This indicates that the thermal expansion coefficient does not only depend on the cation symmetry but is also related to the length of the alkyl substituent [57]. Meanwhile, the behavior of the thermal expansion coefficient is almost similar for all PILs with common cations. Sarkar et al. have also reported a similar variation trend of the thermal expansion coefficient for diethylammonium-based PILs [19]. To conclude, the thermal expansion coefficient can be considered as temperature independent as it shows similar results over the temperature range studied.

Table 2. Thermal expansion coefficients (α_p) of the PILs calculated using Equation (1).

T/K	10^{-4} α/K^{-1}					
	[EHA][C5]	[EHA][C6]	[EHA][C7]	[BEHA][C5]	[BEHA][C6]	[BEHA][C7]
293.15	7.92	7.93	7.79	9.55	9.36	9.27
303.15	7.98	7.99	7.86	9.64	9.45	9.36
313.15	8.05	8.05	7.92	9.73	9.54	9.45
323.15	8.11	8.12	7.95	9.83	9.63	9.54
333.15	8.18	8.18	8.05	9.92	9.72	9.63
343.15	8.25	8.25	8.11	10.02	9.82	9.72
353.15	8.31	8.32	8.18	10.13	9.91	9.82
363.15	8.38	8.39	8.24	10.23	10.01	9.92

The volume occupied by one mole of a compound at a given temperature and pressure is denoted as molar volume, V_m. The molar volume was calculated by using an empirical equation as shown in Equation (2) and utilizing the experimental densities [41,58–61]:

$$V_m = M/(\rho \cdot N_A) \qquad (2)$$

where V_m is the molar volume, M is the molar mass of the ammonium-based PILs, ρ is the density of PILs at 303.15 K and N_A is Avogadro's number.

The calculated molar volume for all ammonium-based PILs are tabulated in Table 3. From the calculated value, the molar volume, V_m, is proportional to the anion alkyl chain length as well as the size of the cation. The molar volume increases with the alkyl chain length of the anion and this behavior is caused by the addition of the CH_2 group in the anion of the PILs. Besides that, PILs with [BEHA] cation exhibit a larger molar volume value compared to PILs with [EHA] cation. This could be explained by the difference in the size of the cations. Similar findings have been observed in other studies [19,37].

Table 3. Molar volume, V_m; standard entropy, $S°$; lattice potential energy, U_{pot} at 303.15 K.

Ionic Liquids	V_m	$S°$	U_{pot}
	nm^3	J·K^{-1}·mol^{-1}	kJ·mol^{-1}
[EHA][C5]	0.4242	558.3	416.0
[EHA][C6]	0.4509	591.5	409.8
[EHA][C7]	0.4771	624.2	404.1
[BEHA][C5]	0.6590	850.9	373.4
[BEHA][C6]	0.6883	887.4	369.5
[BEHA][C7]	0.7156	921.5	366.1

Entropy is the measurement of the randomness of molecules, and generally, entropy increases with molar volume [19]. The relationship between molar volume (V_m) and standard entropy ($S°$) for the ammonium-based PILs in this work can be explored by using the following standard equation that is available in the literature [62]:

$$S° = 1246.5\, V_m + 29.5 \qquad (3)$$

The results presented in Table 3 clearly indicate that the standard entropy increased with the molar volume value for all ammonium-based PILs. The increasing number of carbon atoms in the alkyl chain of carboxylate anion has resulted in the increment of the $S°$ of the ammonium-based PILs. From the calculated values obtained, [BEHA]-based PILs depicted the highest standard entropy due to their larger size compared to [EHA]-based PILs, which causes the least interaction between cation and anion [41]. In this work, the standard entropy of [EHA] and [BEHA] PILs increases in the sequence of [C5] < [C6] < [C7].

In addition, to predict the relative stabilities of ILs, Glasser [62] has also developed a method for calculating lattice potential energies (U_{pot}) of ILs by using Equation (4):

$$U_{pot} = [\gamma \, (\rho/M)^{1/3}] + \delta \qquad (4)$$

where γ and δ are fitting coefficients with values of 1981.9 kJ·mol^{-1} and 103.8 kJ·mol^{-1}, respectively.

The lattice potential energy of the studied PILs was calculated at 303.15 K. The main factor contributing to lattice potential energy is electrostatic or columbic interaction. However, lattice potential energy is inversely related to the volume of ions [19,52,54]. As can be seen in Table 3, lattice potential energy decreases with the addition of the carbon chain length of the carboxylate groups. The addition of methylene group in the alkyl chain of both cation and anion increases the entropy, and consequently reduces packing efficiency in the PILs [63]. As a result, lattice potential energy will decrease with the increase in the alkyl chain length of the PILs.

2.4. Viscosity (η) Measurement

Viscosity is one of the important properties that governs the potential applications of any solvents, and it is largely influenced by intermolecular interactions namely hydrogen bonding, dispersive forces and columbic interactions [64]. The experimental data and the plots with standard errors for viscosity values are available in Table S2 and Figure S14, respectively, in the Supplementary Materials. The viscosity was measured in a temperature range of 293.15 to 363.15 K and graphically shown in Figure 4. The viscosity of all ammonium-based PILs decreased exponentially with an increase in temperature in each PIL as depicted in Figure 4. For example, the viscosity of [EHA][C5] at 293.15K is 45 times larger than at 363.15K. In another study, Liu et al. performed dynamic density measurement on three series of ILs, namely N-alkylpyridinium bis(trifluoromethylsulfonyl)imide ([Cnpy][NTf2], n = 2, 4, 5, 6), N-alkyl-3-methylpyridinium bis(trifluoromethylsulfonyl)imide ([Cn3mpy][NTf2], n = 2, 3, 4, 6) and N-alkyl-4-methylpyridinium bis(trifluoromethylsulfonyl) imide ([Cn4mpy][NTf2], n = 3, 4, 6) within a temperature range of T = (283.15 to 353.15) K [60]. They suggested that the dynamic viscosity increases with the extension of the alkyl side chain of the cation for the three series of pyridinium-based ILs. However, in this work, PILs with [EHA] cation display a higher viscosity value than PILs with [BEHA] cation. Basically, the van der Waals attraction between the aliphatic alkyl chain affects the viscosity values of the PILs [41,53]. However, the water content of the PILs may also affected the observed viscosity results. Furthermore, PILs with [BEHA] cation displayed a marginal increment in the viscosity values as the alkyl chain length of the anion increased.

2.5. Refractive Index (n_D) Measurement

Generally, the refractive index (n_D) describes how fast light travels through material. It estimates the electronic polarizability of the molecules and shows the dielectric response to an external electric field produced by electromagnetic waves (light) [65]. Figure 5 shows the refractive index of ammonium-based PILs that were measured in a temperature range of 293.15 to 333.15 K at atmospheric pressure. The experimental data is tabulated in Table S3 while the plots with standard errors are presented in Figure S15 in the Supplementary Material. From the table, the n_D values were found to be decreasing with increasing temperature. Moreover, the values of the refractive index increased with the increase in cation and anion chain length of PILs. A similar observation was also found in the literature involving PILs in which the n_D values of the studied PILs were in the range of 1.45–1.41 [50]. The increment of refractive index values with increasing alkyl chain length in the cation is influenced by higher intermolecular interaction such as the van der Waals forces of the PILs [52].

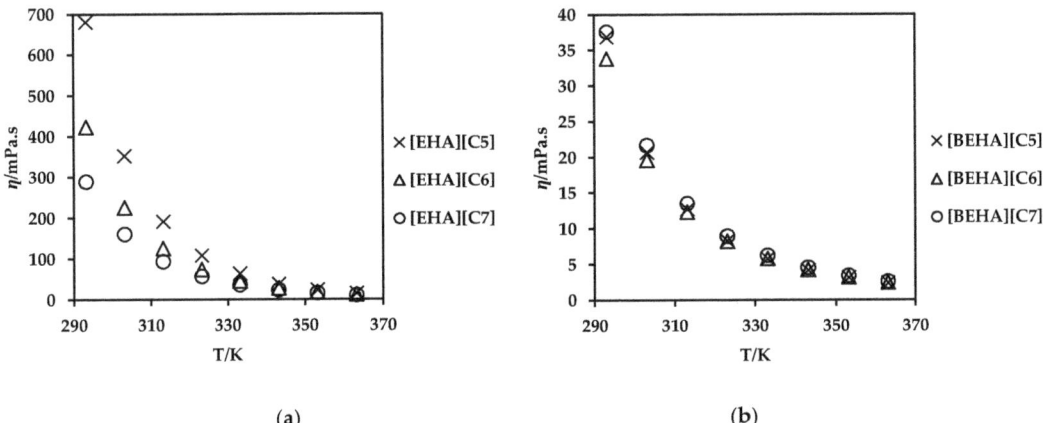

Figure 4. Viscosity (η) values of (**a**) [EHA][C5], [EHA][C6], [EHA][C7], and (**b**) [BEHA][C5], [BEHA][C6], [BEHA][C7] as a function of temperature.

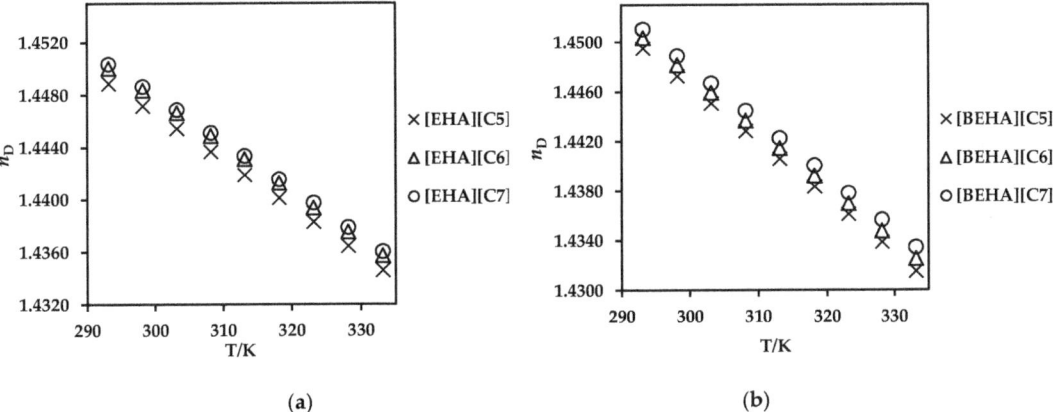

Figure 5. Refractive index (n_D) values of (**a**) [EHA][C5], [EHA][C6] and [EHA][C7], and (**b**) [BEHA][C5], [BEHA][C6] and [BEHA][C7] as a function of temperature.

2.6. Thermophysical Properties Correlations

The density (ρ), dynamic viscosity (η) and refractive index (n_D) experimental values were correlated by using the following equations [53,66]:

$$\rho = A_1 + A_2 T \tag{5}$$

$$\lg \eta = A_3 + A_4/T \tag{6}$$

$$n_D = A_5 + A_6 T \tag{7}$$

where T is the temperature in K, and A_1 through A_6 are correlation coefficients using the least square method. Table 4 represents the estimation of values for correlation coefficients together with the standard deviations, SD which was calculated by using the Equation (8).

Z_{expt} and Z_{calc} are experimental and calculated values, respectively, while n_{DAT} is the number of experimental points.

$$SD = \sqrt{\frac{\sum_{i}^{n_{DAT}} (Z_{expt} - Z_{calc})^2}{n_{DAT}}} \quad (8)$$

Table 4. Fitting parameters of Equation (5) to correlate density (ρ) of PILs and calculated standard deviation (SD_1). Fitting parameters of Equation (6) to correlate viscosity (η) of PILs and calculated standard deviation (SD_2). Fitting parameters of Equation (7) to correlate refractive index (n_D) of PILs and calculated standard deviation (SD_3).

ILs	A_1	A_2	SD_1	A_3	A_4	SD_2	A_5	A_6	SD_3
[EHA][C5]	1.1297	−0.0007	0.0043	2842.7	−8.1026	0.002	1.5535	−0.0004	0.00005
[EHA][C6]	1.1279	−0.0007	0.0004	1772.7	−5.0403	0.006	1.5555	−0.0004	0.00118
[EHA][C7]	1.1091	−0.0007	0.0012	1217.3	−3.4491	0.006	1.5553	−0.0004	0.00146
[BEHA][C5]	1.1161	−0.0008	0.0004	152.1	−0.4278	0.027	1.5809	−0.0004	0.00146
[BEHA][C6]	1.1076	−0.0008	0.0004	139.8	−0.3918	0.024	1.5810	−0.0004	0.00103
[BEHA][C7]	1.1056	−0.0008	0.0003	155.6	−0.4367	0.024	1.5802	−0.0004	0.00115

2.7. CO₂ Absorption Measurement

Carbon dioxide absorption measurements have been performed to investigate the potential ability of the ammonium-based PILs as solvents for CO_2 capture. The measurements were conducted in the pressure range of 1-29 bar at room temperature and the results are plotted in Figures 6 and 7. From the plots, the CO_2 uptake by the ammonium-based PILs shows a trend of polynomial increment with CO_2 pressure. Generally, ammonium-based PILs with [BEHA] cation exhibited marginal difference in CO_2 absorption values than that of ammonium-based PILs with [EHA] cations as shown in Figure 7. At a constant pressure of 29 bar, [BEHA][C7] displayed the highest CO_2 absorption with the value of 0.78 mol fraction when compared to [BEHA][C6] and [BEHA][C5] with the CO_2 mol fractions of 0.68 and 0.64, respectively. This behavior can be explained by using the data reported of density and molar volume of the ammonium-based PILs. The increment in the density value of the PIL increases the molar volume of the PILs which thus in turn causes an increase in the fractional free volume and consequently enhances the CO_2 uptake by the ammonium-based PILs [67,68]. Based on the analysis and comparison of FTIR and ^{13}C NMR, Xu and Oncsik et al. proposed that the mechanism of CO_2 absorption is via the interaction between gas and the basic anion [32,69]. At approximately 20 bar and 25 °C, the CO_2 uptake by the [BEHA][C7] is about 40% higher than that of bis(2-ethylhexyl)ammonium butyrate protic ionic liquid [37]. On the other hand, some researchers have performed investigations on the relationship between the viscosity of PILs and performance of CO_2 absorption by the PILs and found that PIL with low viscosity value has a high absorption capacity of CO_2 [70]. ILs with low viscosities can result in low mass transfer resistance between liquid and gas phases, and this eventually increases the CO_2 absorption rate. The viscosities of [BEHA][C5], [BEHA][C6] and [BEHA][C7] were recorded to have values between 19.64 and 21.70 mPa·s at 30 °C, which are much lower when compared to conventional ILs, for example [Bmim][BF4] with the viscosity value of 68.90 mPa·s at the same temperature [71]. Regardless of the cation, there is an increasing trend of CO_2 absorption in the order of [C5] < [C6] < [C7] anion. As such, both [EHA][C7] and [BEHA][C7] show the highest CO_2 absorption capacity at 29 bar and room temperature with the values of CO_2 mol fractions of 0.77 moles and 0.78 moles, respectively. These results could be considered an indication of the potential ability of the ammonium-based PILs as solvents for CO_2 capture. However, more thorough studies must be conducted for further evaluation before the ammonium-based PILs can be fully used as new solvents in the field of CO_2 removal.

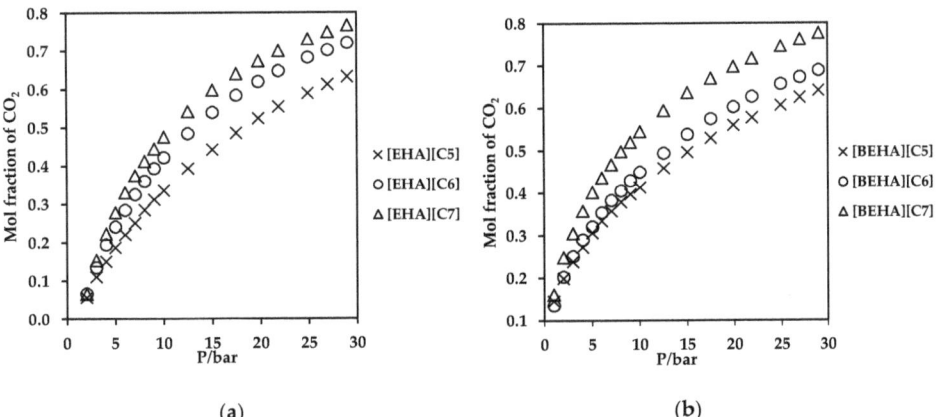

Figure 6. Plot of CO$_2$ absorption in ammonium-based PILs with (**a**) [EHA] cation, and (**b**) [BEHA] cation.

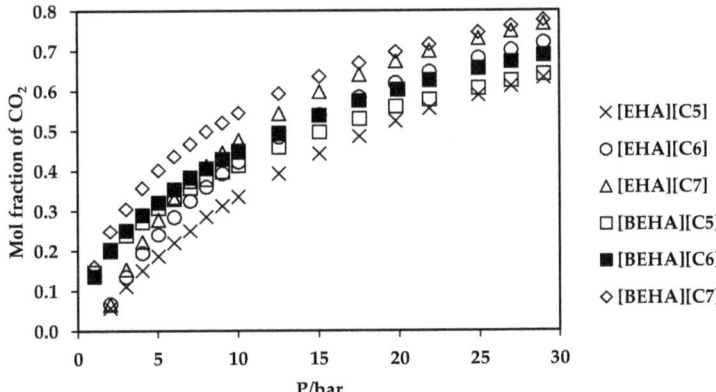

Figure 7. Plot of CO$_2$ absorption in ammonium-based PILs at 298.15 K.

3. Materials and Methods

3.1. Chemicals

To synthesize all six ammonium-based PILs, analytical grade chemicals from Merck, Darmstadt, Germany were used. The CAS numbers, abbreviations, grade percentage, density, viscosity, flash point and melting points of all chemicals are as follows: 2-ethylhexylamine (104-75-6, [EHA], 99.0%, 0.789 g.cm^{-3}, 1.1 cP, 140 °C, −76 °C), bis-(2-ethylhexyl)amine (106-20-7, [BEHA], 99.0%, 0.805 g.cm^{-3}, 3.7 cP, 130 °C, <−20 ° C), pentanoic acid (109-52-4, [C5], 98.0%, 0.939 g.cm^{-3}, 2.3 cP, 96 °C, −34.5 °C), hexanoic acid (142-62-1, [C6], 98.0%, 0.927 g.cm^{-3}, 3.2 cP, 102 °C, −2.78 °C) and heptanoic acid (111-14-8, [C7], 99.0%, 0.917 g cm^{-3}, 3.4 cP, 113 °C, −7.5 °C)).

3.2. Synthesis of PILs

The synthesis of PILs was carried out by using a one-step neutralization reaction and the reaction is written as follows:

$$(R_X)_2NH + HOOC(R_Y) \rightarrow (R_X)_2NH_2^+\ {}^-OOC(R_Y)$$

R$_X$ is 2-ethylhexyl or bis-(2-ethylhexyl and R$_Y$ is either pentyl, hexyl or heptyl. In this work, 2-ethylhexylamine and bis-(2-ethylhexyl)amine, abbreviated as [EHA] and [BEHA], respectively, were the cations while the acids with the variation of alkyl chain length from pentyl [C5], hexyl [C6] and heptyl [C7] were the anions of the PILs.

In a specific procedure, a 1:1 mol ratio of acid was added to the base with continuous stirring at 250 rpm for 24 h at room temperature. The resulting product was dried under vacuum at 80 °C for 6 h to remove any water traces and impurities that might be present resulting from starting reagents as well as surrounding atmosphere. The PILs which were in liquid forms without noticeable solid crystal or precipitation after the purification step were kept in sealed containers until further analysis. The proton transfer reaction had resulted in the formation of six PILs as tabulated in Table 5. Figure 8 depicts the reaction for the synthesis of [EHA][C5].

Table 5. Structures of acids and bases, their names, and abbreviations used.

Structure	Name and Abbreviation
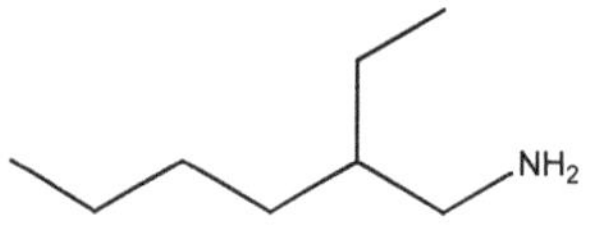	2-ethylhexylamine, [EHA]
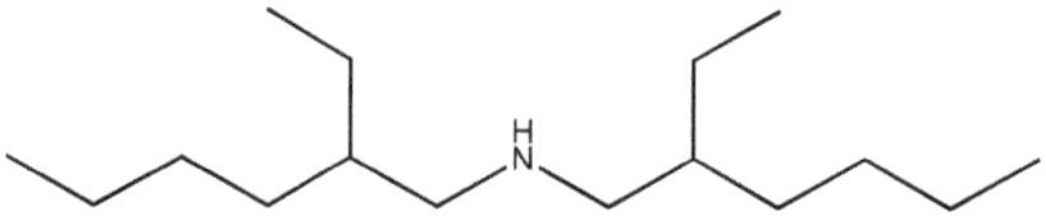	Bis-(2-ethylhexyl)amine, [BEHA]
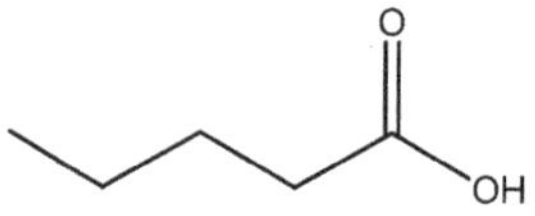	Pentanoic acid, [C5]
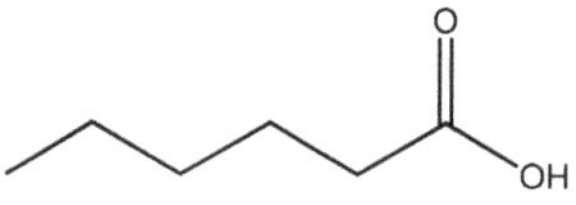	Hexanoic acid, [C6]
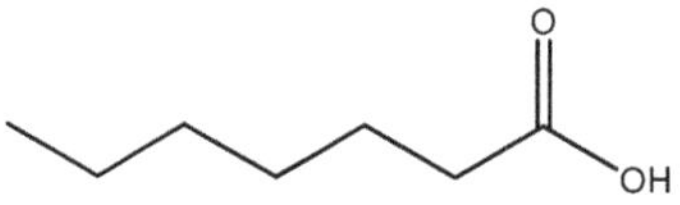	Heptanoic acid, [C7]

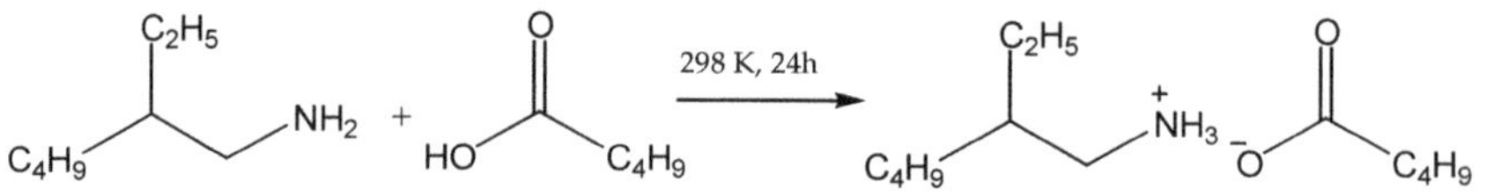

2-ethylhexylamine, [EHA] pentanoic acid, [C5] 2-ethylhexylammonium pentanoate,[EHA][C5]

Figure 8. Synthesis reaction for [EHA][C5].

3.3. *Characterization*

3.3.1. Structural Confirmation and Water Content

Nuclear Magnetic Resonance (NMR) spectroscopy (Bruker Ascend TM 500, Billerica, MA, USA) was used to confirm the structures of the synthesized PILs. In each analysis, a 100 µL PIL sample was dissolved in a 600 µL solvent ($CDCl_3$). The ^1H and ^{13}C spectra are reported in parts per million and the multiplicities, where applicable, are written as *d* (doublet), *t* (triplet) and *m* (multiplet). The water content of each PIL was determined by using Volumetric Karl Fisher V30 Mettler Toledo (Columbus, OH, USA).

3.3.2. TGA Analysis

A Simultaneous Thermal Analyzer (STA) 6000 (Perkin Elmer, Waltham, MA, USA) was used to study the thermal stability of the PILs. The reproducibility for TGA STA 6000 is $<\pm 0.5$ °C with $\pm$ 2% based on metal standard. In each analysis, 10 mg of the PILs sample was placed in a crucible pan and the thermal analysis was conducted in a temperature range of 50–650 °C at a heating rate of 10 °C·min^{-1} under 20 mL/min of nitrogen flow.

3.3.3. DSC Analysis

A Differential Scanning Calorimeter (DSC) 1 Star system (Mettler Toledo, Columbus, OH, USA) was used to investigate the phase transition of the PILs. The reproducibility for DSC is $\pm$ 0.2K with <1% based on indium calibration. In total, 10 mg of samples were sealed in aluminum pans and subject to analysis in a temperature range of 50–150 °C with a heating rate of 10 °C·min^{-1}. The phase transition data were analyzed by using the second heating plot.

3.3.4. Density (p) and Viscosity (η) Measurement

An Anton Paar Stabinger Viscometer (Graz, Austria) was used to simultaneously measure the density and viscosity of the PILs in a temperature range of 293.15–363.15 K with a temperature measurement accuracy of 0.02 K. The reproducibility of the density and viscosity measurements were $\pm 5.10^{-4}$ g.cm^{-3} and 0.35%, respectively. The measurements were repeated several times and the average value was considered for further analysis. Prior to the density and viscosity measurements, the equipment was calibrated using a standard fluid provided by the supplier. A commercial imidazolium IL with known density and viscosity values was also used to validate the equipment.

3.3.5. Refractive Index (n_D) Measurement

An ATAGO RX-5000 Alpha Digital Refractometer (Tokyo, Japan) with a measuring accuracy of $\pm 4.10^{-5}$ was used to determine the refractive index values of the PILs. The measurement was done in a temperature range of 293.15 to 333.15 K. The instrument was also calibrated by using standard organic solvents provided by the supplier. In addition, a commercial imidazolium IL was also used while conducting the validation test and the result was compared with the values available from the literature [50].

3.3.6. CO_2 Absorption Measurement

The CO_2 absorption of the PILs was studied by using a magnetic suspension balance (MSB) from Rubotherm Präzisionsmesstechnik GmbH (Bochum, Germany). In this gravimetric method, the weight change of the PILs upon absorption of CO_2 was measured and calculated in a range of pressure from 1 to 29 bar at room temperature. The sample absorption chamber linked to the microbalance, which has a precision of ± 20 µg, via an electromagnet and a suspension magnet which keeps the balance at ambient conditions during the CO_2 absorption experiments. In a typical CO_2 absorption measurement, approximately 1g of PIL sample was loaded in the sample chamber and the absorption system was evacuated at 10^{-3} mbar (Pfeiffer model DUO5) to remove any impurities until the weight remained constant. Then, the sample chamber was pressurized with CO_2 at a constant temperature by means of an oil circulator (Julabo, model F25-ME, ± 0.1 °C accuracy, Seelbach, Germany) and the weight change due to the absorption of the gas in the PIL was observed and recorded. Once a constant weight reading was recorded, the system was allowed to stand in the condition for an additional 3–4 h to ensure complete equilibration of the binary CO_2–PIL system. The absorption measurement was repeated with different pressure values of CO_2 to yield a series of absorption isotherm. The weight of the CO_2 dissolved in the PILs sample was calculated using Equation (9) available from literature [72,73].

$$\text{wt } CO_2 = [\text{wt} - (\text{wt}_{Sc} + \text{wt}_S)] + [(V_{Sc} + V_S)(\rho CO_2)] \tag{9}$$

where wt (g) is the corrected weight of the balance, $\text{wt}_{Sc} + \text{wt}_S$ (g) are the weights of sample cell and sample, respectively, $V_{Sc} + V_S$ (cm^{-3}) are the volumes of the sample cell and sample, respectively, and ρCO_2 (g.cm^{-3}) is the density of CO_2 at the pressure and temperature during the CO_2 absorption. The results of CO_2 absorption are presented in terms of mole fraction of CO_2 (x) dissolved in the PIL, which was calculated using Equation (10):

$$x = n_{CO2}/(n_{liq} + n_{CO2}) \tag{10}$$

where n_{CO2} is the mole of CO_2 absorbed in the PIL and n_{liq} is the mole of the PIL.

4. Conclusions

In this work, six new ammonium-based PILs have been successfully synthesized through a one-step procedure. The thermophysical properties including density, viscosity, refractive index and thermal stability have been measured. The experimental results revealed the dependency of the experimental values namely the ρ, η, n_D and T_d on the alkyl chain of the anion, size of the cations and the temperature of measurement. The phase transition analysis of the PILs yielded the glass transition temperature (T_g) and melting point (T_m) of the PILs studied. These synthesized ammonium-based PILs have been tested for their ability towards CO_2 absorption in which [BEHA][C7] displayed the highest CO_2 uptake in the experimental conditions signifying its capability to be a potential solvent in the application of CO_2 capture. Future works should include CO_2 desorption studies of the PILs for the purpose of recyclability and sustainability of the absorbents.

Supplementary Materials: The following are available online, Table S1: Density (ρ) values of PILs at temperatures (293.15–363.15) K, Table S2: Dynamic viscosity (η) values of PILs at temperatures (293.15–363.15) K, Table S3: Refractive index (n_D) values of PILs at temperatures (293.15–333.15 K), Figures S1–S12: NMR analysis of the PILs, Figure S13: Plots of density (ρ) values with standard errors of (a) [EHA][C5], [EHA][C6], [EHA][C7], and (b) [BEHA][C5], [BEHA][C6], [BEHA][C7] as a function of temperature, Figure S14: Plots of viscosity (η) values with standard errors of (a) [EHA][C5], [EHA][C6], [EHA][C7], and (b) [BEHA][C5], [BEHA][C6], [BEHA][C7] as a function of temperature, and Figure S15: Plots of refractive index (n_D) values with standard errors of (a) [EHA][C5], [EHA][C6], [EHA][C7], and (b) [BEHA][C5], [BEHA][C6], [BEHA][C7] as a function of temperature.

Author Contributions: Conceptualization, N.M.Y. and N.H.Z.O.Z.; methodology, N.H.Z.O.Z. and N.M.Y. validation, N.H.Z.O.Z. and A.H.A.R.; formal analysis, N.H.Z.O.Z. and N.M.Y.; resources, N.M.Y. and M.A.B.; data curation, N.M.Y.; writing—original draft preparation, N.H.Z.O.Z. and A.H.A.R.; writing—review and editing, N.M.Y.; supervision, N.M.Y. and M.A.B.; project administration, N.M.Y. and M.A.B.; funding acquisition, N.M.Y. All authors have read and agreed to the published version of the manuscript.

Funding: This research was funded by Yayasan Universiti Teknologi PETRONAS-Fundamental Research Grant (YUTP-FRG) (cost centre 015LC0-054) under research project "Design, Synthesis and Evaluation of Protic Ionic Liquids for CO_2 Removal from Natural Gas", and the APC was funded by 015LC0-054.

Institutional Review Board Statement: Not applicable.

Informed Consent Statement: Not applicable.

Data Availability Statement: The data presented in this study is available in this article and Supplementary Materials.

Acknowledgments: The authors would like to acknowledge the financial assistance and support provided by Universiti Teknologi PETRONAS and the Centre of Research in Ionic Liquids (CORIL), UTP.

Conflicts of Interest: The authors declare no conflict of interest.

References

1. Muda, N.; Jin, T. On prediction of depreciation time of fossil fuel in Malaysia. *J. Math Stat.* **2012**, *8*, 136–143. [CrossRef]
2. Mokhatab, S.; Poe, W.A.; Mak, J.Y. (Eds.) Chapter 1—Natural Gas Fundamentals. In *Handbook of Natural Gas Transmission and Processing*, 3rd ed.; Gulf Professional Publishing: Boston, MA, USA, 2015; pp. 1–36. ISBN 978-0-12-801499-8.
3. Guo, B.; Ghalambor, A. (Eds.) Chapter 1—Introduction. In *Natural Gas Engineering Handbook*, 2nd ed.; Gulf Publishing Company: Houston, TX, USA, 2005; pp. 1–11. ISBN 978-1-933762-41-8.
4. Speight, J.G. (Ed.) 5—Recovery, Storage, and Transportation. In *Natural Gas*, 2nd ed.; Gulf Professional Publishing: Boston, MA, USA, 2019; pp. 149–186. ISBN 978-0-12-809570-6.
5. Liang, Z.; Rongwong, W.; Liu, H.; Fu, K.; Gao, H.; Cao, F.; Zhang, R.; Sema, T.; Henni, A.; Sumon, K.; et al. Recent progress and new developments in post-combustion carbon-capture technology with amine based solvents. *Int. J. Greenh. Gas Control.* **2015**, *40*, 26–54. [CrossRef]
6. Gómez-Díaz, D.; Muñiz-Mouro, A.; Navaza, J.M.; Rumbo, A. Diamine versus amines blend for CO_2 chemical absorption. *AIChE J.* **2021**, *67*, e17071. [CrossRef]
7. Huertas, J.I.; Gomez, M.D.; Giraldo, N.; Garzón, J. CO_2 absorbing capacity of MEA. *J. Chem.* **2015**, e965015. [CrossRef]
8. Dubois, L.; Thomas, D. Postcombustion CO_2 capture by chemical absorption: Screening of aqueous amine(s)-based solvents. *Energy Procedia* **2013**, *37*, 1648–1657. [CrossRef]
9. Spigarelli, B.P.; Kawatra, S.K. Opportunities and challenges in carbon dioxide capture. *J. CO2 Util.* **2013**, *1*, 69–87. [CrossRef]
10. Singh, P.; Niederer, J.P.M.; Versteeg, G.F. Structure and activity relationships for amine-based CO_2 absorbents-II. *Chem. Eng. Res. Des.* **2009**, *87*, 135–144. [CrossRef]
11. Ünveren, E.E.; Monkul, B.Ö.; Sarıoğlan, Ş.; Karademir, N.; Alper, E. Solid amine sorbents for CO_2 capture by chemical adsorption: A review. *Petroleum* **2017**, *3*, 37–50. [CrossRef]
12. Lu, Y.-K.; Wang, H.-H.; Hu, Q.-X.; Ma, Y.-Y.; Hou, L.; Wang, Y.-Y. A stable Cd(II)-based MOF with efficient CO_2 capture and conversion, and fluorescence sensing for ronidazole and dimetridazole. *J. Solid State Chem.* **2020**, *295*, 121890. [CrossRef]
13. Choi, S.; Drese, J.; Jones, C. Adsorbent materials for carbon dioxide capture from large anthropogenic point sources. *ChemSusChem* **2009**, *2*, 796–854. [CrossRef]
14. Sayari, A.; Belmabkhout, Y.; Serna-Guerrero, R. Flue gas treatment via CO_2 adsorption. *Chem. Eng. J.* **2011**, *171*, 760–774. [CrossRef]
15. Chen, G.; Chen, G.; Cao, F.; Zhang, R.; Gao, H.; Liang, Z. Mass transfer performance and correlation for CO_2 absorption into aqueous 3-diethylaminopropylamine solution in a hollow fiber membrane contactor. *Chem. Eng. Process* **2020**, *152*, 107932. [CrossRef]
16. Saidi, M. CO_2 absorption intensification using novel DEAB amine-based nanofluids of CNT and SiO_2 in membrane contactor. *Chem. Eng. Process* **2020**, *149*, 107848. [CrossRef]
17. Sohaib, Q.; Vadillo, J.M.; Gómez-Coma, L.; Albo, J.; Druon-Bocquet, S.; Irabien, A.; Sanchez-Marcano, J. CO_2 capture with room temperature ionic liquids; coupled absorption/desorption and single module absorption in membrane contactor. *Chem. Eng. Sci.* **2020**, *223*, 115719. [CrossRef]
18. Cao, F.; Gao, H.; Xiong, Q.; Liang, Z. Experimental studies on mass transfer performance for CO_2 absorption into aqueous N,N-dimethylethanolamine (DMEA) based solutions in a PTFE hollow fiber membrane contactor. *Int. J. Greenh. Gas Control* **2019**, *82*, 210–217. [CrossRef]
19. Sarkar, A.; Sharma, G.; Singh, D.; Gardas, R.L. Effect of anion on thermophysical properties of N,N-diethanolammonium based protic ionic liquids. *J. Mol. Liq.* **2017**, *242*, 249–254. [CrossRef]
20. Bounaceur, R.; Lape, N.; Roizard, D.; Vallieres, C.; Favre, E. Membrane processes for post-combustion carbon dioxide capture: A parametric study. *Energy* **2006**, *31*, 2556–2570. [CrossRef]
21. Endo, T.; Murata, H.; Imanari, M.; Mizushima, N.; Seki, H.; Nishikawa, K. NMR study of cation dynamics in three crystalline states of 1-butyl-3-methylimidazolium hexafluorophosphate exhibiting crystal polymorphism. *J. Phys. Chem. B* **2012**, *116*, 3780–3788. [CrossRef]
22. Smith, J.A.; Webber, G.B.; Warr, G.G.; Atkin, R. Rheology of protic ionic liquids and their mixtures. *J. Phys. Chem. B* **2013**, *117*, 13930–13935. [CrossRef]
23. Patra, R.N.; Gardas, R.L. Effect of nitro groups on desulfurization efficiency of benzyl-substituted imidazolium-based ionic liquids: Experimental and computational approach. *Energy Fuels* **2019**, *33*, 7659–7666. [CrossRef]
24. Singh, V.; Sharma, G.; Gardas, R.L. Thermodynamic and ultrasonic properties of ascorbic acid in aqueous protic ionic liquid solutions. *PLoS ONE* **2015**, *10*, e0126091. [CrossRef]
25. Davis, J.; Rochelle, G. Thermal degradation of monoethanolamine at stripper conditions. *Energy Procedia* **2009**, *1*, 327–333. [CrossRef]
26. Kang, S.; Chung, Y.G.; Kang, J.H.; Song, H. CO_2 absorption characteristics of amino group functionalized imidazolium-based amino acid ionic liquids. *J. Mol. Liq.* **2020**, *297*, 111825. [CrossRef]

27. Tu, Z.; Liu, P.; Zhang, X.; Shi, M.; Zhang, Z.; Luo, S.; Zhang, L.; Wu, Y.; Hu, X. Highly-selective separation of CO_2 from N_2 or CH_4 in task-specific ionic liquid membranes: Facilitated transport and salting-out effect. *Sep. Purif. Technol.* **2021**, *254*, 117621. [CrossRef]
28. Gao, F.; Wang, Z.; Ji, P.; Cheng, J.-P. CO_2 absorption by DBU-based protic ionic liquids: Basicity of anion dictates the absorption capacity and mechanism. *Front. Chem.* **2019**, *6*, 658. [CrossRef] [PubMed]
29. Vijayaraghavan, R.; Oncsik, T.; Mitschke, B.; MacFarlane, D.R. Base-rich diamino protic ionic liquid mixtures for enhanced CO_2 capture. *Sep. Purif. Technol.* **2018**, *196*, 27–31. [CrossRef]
30. Vijayraghavan, R.; Pas, S.; Izgorodina, E.; Macfarlane, D. Diamino protic ionic liquids for CO_2 capture. *Phys. Chem. Chem. Phys.* **2013**, *15*, 19994. [CrossRef]
31. Zheng, W.-T.; Zhang, F.; Wu, Y.-T.; Hu, X.-B. Concentrated aqueous solutions of protic ionic liquids as effective CO_2 absorbents with high absorption capacities. *J. Mol. Liq.* **2017**, *243*, 169–177. [CrossRef]
32. Xu, Y. CO_2 absorption behavior of azole-based protic ionic liquids: Influence of the alkalinity and physicochemical properties. *J. CO2 Util.* **2017**, *19*, 1–8. [CrossRef]
33. Wei, L.; Guo, R.; Tang, Y.; Zhu, J.; Liu, M.; Chen, J.; Xu, Y. Properties of aqueous amine based protic ionic liquids and its application for CO_2 quick capture. *Sep. Purif. Technol.* **2020**, *239*, 116531. [CrossRef]
34. Zhu, X.; Song, M.; Xu, Y. DBU-based protic ionic liquids for CO_2 capture. *ACS Sustain. Chem. Eng.* **2017**, *5*, 8192–8198. [CrossRef]
35. Fredlake, C.P.; Crosthwaite, J.M.; Hert, D.G.; Aki, S.N.V.K.; Brennecke, J.F. Thermophysical properties of imidazolium-based ionic liquids. *J. Chem. Eng. Data* **2004**, *49*, 954–964. [CrossRef]
36. Bhattacharjee, A.; Lopes-da-Silva, J.A.; Freire, M.G.; Coutinho, J.A.P.; Carvalho, P.J. Thermophysical properties of phosphonium-based ionic liquids. *Fluid Phase Equilib.* **2015**, *400*, 103–113. [CrossRef] [PubMed]
37. Yunus, N.M.; Halim, N.H.; Wilfred, C.D.; Murugesan, T.; Lim, J.W.; Show, P.L. Thermophysical properties and CO_2 absorption of ammonium-based protic ionic liquids containing acetate and butyrate anions. *Processes* **2019**, *7*, 820. [CrossRef]
38. Chen, Y.; Cao, Y.; Lu, X.; Zhao, C.; Yan, C.; Mu, T. Water sorption in protic ionic liquids: Correlation between hygroscopicity and polarity. *New J. Chem.* **2013**, *37*, 1959–1967. [CrossRef]
39. Seddon, K.; Stark, A.; Torres, M.-J. Influence of chloride, water, and organic solvents on the physical properties of ionic liquids. *Pure Appl. Chem.* **2000**, *72*, 2275–2287. [CrossRef]
40. Bhattacharyya, S.; Shah, F.U. Thermal stability of choline based amino acid ionic liquids. *J. Mol. Liq.* **2018**, *266*, 597–602. [CrossRef]
41. Keshapolla, D.; Srinivasarao, K.; Gardas, R.L. Influence of temperature and alkyl chain length on physicochemical properties of trihexyl- and trioctylammonium based protic ionic liquids. *J. Chem. Thermodyn.* **2019**, *133*, 170–180. [CrossRef]
42. Bandrés, I.; Royo, F.M.; Gascón, I.; Castro, M.; Lafuente, C. Anion influence on thermophysical properties of ionic liquids: 1-butylpyridinium tetrafluoroborate and 1-butylpyridinium Ttriflate. *J. Phys. Chem. B* **2010**, *114*, 3601–3607. [CrossRef]
43. Cai, G.; Yang, S.; Zhou, Q.; Liu, L.; Lu, X.; Xu, J.; Zhang, S. Physicochemical properties of various 2-hydroxyethylammonium sulfonate -based Pprotic ionic liquids and their potential application in hydrodeoxygenation. *Front. Chem.* **2019**, *7*, 196. [CrossRef]
44. Del Sesto, R.E.; Corley, C.; Robertson, A.; Wilkes, J.S. Tetraalkylphosphonium-based ionic liquids. *J. Organomet. Chem.* **2005**, *690*, 2536–2542. [CrossRef]
45. Kulkarni, P.S.; Branco, L.C.; Crespo, J.G.; Nunes, M.C.; Raymundo, A.; Afonso, C.A.M. Comparison of physicochemical properties of new ionic liquids based on imidazolium, quaternary ammonium, and guanidinium cations. *Chem. Eur. J.* **2007**, *13*, 8478–8488. [CrossRef] [PubMed]
46. Xu, C.; Cheng, Z. Thermal stability of ionic liquids: Current status and prospects for future development. *Processes* **2021**, *9*, 337. [CrossRef]
47. Greaves, T.L.; Weerawardena, A.; Fong, C.; Krodkiewska, I.; Drummond, C.J. Protic ionic liquids: Solvents with tunable phase behavior and physicochemical properties. *J. Phys. Chem. B* **2006**, *110*, 22479–22487. [CrossRef] [PubMed]
48. Nancarrow, P.; Mohammed, H. Ionic liquids in space technology-current and future trends. *ChemBioEng Rev.* **2017**, *4*, 106–119. [CrossRef]
49. Shen, Y.; Kennedy, D.F.; Greaves, T.L.; Weerawardena, A.; Mulder, R.J.; Kirby, N.; Song, G.; Drummond, C.J. Protic ionic liquids with fluorous anions: Physicochemical properties and self-assembly nanostructure. *Phys. Chem. Chem. Phys.* **2012**, *14*, 7981–7992. [CrossRef]
50. Othman Zailani, N.H.Z.; Yunus, N.M.; Ab Rahim, A.H.; Bustam, M.A. Thermophysical properties of newly synthesized ammonium-based protic ionic liquids: Effect of temperature, anion and alkyl chain length. *Processes* **2020**, *8*, 742. [CrossRef]
51. Perumal, M.; Balraj, A.; Jayaraman, D.; Krishnan, J. Experimental investigation of density, viscosity, and surface tension of aqueous tetrabutylammonium-based ionic liquids. *Environ. Sci. Pollut. Res.* **2021**, *28*, 63599–63613. [CrossRef]
52. Gusain, R.; Panda, S.; Bakshi, P.S.; Gardas, R.L.; Khatri, O.P. Thermophysical properties of trioctylalkylammonium bis(salicylato)borate ionic liquids: Effect of alkyl chain length. *J. Mol. Liq.* **2018**, *269*, 540–546. [CrossRef]
53. Yunus, N.M.; Abdul Mutalib, M.I.; Man, Z.; Bustam, M.A.; Murugesan, T. Thermophysical properties of 1-alkylpyridinum bis(trifluoromethylsulfonyl)imide ionic liquids. *J. Chem. Thermodyn.* **2010**, *42*, 491–495. [CrossRef]
54. Wu, B.; Yamashita, Y.; Endo, T.; Takahashi, K.; Castner, E.W. Structure and dynamics of ionic liquids: Trimethylsilylpropyl-substituted cations and bis(sulfonyl)amide anions. *J. Chem. Phys.* **2016**, *145*, 244506. [CrossRef] [PubMed]
55. Pinto, R.R.; Mattedi, S.; Aznar, M. Synthesis and physical properties of three protic ionic liquids with the ethylammonium cation. *Chem. Eng. Trans.* **2015**, *43*, 1165–1170. [CrossRef]

56. Singh, D.; Gardas, R.L. Influence of cation size on the ionicity, fluidity, and physiochemical properties of 1,2,4-triazolium based ionic liquids. *J. Phys. Chem. B* **2016**, *120*, 4834–4842. [CrossRef] [PubMed]
57. Zec, N.; Vraneš, M.; Bešter-Rogač, M.; Trtić-Petrović, T.; Dimitrijević, A.; Čobanov, I.; Gadžurić, S. Influence of the alkyl chain length on densities and volumetric properties of 1,3-dialkylimidazolium bromide ionic liquids and their aqueous solutions. *J. Chem. Thermodyn.* **2018**, *121*, 72–78. [CrossRef]
58. Tariq, M.; Forte, P.A.S.; Gomes, M.F.C.; Lopes, J.N.C.; Rebelo, L.P.N. Densities and refractive indices of imidazolium- and phosphonium-based ionic liquids: Effect of temperature, alkyl chain length, and anion. *J. Chem. Thermodyn.* **2009**, *41*, 790–798. [CrossRef]
59. Liu, Q.-S.; Yang, M.; Li, P.-P.; Sun, S.-S.; Welz-Biermann, U.; Tan, Z.-C.; Zhang, Q.-G. Physicochemical properties of ionic liquids [C$_3$py][NTf$_2$] and [C$_6$py][NTf$_2$]. *J. Chem. Eng. Data* **2011**, *56*, 4094–4101. [CrossRef]
60. Liu, Q.-S.; Li, P.-P.; Welz-Biermann, U.; Chen, J.; Liu, X.-X. Density, dynamic viscosity, and electrical conductivity of pyridinium-based hydrophobic ionic liquids. *J. Chem. Thermodyn.* **2013**, *66*, 88–94. [CrossRef]
61. Liu, Q.-S.; Li, Z.; Welz-Biermann, U.; Li, C.-P.; Liu, X.-X. Thermodynamic properties of a new hydrophobic amide-based task-specific ionic liquid [EimCH$_2$CONHBu][NTf$_2$]. *J. Chem. Eng. Data* **2013**, *58*, 93–98. [CrossRef]
62. Glasser, L. Lattice and phase transition thermodynamics of ionic liquids. *Thermochim. Acta* **2004**, *421*, 87–93. [CrossRef]
63. Er, H.; Wang, H. Properties of protic ionic liquids composed of N-alkyl (=hexyl, octyl and 2-ethylhexyl) ethylenediaminum cations with trifluoromethanesulfonate and trifluoroacetate anion. *J. Mol. Liq.* **2016**, *220*, 649–656. [CrossRef]
64. Chennuri, B.K.; Gardas, R.L. Measurement and correlation for the thermophysical properties of hydroxyethyl ammonium based protic ionic liquids: Effect of temperature and alkyl chain length on anion. *Fluid Phase Equilib.* **2016**, *427*, 282–290. [CrossRef]
65. Zhang, X.U.; Faber, D.J.; Post, A.L.; van Leeuwen, T.G.; Sterenborg, H.J.C.M. Refractive index measurement using single fiber reflectance spectroscopy. *J. Biophotonics* **2019**, *12*, e201900019. [CrossRef] [PubMed]
66. Soave, G. Equilibrium constants from a modified Redlich-Kwong equation of state. *Chem. Eng. Sci.* **1972**, *27*, 1197–1203. [CrossRef]
67. Shaikh, A.R.; Karkhanechi, H.; Kamio, E.; Yoshioka, T.; Matsuyama, H. Quantum mechanical and molecular dynamics simulations of dual-amino-acid ionic liquids for CO$_2$ capture. *J. Phys. Chem. C* **2016**, *120*, 27734–27745. [CrossRef]
68. Gupta, K. Tetracyanoborate based ionic liquids for CO$_2$ capture: From ab initio calculations to molecular simulations. *Fluid Phase Equilib.* **2016**, *415*, 34–41. [CrossRef]
69. Oncsik, T.; Vijayaraghavan, R.; MacFarlane, D.R. High CO$_2$ absorption by diamino protic ionic liquids using azolide anions. *Chem. Commun.* **2018**, *54*, 2106–2109. [CrossRef]
70. Li, F.; Bai, Y.; Zeng, S.; Liang, X.; Wang, H.; Huo, F.; Zhang, X. Protic ionic liquids with low viscosity for efficient and reversible capture of carbon dioxide. *Int. J. Greenh. Gas Control* **2019**, *90*, 102801. [CrossRef]
71. Zhao, Y.; Zhang, X.; Zeng, S.; Zhou, Q.; Dong, H.; Tian, X.; Zhang, S. Density, viscosity, and performances of carbon dioxide capture in 16 absorbents of amine + ionic liquid + H$_2$O, ionic liquid + H$_2$O, and amine + H$_2$O systems. *J. Chem. Eng. Data* **2010**, *55*, 3513–3519. [CrossRef]
72. Sato, Y.; Takikawa, T.; Takishima, S.; Masuoka, H. Solubilities and diffusion coefficients of carbon dioxide in poly(vinyl acetate) and polystyrene. *J. Supercrit. Fluids* **2001**, *19*, 187–198. [CrossRef]
73. Yunus, N.M.; Abdul Mutalib, M.I.; Man, Z.; Bustam, M.A.; Murugesan, T. Solubility of CO$_2$ in pyridinium based ionic liquids. *Chem. Eng. J.* **2012**, *189–190*, 94–100. [CrossRef]

Article

Novel Binary Mixtures of Alkanolamine Based Deep Eutectic Solvents with Water—Thermodynamic Calculation and Correlation of Crucial Physicochemical Properties

Bartosz Nowosielski [1], Marzena Jamrógiewicz [2], Justyna Łuczak [3] and Dorota Warmińska [1,*]

1. Department of Physical Chemistry, Faculty of Chemistry, Gdańsk University of Technology, ul. Narutowicza 11/12, 80-233 Gdańsk, Poland; bartosz.nowosielski@pg.edu.pl
2. Department of Physical Chemistry, Faculty of Pharmacy, Medical University of Gdańsk, Al. Gen. Hallera 107, 80-416 Gdańsk, Poland; marzena.jamrogiewicz@gumed.edu.pl
3. Department of Process Engineering and Chemical Technology, Faculty of Chemistry, Gdańsk University of Technology, ul. Narutowicza 11/12, 80-233 Gdańsk, Poland; justyna.luczak@pg.edu.pl
* Correspondence: dorwarmi@pg.edu.pl; Tel.: +48-583471410

Abstract: This paper demonstrates the assessment of physicochemical and thermodynamic properties of aqueous solutions of novel deep eutectic solvent (DES) built of tetrabutylammonium chloride and 3-amino-1-propanol or tetrabutylammonium bromide and 3-amino-1-propanol or 2-(methylamino)ethanol or 2-(butylamino)ethanol. Densities, speeds of sound, refractive indices, and viscosities for both pure and aqueous mixtures of DES were investigated over the entire range of compositions at atmospheric pressure and T = (293.15 - 313.15) K. It was concluded that the experimental data were successfully fitted using the Jouyban–Acree model with respect to the concentration. Obtained results showed that this mathematical equation is an accurate correlation for the prediction of aqueous DES properties. Key physicochemical properties of the mixtures—such as excess molar volumes, excess isentropic compressibilities, deviations in viscosity, and deviations in refractive indices—were calculated and correlated by the Redlich–Kister equation with temperature-dependent parameters. The non-ideal behavior of the studied systems were also evaluated by using the Prigogine−Flory−Patterson theory and the results were interpreted in terms of interactions between the mixture components.

Keywords: DES; deep eutectic solvents; aqueous mixtures; excess properties; JAM; PFP

1. Introduction

Deep eutectic solvents (DES) are very important and well-known components or materials used often in chemistry but also in other industries as pharmacy, chemical technology as an inexpensive solvent/component being sustainable alternative to the conventional organic solvents which are non-ecological. Currently, the advancement of different DES had enabled the production of materials for unique purpose for reaction medium, biodiesel processes, metal electrodeposition, nanotechnology, and others [1]. Global research is still focused on the developing new innovative DES which will probably replace non-ecological classic solvents as well [2,3].

Deep eutectic solvents, based on urea and quaternary ammonium salts, were first reported by Abbott et al. [4]. Since then many derivatives have been invented and applied. Generally, DESs are built from hydrogen bond acceptor (HBA) and hydrogen bond donor (HBD) in the appropriate molar ratio forming complexes through hydrogen bonds. As a result, deep eutectic solvents have a lower melting point than their components [5].

DESs share many properties with room temperature ionic liquid (RTIL). They are practically non-volatile and non-flammable, and exhibit high thermal and electro-chemical stability, but are definitely cheaper, less toxic, and often biodegradable in comparison with RTIL [6]. Additionally, since the physical properties of DES are dependent on the composition and proportions of the components making up a given eutectic mixture, it is possible to propose a particular composition whose properties should be applied specifically [1,7].

Deep eutectic solvents are highly hygroscopic liquids, so trace amounts of water are often unavoidable as impurity [8]. However, depending on the application, water can be deliberately added to the DES to modulate the physicochemical properties of the solvent, especially mass transfer properties such as viscosity.

Aqueous mixtures of DES have already been used in nanoparticle synthesis [9], carbon dioxide absorption [10,11], and as media in chemical reactions [12]. Some conflicting information is found in the literature regarding the impact of water in DES on processes of CO_2 absorption. Su et al. [10] discovered that even a small addition of water to DES composed of choline chloride (ChCl) and urea reduces the solubility of carbon dioxide, while Li et al. [11] concluded that the solubility of CO_2 increases in tetramethylammonium chloride and monoethanolamine DES after the addition of H_2O, and reaches a maximum at 10% water content. Therefore, since the water content has a great influence on the physicochemical properties of deep eutectic solvents, it is crucial to control it in aqueous DES solutions from a technological point of view and in further application. A significant part of the research carried out so far has been devoted to the properties of pseudo-pure DES [13–16]. Thermodynamic properties of aqueous DESs are also rarely published [17]. Kuddushi et al. measured densities of aqueous solution of DES composed of ChCl with glutaric acid and malonic acid [18]. They calculated excess molar volume (V_m^E) for those systems that found to be negative, indicating strong interactions between water and DES molecules. Additionally, volumetric properties were described for such aqueous DES systems as: allyltriphenyl phosphonium bromide (ATPPB) with diethylene glycol (DEG) [19] or triethylene glycol (TEG) [20], ChCl with ethylene glycol or glycerol [21] and ChCl with lactic acid [22].

Studies that consider other properties of aqueous DES solutions—such as acoustic properties [18], refractive indices [21,23], and viscosities [23,24]—are still rare and not popular. All of them confirm that the interactions between DES and water molecules significantly influence on thermodynamic properties of aqueous DES solutions.

In general, most of the research that has been carried out to date on deep eutectic solvents as potential CO_2 absorbers are devoted to DESs in which physical absorption of carbon dioxide occurs. The carbon dioxide absorption capacity in those DESs is lower than that in commercially used absorbers therefore their potential application in industry is significantly limited. Therefore, the aim of this work is to characterize and better understand the water solutions of deep eutectic solvents based on alkanolamines with chemical absorption capacity in terms of their suitability for the effective separation of carbon dioxide from gas streams at relatively low pressure. It is known from the literature that carbon dioxide capacity, apart from others factors, depends also on the strength of intermolecular interactions between DES components [25,26]. As the interactions between HBA and HBD increase, the CO_2 solubility decreases. Similar effects can be expected for aqueous solutions of DESs, where increasing strength of the DES-H2O interactions might result in a decrease of carbon dioxide capacity due to the weaker interactions between CO_2 and DES [27]. The final effect of the presence of water in DES on the solubility of CO_2 will also depend on which of the components (DES or water) will have the greater affinity for the gas absorbed. Herein, we prepared deep eutectic solvents built of tetrabutylammonium bromide with 3-amino-1-propanol (AP), 2-(methylamino)ethanol (MAE), or 2-(butylamino)ethanol (BAE) and tetrabutylammonium chloride (TBAC) with AP, at 1:6 molar ratios. Then, the physicochemical properties of pure and aqueous solutions of DESs as density, speed of sound, viscosity, and refractive index were measured, and Jouyban–Acree predictive model (JAM) was used to correlate the experimental data. Several mathematical models for the correlation of physical properties of binary mixtures can be found in the literature [28]. However, these models were used mainly to correlate the density of mixtures. Thus, we decided to use the JAM equation, as so far it has also been used to predict the viscosity of two-component mixtures of classical solvents [29]. Thermodynamic excess properties—including excess molar volume, excess isobaric thermal expansion, excess isentropic compressibility, deviation in refractive indices, deviation in viscosities, and excess Gibbs energy of activation for viscous flow—were calculated and correlated by the Redlich–Kister-type polynomial equation considered as the most common and accurate mathematical model for this purpose. Prigogine–Flory–Patterson Theory (PFP) was used to correlate the experimental excess molar volume as the most accepted theory to interpret the behavior of non-ideal liquid solution, which has been applied to many mixtures of classical solvents as well as to systems containing ionic liquids and some aqueous solutions of DESs. The effect of the HBA anion type of obtained DES solutions was evaluated as well as the order and length of the alkyl chain of each alkanolamine were discussed. The influence of temperature on thermodynamic properties of DES solutions was also involved in this study.

2. Materials and Methods

2.1. Chemicals

The chemicals used in this study—3-amino-1-propanol (AP), 2-(methylamino)ethanol (MAE), 2-(butylamino)ethanol (BEA), tetrabutylammonium bromide (TBAB), and tetrabutylammonium chloride (TBAC)—were purchased from Sigma-Aldrich. TBAC was purified by double crystallization from acetone by adding diethyl ether. All salts were dried under reduced pressure before use, TBAB at 323 K for 48 h while TBAC at 298.15 K for several days. The corresponding information and the chemical structures of the DESs components are presented in Table 1.

Table 1. Provenance, mass fraction purity, and chemical structures of the compounds studied.

Chemical Name	Source	CAS Number	Molecular Weight $M/(g \cdot mol^{-1})$	Mass Fraction Purity	Chemical Structure
3-amino-1-propanol (AP)	Sigma Aldrich	156-87-6	75.11	0.99 [a]	
2-(methylamino)-ethanol (MEA)	Sigma Aldrich	109-83-1	75.11	≥98 [a]	
2-(butylamino)-ethanol (BEA)	Sigma Aldrich	111-75-1	117.19	≥98 [a]	
tetrabutylammonium bromide (TBAB)	Sigma Aldrich	1643-19-2	322.37	≥0.99 [a]	
tetrabutylammonium chloride (TBAC)	Sigma Aldrich	1112-67-0	277.92	≥0.98 [b]	

[a] As stated by supplier. [b] After crystalization, determined by potentiometric titration.

2.2. Preparation of DESs and Their Aqueous Solutions

DESs were prepared by mass with the same molar ratio of 1:6 salt to amino alcohol. The weighing was done using an analytical balance (Mettler Toledo) with the precision of 0.1 mg. The standard uncertainty in the mass fraction was estimated to be less than $\pm 1 \times 10^{-4}$. The combinations of the quaternary ammonium salts and AP/MAE/BEA were mixed at 353.15 K for 1 h using a magnetic stirrer in a fume hood until a homogeneous and uniform liquid without any precipitate was formed. Water content of DESs was measured using a Mettler Toledo Coulometric Karl–Fischer titrator (899 Coulometer apparatus from Metrohm) and it was found to be less than 0.0016 mass fraction. Table 2 displays the abbreviation of chemicals and DESs along with their molar mass, molar ratio, mass fraction, and water content.

Table 2. Abbreviations of chemicals and DESs along with their molar mass, molar ratio, mass fraction, and water content.

Symbol	HBA	HBD	Molar Ratio	M_{DES}/ (g·mol^{-1})	Mass Fraction of HBA [a]	Water Content [b]
DES1	TBAB	AP	1:6	110.449	0.4172	0.00059
DES2	TBAC	AP	1:6	104.091	0.3815	0.00121
DES3	TBAB	MAE	1:6	110.433	0.4170	0.00118
DES4	TBAB	BAE	1:6	146.519	0.3145	0.00159

[a] The standard uncertainty of DES mass fraction composition is 0.0001. [b] Water content of DESs in mass fraction determined by Karl Fisher titration with the standard uncertainty ±0.0001.

Deionized, double distilled, degassed water with a specific conductance of 1.15×10^{-6} S·cm^{-1} was used for the preparation of aqueous mixtures of DESs. The water contents in DES was accounted for upon solution preparation.

2.3. Physical Properties Measurements

2.3.1. Density and Speed of Sound

Measurements of density and speed of sound were performed by using a digital vibration-tube analyzer (Anton Paar DSA 5000, Graz, Austria) with proportional temperature control that kept the samples at working temperature with an accuracy of 0.01 K. Experimental frequency for the measurements of the ultrasonic speed was equal to 3 MHz. The apparatus was calibrated with double distilled deionized and degassed water and dry air at atmospheric pressure according to the apparatus catalog procedure. The experimental uncertainty of density and ultrasonic velocity measurements was better than 35×10^{-3} kg m^{-3} and 2×10^{-1} m s^{-1}, respectively.

2.3.2. Viscosity

Viscosities of the solvents were determined using LVDV-III Programmable Rheometer (cone-plate viscometer; Brookfield Engineering Laboratory, USA), controlled by a computer. The temperature of the samples was controlled within ± 0.01 K using a thermostatic water bath (PolyScience 9106, Niles, IL, USA). The display of the viscosimeter was verified with certified viscosity standard N100 and S3 provided by Cannon at 298.15 ± 0.01 K. The standard uncertainty of viscosity measurement was better than 1%.

2.3.3. Refractive Index

The refractive indices were measured using an Abbe refractometer (RL-3, Warsaw, Poland) equipped with a thermostat for controlling the cell temperature with an accuracy of ± 0.1 K. The standard uncertainty of refractive index measurement on the n_D scale was 0.0002. At least three independent measurements were taken for each sample at each temperature to assure reproducibility of the measurement.

2.3.4. Isobaric Heat Capacity

A Mettler Toledo Star One differential scanning calorimeter (DSC), STAR-1 System (Mettler Toledo, Greifensee, Switzerland), was used to measure specific heat capacities of novel DESs. The DSC instrument was calibrated by the indium standard prior to sample measurements. During the measurement, an inert atmosphere was created under a nitrogen flow of 60 mL min^{-1}. The sapphire method for cp determination was used [30]. A 'baseline' or blank measurement was performed for heating rate 10 K min^{-1}. All of the results obtained were blank curve corrected and performed twice. The test material and the reference were placed into individual aluminium crucibles which were then sealed with pierced lids. The data from the DSC were recorded and then analyzed to obtain the Cp from the data.

3. Results and Discussion
3.1. Physical Properties of Binary Mixtures

The experimental values of density, speed of sound, viscosity, and refractive indices for aqueous solutions of DESs consisting of tetrabutylammonium bromide and 3-amino-1-propanol or 2-(methylamino)ethanol or 2-(butylamino)ethanol and for the aqueous solutions of DES built of tetrabutylammonium chloride and 3-amino-1-propanol over the entire range of compositions at temperatures ranging from 293.15 K to 313.15 K are reported in Tables S1–S4. Moreover, in Figure 1 the physical properties are plotted as a function of the DES molar fraction at 298.15 K for all systems studied. As it can be observed, depending on the properties, its dependence on the deep eutectic solvent content varies. Moreover, all trends are nonlinear, indicating deviation from the ideal course.

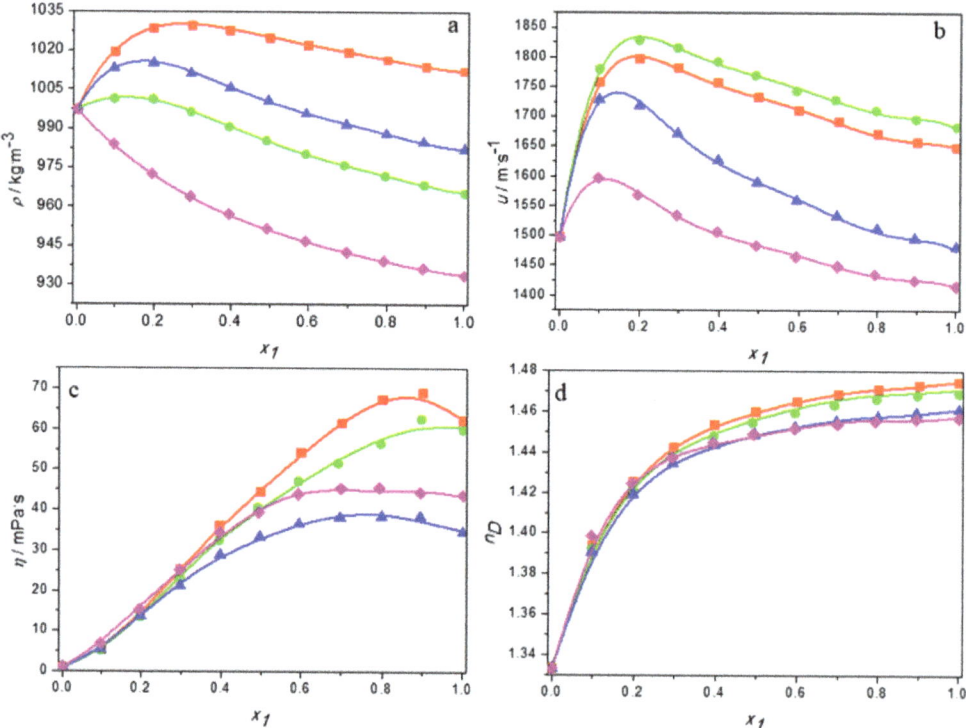

Figure 1. The dependence of the physical properties of aqueous solutions of DESs on molar fraction of deep eutectic solvent at 298.15K: (**a**) density; (**b**) speed of sound; (**c**) viscosity; (**d**) refractive index. ■ DES1; ● DES2; ▲ DES3; ♦ DES4; —, Equation (1).

Taking into account the density, its values decrease in the whole range of DES concentrations only for aqueous TBAB:BAE (DES4) solutions for which a negative deviation from ideal behavior is observed. For the other systems, the density increases with increasing DES concentration at low deep eutectic solvent content, reaches its maximum value at certain molar fraction of DES, and afterwards begins to decrease. The composition of the solution with the highest density depends on both the amino alcohol and the salt.

The dependence of the sound velocity and viscosity of aqueous DES solutions on the molar fraction of DES also shows a maximum. However, in the case of viscosity, unlike the speed of sound and the density, it occurs at high DES content. The refractive index increases monotonically in the whole range of DES concentrations for all systems.

When the temperature dependence is considered, one can observe that all properties decrease with the increase of temperature as the result of thermal expansion.

The experimental values of of density, speed of sound, viscosity, and refractive indices of binary mixtures were correlated by using of Jouyban–Acree model [29,31–33]. This mathematical model uses the physicochemical properties of the individual solvents as input data and a number of curve-fitting parameters represent the effects of solvent–solvent interactions in the solution

$$lny = x_1 lny_1 + x_2 lny_2 + \frac{x_1 x_2}{T} \sum_{i=0}^{i=n} J_i (x_1 - x_2)^2 \qquad (1)$$

The y, y_1, and y_2 are the physical properties of the mixture, deep eutectic solvent and water, at specific temperature. The x_1 and x_2 are mole fractions of DES and water, respectively. The J_i terms are coefficients of the model computed by using a zero-intercept regression analysis

$$lny - x_1 lny_1 - x_2 lny_2 = \frac{x_1 x_2}{T} \sum_{i=0}^{i=n} J_i (x_1 - x_2)^2 \qquad (2)$$

Root mean square deviation of fit (RMSD) and the average deviation (ARD %) were calculated according to the following equations

$$RMSD = \left[\frac{\sum (Y_{exp} - Y_{pred})^2}{n - k} \right]^{1/2} \qquad (3)$$

$$ARD \% = \frac{100}{n} \sum \frac{|y_{exp} - y_{pred}|}{y_{pred}} \qquad (4)$$

where n is the number of experimental data, k is the number of parameters of model and Y is equal to lny.

Parameters J_i of Equation (1), root mean square deviation of fit (RMSD) and the corresponding average relative deviation (ARD %) for the systems studied at $T = (293.15$ to $313.15)$ K are presented in Table S5 (Supplementary Materials).

Moreover, the values of density, speed of sound, viscosity, and refractive index obtained by JAM are depicted as the smoothed solid lines in Figure 1. As can be seen, the Jouyban–Acree model correlates the experimental physical properties satisfactorily, especially for density and refractive index, for which average relative deviations are the same order as the experimental uncertainty. Thus, the JAM can be considered a reliable model for predicting the densities and refractive indices as well as it can be used for estimation of the speeds of sound and viscosity of aqueous DES solutions, for which, however, higher ARD % are observed.

3.2. Volumetric Properties

3.2.1. Excess Molar Volume

The excess molar volumes (V^E) were calculated using the experimental density data according to the following equation

$$V^E = \frac{x_1 M_1 + x_2 M_2}{\rho} - \frac{x_1 M_1}{\rho_1} - \frac{x_2 M_2}{\rho_2} \quad (5)$$

where d is the density of the mixture and x_i, M_i, and ρ_i are: the mole fraction, the molar mass and density of DES ($i = 1$) and water ($i = 2$), respectively.

The obtained values of V^E are presented in Tables S1–S4 and plotted as a function of the DES mole fraction, x_1, in Figure 2. Figure 2a shows the plots of V^E against mole fraction of DES for all studied mixtures at 298.15 K and Figure 2b depicts the temperature dependence of excess molar volumes for the system (TBAB:BAE + water) as an example. It can be observed in these figures that the curves of V^E are asymmetrical and their values are negative over the whole composition and temperature range. The minimum was found between mol fraction of 0.35 and 0.4 of DES equal to −0.98 for DES1, −1.00 for DES2, −1.13 for DES3, −1.02 for DES4. The asymmetry of the curves is due to the difference between the molar volumes of the components mixture.

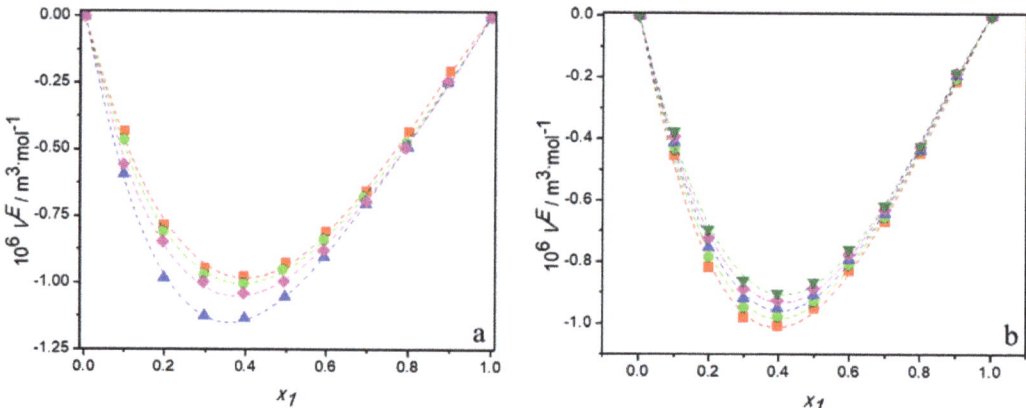

Figure 2. Dependence of the excess molar volume of aqueous solutions of DESs on molar fraction of deep eutectic solvent: (a) at 298.15 K for ■ DES1; • DES2; ▲ DES3; ♦ DES4; (b) for DES1 at 298.15 K (■); 298.15 K (•); 303.15 K (▲); 308.15 K (♦); 313.15 K (▼); —, Equation (6).

In Figure 2, the dashed lines represent the correlated values according to the Redlich–Kister polynomial [34]

$$V^E = x_1 x_2 \sum_{i=0}^{2} A_i (x_1 - x_2)^i \quad (6)$$

where the A_i values are adjustable parameters.

As it can be seen, the calculated values agree very well with the experimental data. The A_i values were determined using the least squares method and they are listed in Tables S6–S9, along with their RMSD. For all systems, excess molar volumes were correlated using three-parameter Redlich–Kister polynomial equation.

The negative values of excess molar volumes can be explained based on the strength of the specific interactions, size, and shape of molecules. When DES is added to water, the intra-molecular interactions between DES or water molecules are disrupted and new hydrogen bonding interactions between water and chloride/bromide anion of HBA and between water and -OH group and -NH$_2$ group or –NH of amino alcohol are forming. Moreover, water molecules—as much smaller than the deep eutectic solvent one—may fit into the interstices of the DES. Therefore, the filling effect of water in the interstices of DES, and the strong hydrogen bonding interactions between the unlike components of the systems, all lead to the negative values of the excess molar volumes.

The temperature dependence of the excess molar volumes can determine what kind of effect—i.e., the packing phenomenon or the strong forces between the components—is responsible for the negative values of V^E. In general, as temperature increases, the specific interactions break down and due to the increased thermal fluctuation, more holes of sufficient size for the accommodation of the unlike component are formed. These effects influence the excess molar volume in a reverse manner. A decrease of specific interactions causes an increase in V^E values, while a loosening of the DES structure leads to a decrease of excess molar volume with temperature. Thus, the observed increase of V^E with rising temperature for all systems investigated suggests that specific interactions determine the volumetric behavior of aqueous solutions of deep eutectic solvents based on alcohol amine. A similar phenomenon was observed by other researchers for aqueous solutions of DES based on choline chloride [18,21,22,24,35–38] or allyltriphenylphosphonium bromide [19,20]. What is interesting, for the (DES + alcohol) systems, due to the decrease in the excess molar volume with temperature, the dominance of the packing effect was postulated [39,40].

The dominance of specific interactions in the aqueous solutions of the DESs studied can be confirmed by the Prigogine–Flory–Patterson (PFP) theory [41–45]. This theory has been originally used in interpreting the values of the excess molar volumes of binary systems formed by polar compounds which do not form strong electrostatic or hydrogen bond interactions. Over time, however, it has emerged that the use of the Flory formalism can still provide an interesting correlation between the excess volumes of more complex mixtures. So far, the PFP theory has been successfully applied to predict and model the excess molar volumes of many mixtures containing ionic liquids [46–50] and some aqueous systems with deep eutectic solvents [18,51,52].

According to the PFP theory, the excess molar volume contains three contributions: an interactional contribution, a free volume contribution, and a pressure contribution. The expression for V^E is given as

$$\frac{V^E}{x_1 V_1^* + x_2 V_2^*} = \frac{\left(\widetilde{V}^{1/3} - 1\right)\widetilde{V}^{2/3}\psi_1\Theta_2\chi_{12}}{\left[(4/3)\widetilde{V}^{-1/3} - 1\right]P_1^*} + \frac{-\left(\widetilde{V}_1 - \widetilde{V}_2\right)^2\left[(14/9)\widetilde{V}^{-1/3} - 1\right]\psi_1\psi_2}{\left[(4/3)\widetilde{V}^{-1/3} - 1\right]\widetilde{V}} + \frac{\left(\widetilde{V}_1 - \widetilde{V}_2\right)\left(P_1^* - P_2^*\right)\psi_1\psi_2}{P_2^*\psi_1 + P_1^*\psi_2} \quad (7)$$

where V^E is excess molar volume, x mole fraction, V^* characteristic volume, P^* characteristic pressure, ψ molecular contact energy fraction, θ molecular surface fraction, $\widetilde{V}$ reduced volume, and χ_{12} interactional parameter.

The reduced volume for pure substance i is defined in terms of the thermal expansion coefficients, α_i, as

$$\widetilde{V}_i = \left(\frac{1 + \frac{4}{3}\alpha_i T}{1 + \alpha_i T}\right)^3 \quad (8)$$

The reduced volume of mixture, $\widetilde{V}$, is calculated from

$$\widetilde{V} = \psi_1 \widetilde{V}_1 + \psi_2 \widetilde{V}_2 \quad (9)$$

where the molecular contact energy fraction, ψ, is expressed by: $\psi_1 = 1 - \psi_2 = \frac{\phi_1 p_1^*}{\phi_1 p_1^* + \phi_2 p_2^*}$ with the hardcore volume fraction, ϕ, calculated from $\phi_1 = 1 - \phi_2 = \frac{x_1 V_1^*}{x_1 V_1^* + x_2 V_2^*}$.

The characteristic volume, V_i^*, is calculated from the molar volume from the expression $V_i^* = \frac{V_i^0}{\widetilde{V}_i}$ and the characteristic pressure is expressed by

$$p_i^* = \frac{\alpha_i}{\kappa_{Ti}} T \widetilde{V}_i^2 \quad (10)$$

where κ_{Ti} is the isothermal compressibility obtained from the isentropic compressibility from the thermodynamic relation

$$\kappa_{Ti} = \kappa_{Si} + \frac{V_i^0 \alpha_i^2 T}{C_{pi}} \quad (11)$$

with the isobaric heat capacity, C_{pi}.

The molecular surface fraction of component 2 is given by: $\Theta_2 = \frac{\phi_2}{\phi_1 \frac{s_1}{s_2} + \phi_2}$, in which the ratio of the surface contact sites per segment is given by

$$\frac{s_1}{s_2} = \left(\frac{v_2^*}{v_1^*}\right)^{\frac{1}{3}} \quad (12)$$

In present study, the thermal expansion coefficient, α_p, defined as: $\alpha_p = \frac{1}{V}\left(\frac{\partial V}{\partial T}\right)_p$, was calculated by use the temperature dependence of density, which was found to be the second-order polynomial equation.

The isobaric heat capacity of DESs was determined experimentally. Table S10 shows, for all pure DESs used in this work, the thermal expansion coefficient, the isobaric heat capacity and the Flory parameters necessary for the application of the PFP theory. The results of our experiments compare the Cp of DES1–DES4. Quite noticeable differences are observed. Generally, the Cp values [J·mol^{-1} K^{-1}] are arranged in order: DES 3 < DES 2 < DES 1 < DES 4 at set temperature points.

According to the PFP theory, for the separation of the values of excess molar volume into three contributions, the interactional parameter χ_{12} must be found. In the present study, it has been done by minimalization of the objective function, considering deviations in the prediction of the excess volume, defined as

$$OF = \sum_{i=1}^{n}\left(V_{exp}^E - V_{calc}^E\right)^2 \quad (13)$$

In calculations, the value of the interactional parameter χ_{12} was assumed to be independent on the composition of mixture. Figure 3 shows the composition dependence of calculated excess molar volume, together with the three contributions (V_{int}^E, V_{fv}^E, $V_{P^*}^E$), compared with the experimental V^E data for each system studied at 298.15 K.

Moreover, Table 3 reports the adjusted values of interactional parameter χ_{12} and calculated three contributions to excess molar volume for the binary mixtures of deep eutectic solvents with water at all temperatures investigated and at x_1 = 0.4 together with RMSD.

Study of the data presented in Table 3 as well as an analysis of Figure 3 reveals that the interactional contribution is always negative and it seems to be the most important to explain the values of the excess molar volume. It decides about the sign and magnitude of the V^E due to its greater value compared to the other two contributions for all investigated systems at all temperatures. The free volume contribution, which is a measure of geometrical accommodation, is negative but its magnitude is much smaller than for the interactional contributions. Therefore, it can be said that the PFP model confirms the conclusions resulting from the dependence of excess molar volume on the temperature, postulating little significance of the packing effect for the systems studied. The third contribution is the result of differences in internal pressure and in the reduced volumes of the components. It is positive, its magnitude is smaller than the interactional contribution and decreases distinctly with temperature.

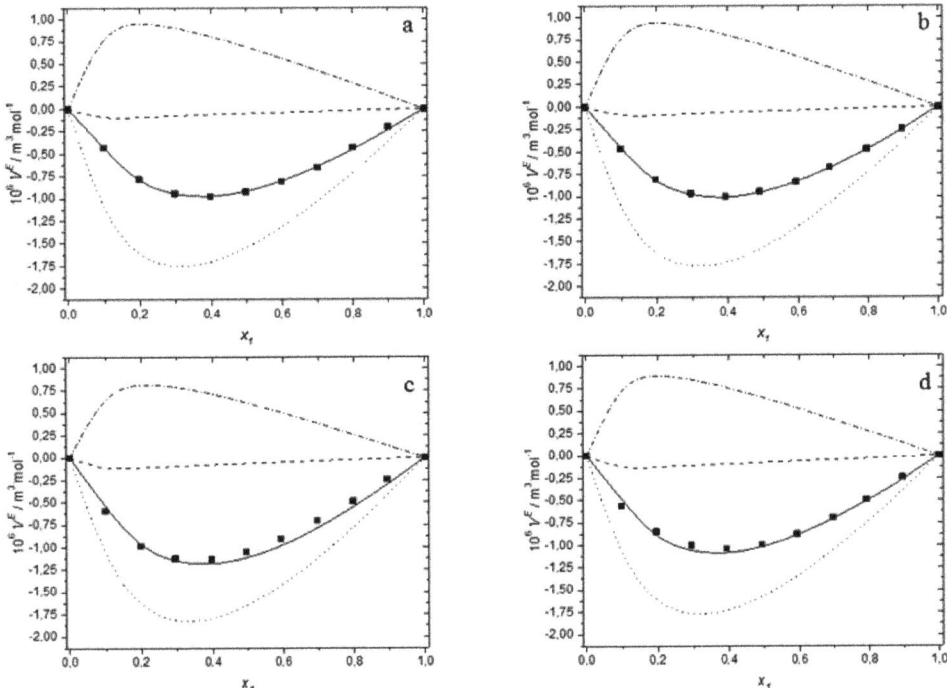

Figure 3. The dependence of the excess molar volume of aqueous solutions of DESs on molar fraction of deep eutectic solvent at 298.15 K: (**a**) for DES1; (**b**) for DES2; (**c**) for DES3; (**d**) for DES4; (■) experimental data; (—) calculated using the PFP model; (········) interactional contribution (V_{int}^E); (– – –), free volume contribution (V_{fv}^E); (– · –), characteristic pressure contribution (V_{p*}^E).

Table 3. Calculated values of the interactional parameter χ_{12}, root mean square deviation of fit RMSD, and the three contributions (V_{int}^E, V_{fv}^E, V_{p*}^E) from the PFP theory to the excess molar volumes for the binary mixtures of deep eutectic solvents with water at $x_1 = 0.4$ and T = (293.15 − 313.15) K.

T/K	293.15	298.15	303.15	308.15	313.15
		DES$_1$ (1) + water (2)			
$10^{-6}\, \chi_{12}$/J·m^{-3}	−393.3	−354.2	−317.6	−283.2	−252.1
$10^{6}\, V_{int}^E$/m^3·mol^{-1}	−1.932	−1.726	−1.534	−1.359	−1.205
$10^{6}\, V_{p*}^E$/m^3·mol^{-1}	0.979	0.814	0.661	0.522	0.398
$10^{6}\, V_{fv}^E$/m^3·mol^{-1}	−0.062	−0.064	−0.066	−0.059	−0.053
RMSD	0.018	0.016	0.022	0.033	0.046
		DES$_2$ (1) + water (2)			
$10^{-6}\, \chi_{12}$/J·m^{-3}	−397.7	−358.5	−321.7	−287.2	−256.0
$10^{6}\, V_{int}^E$/m^3·mol^{-1}	−1.960	−1.752	−1.559	−1.384	−1.230
$10^{6}\, V_{p*}^E$/m^3·mol^{-1}	0.964	0.806	0.652	0.513	0.390
$10^{6}\, V_{fv}^E$/m^3·mol^{-1}	−0.062	−0.065	−0.064	−0.060	−0.053
RMSD	0.020	0.006	0.008	0.020	0.033
		DES$_3$ (1) + water (2)			
$10^{-6}\, \chi_{12}$/J·m^{-3}	−324.1	−294.1	−266.0	−241.2	−218.9
$10^{6}\, V_{int}^E$/m^3·mol^{-1}	−2.031	−1.818	−1.624	−1.454	−1.305
$10^{6}\, V_{p*}^E$/m^3·mol^{-1}	0.888	0.716	0.559	0.421	0.301
$10^{6}\, V_{fv}^E$/m^3·mol^{-1}	−0.078	−0.081	−0.079	−0.074	−0.067
RMSD	0.080	0.068	0.054	0.040	0.029

Table 3. Cont.

T/K	293.15	298.15	303.15	308.15	313.15
			DES$_4$ (1) + water (2)		
$10^{-6}\,\chi_{12}/\text{J}\cdot\text{m}^{-3}$	−245.4	−217.4	−190.9	−166.9	−145.5
$10^{6}\,V^{E}_{int}/\text{m}^{3}\cdot\text{mol}^{-1}$	−1.989	−1.750	−1.529	−1.332	−1.158
$10^{6}\,V^{E}_{p*}/\text{m}^{3}\cdot\text{mol}^{-1}$	0.946	0.761	0.591	0.438	0.302
$10^{6}\,V^{E}_{fv}/\text{m}^{3}\cdot\text{mol}^{-1}$	−0.091	−0.096	−0.097	−0.094	−0.087
RMSD	0.062	0.045	0.031	0.027	0.033

For deeper analysis of the obtained results, the percentage of the three contributions in excess molar volume was calculated. Figure 4 presents the obtained results at 298.15 K.

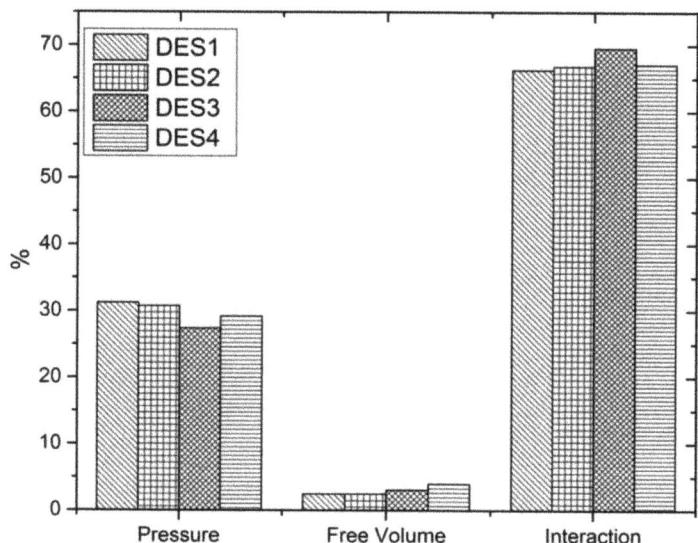

Figure 4. Percentages of the three contributions in excess molar volume of aqueous solutions of DESs for $x_1 = 0.4$ at 298.15 K.

As can be seen, the interactional contribution and the characteristic pressure contribution determine the order of the excess molar volume observed for the studied systems, which is as follows: TBAB:MAE (DES 3) < TBAB:BAE (DES 4) < TBAC:AP (DES 2) ≈ TBAB:AP (DES 1). The free volume fraction has practically no effect on the excess molar volume, and its absolute value increases with the length of the alkyl chain in the amino alcohol.

Moreover, the results show that the anion of the salt in DES does not practically effect on the value of excess molar volume. As depicted in Figures 3 and 4, almost identical interactional contribution, free volume contribution and characteristic pressure contribution are observed for aqueous solutions of TBAC:AP (DES 2) and TBAB:AP (DES 1).

Further analysis of Table 3 shows that the interactional contribution is mainly responsible for the increase of excess molar volume with increasing temperature. The decrease in its absolute value is greater than the decrease of the positive characteristic pressure contribution, and consequently the excess volumes of the studied systems increase with temperature. The percentage of free volume contribution increases with increasing temperature, but due to the low absolute values V^{E}_{int}, it has no influence on the excess molar volume or its dependence on temperature.

Summing up, it is evident from Figure 3 that the PFP theory predicts the experimental data satisfactorily. Thus, while the PFP theory does not take into account the strong interactions between components—such as electrostatic, hydrogen bonding, and complex formation—we can infer that the PFP model reproduces the main characteristics of the experimental data by using only one fitted parameter to describe excess molar volume.

3.2.2. Excess of Thermal Expansion

As the temperature dependence of density was found to be second order polynomial, type: $\ln(\rho) = at^2 + bt + c$, the isobaric thermal expansion coefficients at different temperatures were derived according to the equation

$$\alpha_p = \frac{1}{V}\left(\frac{\delta V}{\delta T}\right)_P = -\frac{1}{\rho}\left(\frac{\delta \rho}{\delta T}\right)_P = -\left(\frac{\delta \ln \rho}{\delta T}\right)_P = -(2a+b) \quad (14)$$

Then, excess thermal expansion, $\Delta \alpha_p$, was calculated using the equation

$$\Delta \alpha_p = \alpha_p - \sum_{i=1}^{n} \Phi_i \alpha_{p,i} \quad (15)$$

where Φ_i is the volume fraction of pure component i, defined as $\phi_i = x_i V_i / \sum_i x_i V_{i,}$. The values of α_p and $\Delta \alpha_p$ are given in Tables S1–S6 in Supporting Material and variation of the excess thermal expansion with DES mole fraction, x_1, is plotted in Figure 5.

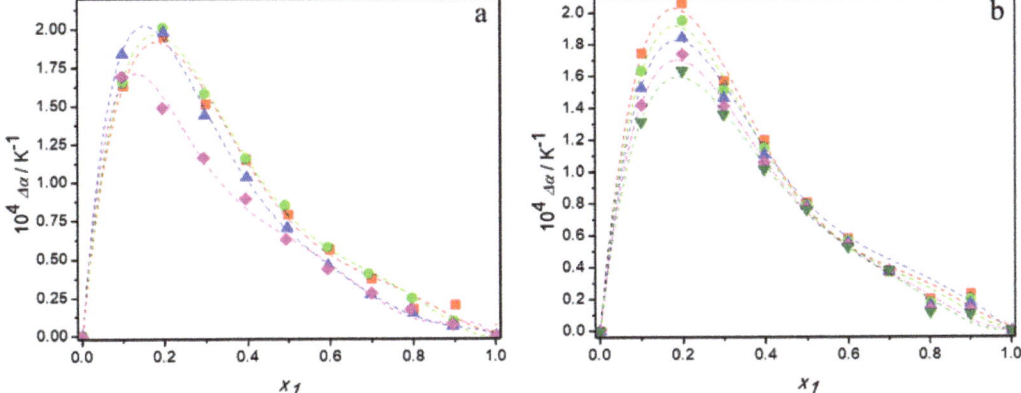

Figure 5. Dependence of the excess thermal expansion of aqueous solutions of DESs on molar fraction of deep eutectic solvent: (**a**) at 298.15 K for ■ DES1; • DES2; ▲ DES3; ♦ DES4; (**b**) for DES1 at 293.15 K (■); 298.15 K (•); 303.15 K (▲); 308.15 K (♦); 313.15 K (▼); —, Equation (6).

It can be seen that the values of excess thermal expansion are positive in the entire composition range for all systems studied, regardless of temperature. Since positive $\Delta \alpha_p$ are typical for the systems containing molecules capable to self-associate, the obtained results confirm strong hydrogen bonds between water molecules or between molecules of deep eutectic solvents, the strength of which decreases with temperature, as indicated by a reduction in excess thermal expansion with temperature [53]. Moreover, the less positive values of excess thermal expansion obtained for the (TBAB:BAE (DES 4) + water) system indicate the weakest hydrogen bond interactions between molecules of this DES compared to the others.

3.2.3. Partial Molar Volumes

The partial molar volumes of the studied DESs and water in their binary mixtures, $\overline{V}_1$ and $\overline{V}_2$, were calculated from the Equations (16) and (17), using the parameters of Redlich Kister equation (Tables S6–S9) and the molar volumes of the pure components, V_1^o and V_2^o

$$\overline{V}_1 = V_1^o + (x_1 - 1)^2 \sum_{i=0}^{j} A_i(2x_1 - 1)^i + 2x_1(1 - x_1)^2 \sum_{i=0}^{j} A_i i(2x_1 - 1)^{i-1} \quad (16)$$

$$\overline{V}_2 = V_2^o + x_1^2 \sum_{i=0}^{j} A_i(2x_1 - 1)^i - 2x_1^2(1 - x_1) \sum_{i=0}^{j} A_i i(2x_1 - 1)^{i-1} \quad (17)$$

The obtained values of the partial molar volumes together with the molar volumes of pure components are presented in Table S11 in Supplementary Material. As can be seen, at all studied temperatures, the molar volumes of the interacting compounds in the pure state were higher than their corresponding values in the mixture, indicating the reduction in volume upon adding a deep eutectic solvent to water. Figure 6 shows the excess partial molar volumes of the components at 298.15 K. These properties were calculated from definition as: $\overline{V}_i^E = \overline{V}_i - V_i^o$ and their values are negative over the whole composition range. In general, the negative $\overline{V}_1^E$ and $\overline{V}_2^E$ values indicate the presence of significant solute–solvent interactions between unlike molecules, whereas the positive $\overline{V}_1^E$ and $\overline{V}_2^E$ values indicate the presence of solute–solute or solvent–solvent interactions between like molecules in the mixture [54]. In the present work, the negative excess partial molar volumes of the components indicate that the DES—water interactions are stronger than the DES—DES or the water–water interactions, what is consistent with the conclusions from excess molar volumes.

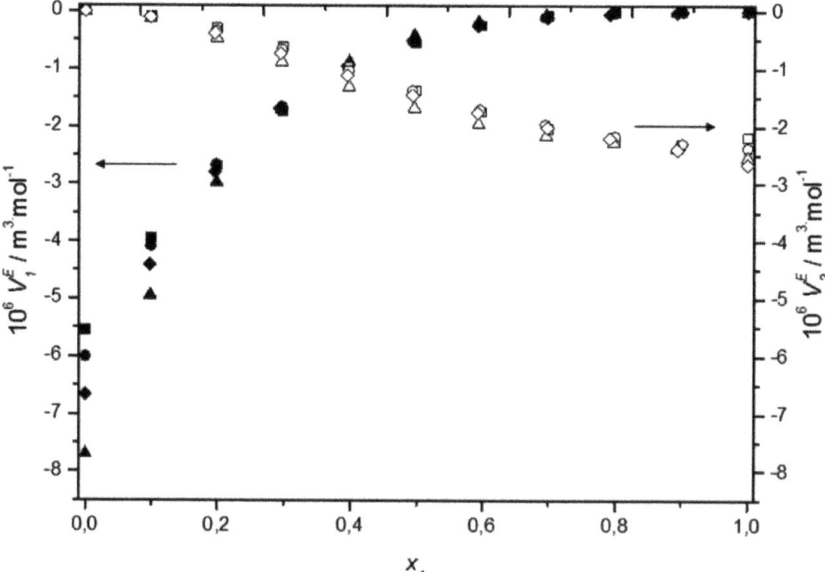

Figure 6. Dependence of the excess partial molar volume $\overline{V}_1^E$ of ■ DES1, ● DES2, ▲ DES3, ◆ DES4, and water $\overline{V}_2^E$ in □ DES1, ○ DES2, △ DES3, ◊ DES4 on molar fraction of deep eutectic solvent at 298.15 K.

Since the partial molar properties at infinite dilution provide useful information about the interactions between components of a mixture that are independent on composition, their values for DES and water were calculated.

The partial molar volumes at infinite dilution of DES were obtained by setting $x_1 = 0$ in Equation (18) as

$$\overline{V}_1^\infty = V_1^o + \sum_{i=0}^{j} A_i(-1)^i \tag{18}$$

Similarly, setting $x_1 = 1$ in Equation (19) allowed to estimate the partial molar volumes at infinite dilution of water

$$\overline{V}_2^\infty = V_2^o + \sum_{i=0}^{j} A_i \tag{19}$$

Table 4 presents the obtained values of partial molar volumes at infinite dilutions of DES and water in their binary systems.

Table 4. Partial molar volumes at infinite dilution of DES and water in their binary mixtures at T = (293.15 to 313.15) K and at atmospheric pressure (0.1 MPa).

T/K	$10^6\,\overline{V}_1^\infty/$ m^3·mol^{-1}	$10^6\,\overline{V}_1^{E\infty}/$ m^3·mol^{-1}	$10^6\,\overline{V}_2^\infty/$ m^3·mol^{-1}	$10^6\,\overline{V}_2^{E\infty}/$ m^3·mol^{-1}	$10^6\,\overline{V}_1^\infty/$ m^3·mol^{-1}	$10^6\,\overline{V}_1^{E\infty}/$ m^3·mol^{-1}	$10^6\,\overline{V}_2^\infty/$ m^3·mol^{-1}	$10^6\,\overline{V}_2^{E\infty}/$ m^3·mol^{-1}
	DES 1 + water				DES 2 + water			
293.15	102.91	−5.83	15.94	−2.11	101.10	−6.34	15.58	−2.47
298.15	103.60	−5.54	15.88	−2.19	101.82	−6.01	15.69	−2.38
303.15	104.32	−5.22	16.13	−1.96	102.51	−5.71	15.79	−2.30
308.15	105.00	−4.95	16.24	−1.88	103.18	−5.44	15.89	−2.24
313.15	105.65	−4.70	16.31	−1.85	103.84	−5.19	15.98	−2.17
	DES 3 + water				DES 4 + water			
293.15	104.00	−8.04	15.45	−2.6	149.34	−6.98	15.29	−2.75
298.15	104.75	−7.70	15.52	−2.54	150.27	−6.66	15.40	−2.67
303.15	105.47	−7.39	15.60	−2.49	151.17	−6.36	15.50	−2.60
308.15	106.18	−7.10	15.69	−2.43	152.06	−6.09	15.59	−2.53
313.15	106.88	−6.83	15.77	−2.39	152.94	−5.83	15.68	−2.47

As can be seen, both the partial molar volumes and the excess partial molar volumes at infinite dilution increase with the increasing temperature. Such results seem to indicate that the weakening of hydrogen bond interactions between DES and water molecules with increase in temperature is the most important factor controlling the properties of the systems and it dominates over the packing effect. Moreover, the excess partial molar volumes at infinite dilution of DES change in the order TBAB:MAE (DES 3) < TBAB:BAE (DES 4) < TBAC:AP (DES 2) < TBAB:AP (DES 1) confirming the conclusions obtained on the basis of excess molar volumes. The dependence of excess partial molar volumes at infinite dilution of water on the DES: TBAB:BAE (DES 4) ≤ TBAB:MAE (DES 3) < TBAC:AP (DES 2) < TBAB:AP (DES 1) is similar and only the reordering of DES4 and DES3 takes place. However, taking into an account the uncertainty of partial molar volume at infinite dilution, it can be said, that the $\overline{V}_2^{E\infty}$ for systems (DES4+water) and (DES3+water) are practically equal.

3.3. Excess Isentropic Compressibility

The isentropic compressibilities of aqueous solutions of DESs were estimated using experimental values of densities and sound velocities by the Laplace equation

$$\kappa_S = -\frac{1}{V_m}\left(\frac{\partial V_m}{\partial P}\right)_S = \frac{1}{u^2 \cdot \rho} \tag{20}$$

providing the link between thermodynamics and acoustics.

Then, according to the approach developed by Benson et al. [55], the excess isentropic compressibility was calculated as

$$\kappa_S^E = \kappa_S - \sum_i \Phi_i \kappa_{S,i} - T\sum_i \Phi_i V_i \alpha_{p,i}^2 / C_{p,i} + T\sum_i x_i V_i \left(\sum_i \Phi_i \alpha_{p,i}\right)^2 / \sum_i x_i C_{p,i} \qquad (21)$$

Tables S1–S4 in Supplementary Material present the experimental values of sound velocity and the calculated values of excess isentropic compressibility for aqueous solutions of DESs made of tetrabutylammonium bromide and 3-amino-1-propanol or 2-(methylamino)ethanol or 2-(butylamino)ethanol and for the aqueous solutions of DES built of tetrabutylammonium chloride and 3-amino-1-propanol over the entire range of compositions at temperatures ranging from 293.15 K to 313.15 K. Figure 7a shows the plots of κ_S^E against mole fraction of DES for the all studied mixtures at 298.15 K and Figure 7b depicts the temperature dependence of excess isentropic compressibility for the system (TBAB:AP +water) as an example. It is evident that the curves are remarkably asymmetric, with their minima shifted towards a rich mole fraction of water, even more than the excess molar volume curves are.

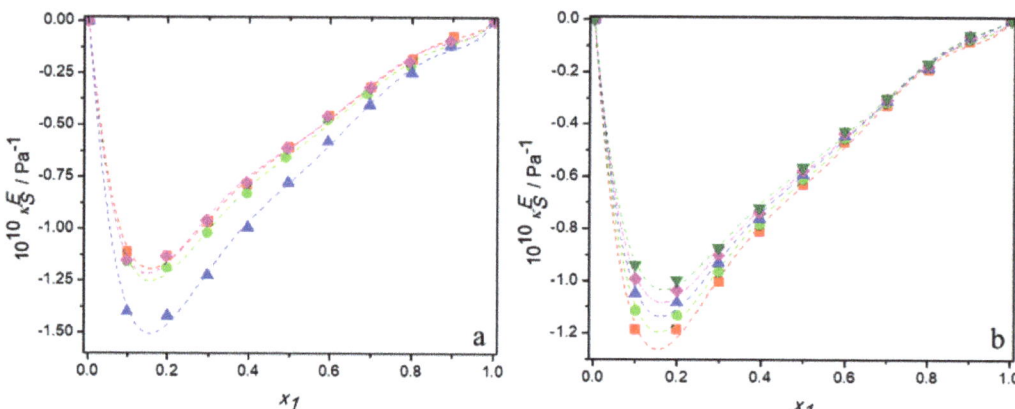

Figure 7. Dependence of the excess isentropic compressibility of aqueous solutions of DESs on molar fraction of deep eutectic solvent: (a) at 298.15 K for ■ DES1; • DES2; ▲ DES3; ♦ DES4; (b) for DES1 at 298.15 K (■); 298.15 K (•); 303.15 K (▲); 308.15 K (♦); 313.15 K (▼); —, Equation (6).

In these figures, it can be also seen that the values of κ_S^E are negative for all systems over the entire range of the mole fraction as well as the temperature range. This indicates that the mixtures might be less compressible than the corresponding ideal mixtures due to a closer approach and stronger interactions between the unlike molecules of the mixtures. The negative values of excess isentropic compressibility for the binary systems follow the order: TBAB:MAE (DES 3) < TBAB:BAE (DES 4) ≈ TBAC:AP (DES 2) ≈ TBAB:AP (DES 1). Thus, the behavior of the excess isentropic compressibility seems to be consistent with the obtained values of the excess molar volume, which suggests that the interactions and the packing effect dominate in aqueous solutions of TBAB:MAE and do not practically depend on the anion of the salt in DES.

Moreover, as can be seen in Figure 7b, the values of κ_S^E become less negative with increasing temperature for all systems at a fixed composition. It is due to the reduction of interactions between unlike molecules, what has already been suggested by volumetric properties. Indeed, the increase in temperature also increases the thermal motion of the molecules and enlargement of interstices. However, the decrease in hydrogen bonding is greater and, in the result, the excess molar compressibility of all aqueous solutions of DES decreases with increasing temperature.

3.4. Deviations in Refractive Index

From refractive indices the deviations in refractive index, Δn_D, were calculated using the equation

$$\Delta n_D = n_D - \sum_{i=1}^{n} \Phi_i n_{Di} \qquad (22)$$

where n_D and n_{Di} are the refractive index of a mixture and a pure component i, respectively and Φ_i denotes the volume fraction of a pure component i.

Tables S1–S4 in Supplementary Material present the experimental values of refractive index and calculated values of the deviations in refractive index for aqueous solutions of DESs over the entire range of compositions at temperatures ranging from 293.15 K to 313.15 K. Figure 8a shows the plots of Δn_D against mole fraction of DES for the all studied mixtures at 298.15 K and Figure 8b depicts the temperature dependence of deviations in refractive index for the system (TBAB:AP +water) as an example.

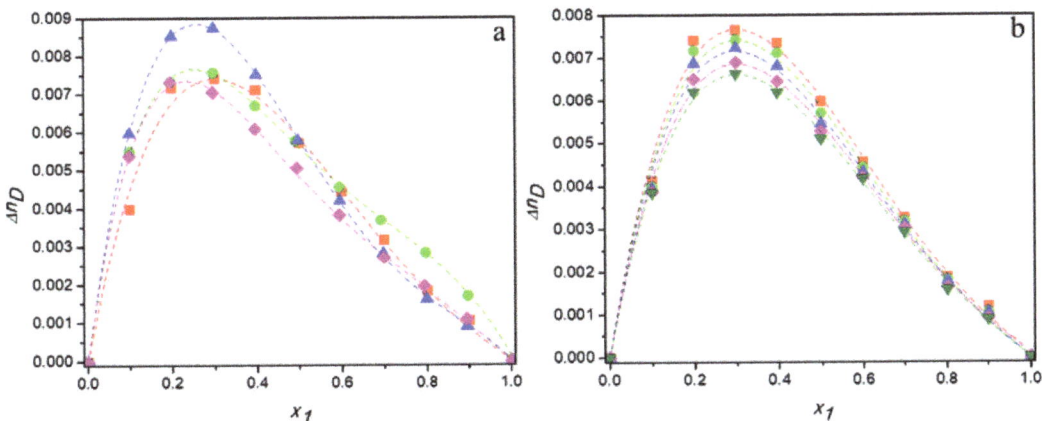

Figure 8. Dependence of the deviations in refractive of aqueous solutions of DESs on molar fraction of deep eutectic solvent: (**a**) 298. 15 K for ■ DES1; • DES2; ▲ DES3; ♦ DES4; (**b**) for DES1 at 298.15 K (■); 298.15 K (•); 303.15 K (▲); 308.15 K (♦); 313.15 K (▼); —, Equation (6).

As can be seen from Figure 8, the values of deviations in refractive index are positive over the whole composition range of binary mixtures and their dependences on mole fraction of DES are asymmetrical.

It is known that deviations of refractive index are negatively correlated with excess molar volumes [56]. If excess molar volume is negative, then there will be less free volume available than in an ideal mixture and the photons will interact more strongly with the components of the solution. As a result, light will travel with a weaker velocity in the mixture and its refractive index will be higher than in an ideal solution. Thus, positive deviations of refractive index will be observed.

This phenomenon occurs in all systems investigated in the present study and it is the strongest for TBAB:MAE (DES 3). Therefore, the obtained deviations of refractive index confirm the conclusion regarding the strongest interactions between this deep eutectic solvent and water molecules compared to other DESs studied. Moreover, as the values Δn_D increase with decreasing temperature, they indicate an increase in the number of hydrogen bonds at lower temperatures, which corresponds with the results of densitometric and acoustic research.

3.5. Deviations in Viscosity and Excess Gibbs Energy of Activation for Viscous Flow

Based on the viscosities of the mixtures, the viscosity deviations $\Delta \eta$ were obtained according to the equation

$$\Delta \eta = \eta - \exp(x_1 ln \eta_1 + x_2 ln \eta_2) \tag{23}$$

which uses the viscosity of the ideal mixture as suggested by Arrhenius.

The excess Gibbs energy of activation for viscous flow ΔG^E, was calculated using the equation

$$\Delta G^E = RT[\ln(\eta V_m) - (x_1 \ln(\eta_1 V_1) + x_2 \ln(\eta_2 V_2)] \tag{24}$$

where R is the gas constant and T is the absolute temperature. The symbols η_1, η_2, V_1, and V_2 represent viscosity of DES, viscosity of water, molar volume of DES, and molar volume of water, respectively.

Tables S1–S4 in Supplementary Material present the experimental values of viscosity, calculated values of the viscosity deviations and the excess Gibbs energy of activation for viscous flow for aqueous solutions of DESs over the entire range of compositions at temperatures ranging from 293.15 K to 313.15 K. Figure 9a,b show the plots of $\Delta \eta$ and ΔG^E against mole fraction of DES for the all studied mixtures at 298.15 K. Figure 9c,d depict the temperature dependence of viscosity deviations and the temperature dependence of the excess Gibbs energy of activation for viscous flow for the system (TBAB:BAE +water) as an example.

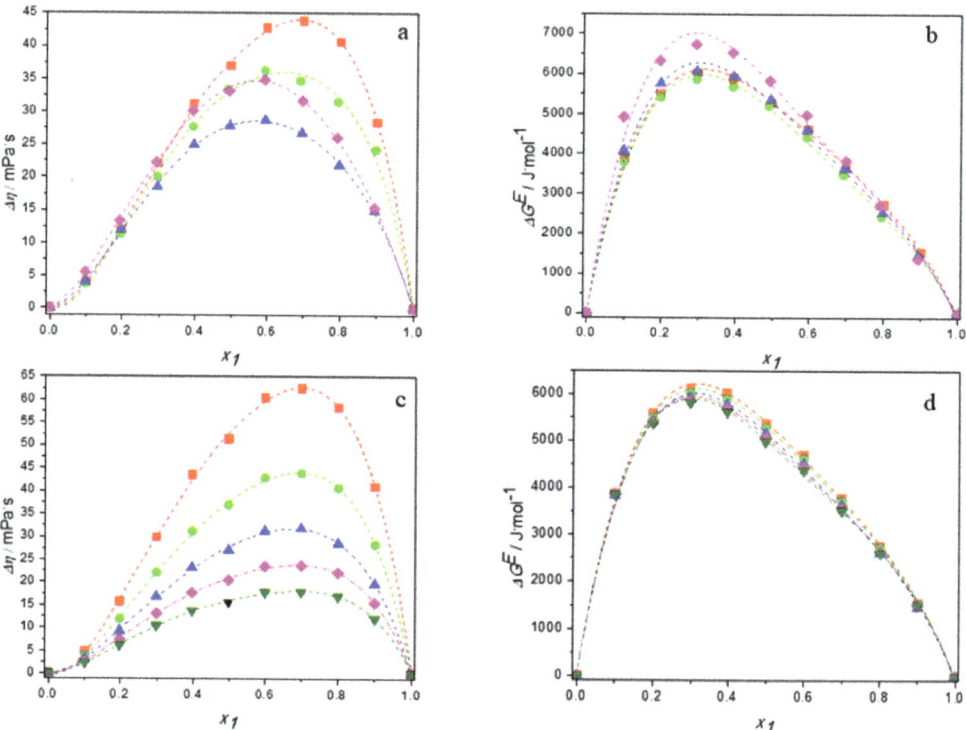

Figure 9. Dependence of the deviations in viscosity and the excess Gibbs energy of activation of aqueous solutions of DESs on molar fraction of deep eutectic solvent: (**a**,**b**) at 298.15 K for ■ DES1; ● DES2; ▲ DES3; ♦ DES4; (**c**,**d**) for DES1 at 298.15 K (■); 298.15 K (●); 303.15 K (▲); 308.15 K (♦); 313.15 K (▼); —, Equation (6).

It is clearly visible that the viscosity deviations of all mixtures are positive in the whole range of composition. It is known that the viscosity of a mixture is related to the liquid structure [57]. Therefore, the viscosity deviation depends on molecular interactions as well as on the size and shape of the molecules forming the solution. The positive viscosity deviations are observed in mixtures with strong specific interactions like hydrogen bonding interactions, whereas the interstitial accommodation of one component with the other within the mixture leads to negative $\Delta\eta$ values [58]. For our DES systems, the predominant effect is the hydrogen bonding, that leads to positive $\Delta\eta$ values.

The order of viscosity deviations is the same as for the viscosity of the studied systems and is as follows: TBAB:AP (DES 1) > TBAC:AP (DES 2) > TBAB:BAE (DES 4) > TBAB:MAE (DES 3). It is different from those obtained for excess molar volume or excess compressibility. Thus, it can concluded that the values of viscosity deviations are determined not only by the interactions between unlike molecules, but also by other effects as shape of the molecules.

Estimation of the results obtained and presented in Figure 9 confirm the temperature dependence of the viscosity deviations in the studied systems because the values of $\Delta\eta$ become less positive with increasing temperature. This is due to the weakening of the interactions between the molecules present in the solution, which seem to dominate over the penetration phenomenon that obviously increases with temperature.

The positive values of excess Gibbs energy of activation presented in Figure 9a,b once again approve the dominance of specific interactions—i.e., hydrogen bonding between DES and water molecules occurring in the studied systems—which become weaker as temperature increases [58].

3.6. Correlations of Excess Properties

Similarly, as in a case of excess molar volumes, in order to correlate the calculated excess thermal expansions, excess isentropic compressibilities, deviations in refractive index, deviations in viscosity, and excess Gibbs energy of activation for viscous flow with the composition, the Redlich–Kister polynomial equation was applied [34]

$$Y^E = x_1 x_2 \sum_{i=0}^{2} A_i (x_1 - x_2)^i \tag{25}$$

where the A_i values are adjustable parameters. They were determined using the least squares method and their values are listed in Tables S6–S9 along with root mean square deviations of fit (RMSD). In order to obtain RMSD close to the experimental uncertainty, a different degree of the polynomial equation was chosen depending on the property and DES. For almost all systems, excess molar volumes, deviations in refractive index, and excess Gibbs energy of activation for viscous flow with composition were correlated using free-parameter Redlich–Kister polynomial equation. For viscosity deviations and excess isentropic compressibilities, a better fit was obtained for four-parameter and for five-parameter Redlich–Kister polynomial equation, respectively. In case of excess thermal expansions four-parameter for DES1 and DES 2 and five-parameter Redlich–Kister polynomial for DES3 and DES5 were chosen.

In Figures 2, 5 and 7–9, the dashed lines represent the correlated values according to the Redlich–Kister equation. As it can be seen, the calculated values agree very well with the experimental data. Thus, the Redlich–Kister equation perfectly represents the data over the experimental temperature range for the novel DES + water binary systems studied in this work.

4. Conclusions

The presented novel DESs built of tetrabutylammonium chloride and 3-amino-1-propanol or tetrabutylammonium bromide and 3-amino-1-propanol or 2-(methylamino)ethanol or 2-(butylamino)ethanol were found to be attractive in their properties, mostly for further evaluation and optimization during development of inexpensive eco-solvents or other valuable material. Most important physicochemical properties have been demonstrated in details such as density, speed of sound, refractive index, and viscosity which were measured for their aqueous solutions over the entire range of compositions at atmospheric pressure and T = (293.15 − 313.15). The chosen Jouyban–Acree model was successfully used to correlate the experimental physical properties with respect to the concentration, and the results showed that this mathematical equation is an accurate correlation for the prediction of aqueous DES properties.

Excess molar volumes, excess isentropic compressibilities, deviations in viscosity, and deviations in refractive indices were calculated to study nonideal behavior of binary mixtures and they were correlated by the Redlich–Kister equation with temperature-dependent parameters. Excess molar volumes and excess compressiblities were negative and deviations in viscosity and deviations in refractive index were positive over the entire range of composition and temperature, suggesting strong intermolecular interactions among unlike molecules. Moreover, the temperature dependences of the excess molar volumes and compressibilities indicate that, in the studied systems, hydrogen bonding prevails over the packing effect (non-specific interactions).

The dominance of specific interactions in the aqueous solutions of the DESs also was confirmed by the Prigogine–Flory–Patterson (PFP) theory, which was applied to excess molar volumes.

The calculated negative values of the excess partial molar volumes of DESs and water demonstrated sufficient DES—water interactions which are stronger than the DES—DES or the water–water ones will probably facilitate the efficient utilization of DES.

In terms of the suitability of the water mixtures of the studied DES for the effective separation of carbon dioxide from gas streams at relatively low pressure, the obtained values of the excess properties allow us to assume that the best absorbent would be TBAB:AP, and the worst of TBAB:MAE.

Supplementary Materials: The following supporting information can be found online: Table S1: Densities ρ, excess molar volumes V^E, isobaric thermal expansion coefficients α_p, excess thermal expansion $\Delta\alpha_p$, speeds of sound u, excess isentropic compressibilities κ_S^E, viscosities η, viscosity deviations $\Delta\eta$, excess Gibbs free energy of activation of viscous flow ΔG^E, refractive indices n_D, refractive index deviations Δn_D as functions of mole fraction, x_1 of DES for TBAB:AP (DES1) + water mixtures at the temperatures (293.15 to 303.15) K and atmospheric pressure; Table S2: Densities ρ, excess molar volumes V^E, isobaric thermal expansion coefficients α_p, excess thermal expansion $\Delta\alpha_p$, speeds of sound u, excess isentropic compressibilities κ_S^E, viscosities η, viscosity deviations $\Delta\eta$, excess Gibbs free energy of activation of viscous flow ΔG^E, refractive indices n_D, refractive index deviations Δn_D as functions of mole fraction, x_1 of DES for TBAC:AP (DES2) + water mixtures at the temperatures (293.15 to 303.15) K and atmospheric pressure; Table S3: Densities ρ, excess molar volumes V^E, isobaric thermal expansion coefficients α_p, excess thermal expansion $\Delta\alpha_p$, speed of sounds u, excess isentropic compressibilities κ_S^E, viscosities η, viscosity deviations $\Delta\eta$, excess Gibbs free energy of activation of viscous flow ΔG^E, refractive indices n_D, refractive index deviations Δn_D as functions of mole fraction, x_1 of DES for TBAB:MAE (DES3) + water mixtures at the temperatures (293.15 to 303.15) K and atmospheric pressure; Table S4: Densities ρ, excess molar volumes V^E, isobaric thermal expansion coefficients α_p, excess thermal expansion $\Delta\alpha_p$, speed of sound u, excess isentropic compressibilities κ_S^E, viscosities η, viscosity deviations $\Delta\eta$, excess Gibbs free energy of activation of viscous flow ΔG^E, refractive indices n_D, refractive index deviations Δn_D as functions of mole fraction, x_1 of DES for TBAB:BAE (DES4) + water mixtures at the temperatures (293.15 to 303.15) K and atmospheric pressure; Table S5: Parameters of the JAM equation, together with RMSD and ARD% for density, speed of sound, viscosity and refractive index of DES (1) + water (2) systems at different temperatures; Table S6: Parameters A_i of Equation (6) and the corresponding RSMD f for TBAB:AP (DES1) + water mixtures at the temperatures (293.15 to 303.15) K and atmospheric pressure; Table S7. Parameters A_i of Equation (6) and the corresponding RSMD for TBAC:AP (DES2) + water mixtures at the temperatures (293.15 to 303.15) K and atmospheric pressure; Table S8: Parameters A_i of Equation (6) and the corresponding RSMD for TBAB:MAE (DES3) + water mixtures at the temperatures (293.15 to 303.15) K and atmospheric pressure; Table S9: Parameters A_i of Equation (6) and the corresponding RSMD for TBAB:BAE (DES4) + water mixtures at the temperatures (293.15 to

303.15) K and atmospheric pressure; Table S10: Isobaric thermal expansion coefficient ($α_p$), isochoric molar heat capacity (C_P), Flory theory parameters: characteristic volume (V*), reduce volume ($\tilde{V}$), characteristic pressure (P*), and ratio of molecular surface to volume ratio (S1/S2) of DES to water; Table S11: Partial molar volumes of DESs and water in their binary mixtures at T = (293.15 to 313.15) K and at atmospheric pressure (0.1 MPa).

Author Contributions: Conceptualization, D.W. and B.N.; Methodology, D.W., B.N. and J.Ł.; Investigation, B.N. and M.J.; Data curation, D.W. and B.N.; Writing—original draft preparation, D.W. and B.N.; Writing—review and editing, J.Ł. and M.J.; Supervision, D.W. All authors have read and agreed to the published version of the manuscript.

Funding: This research received no external funding.

Institutional Review Board Statement: Not applicable.

Informed Consent Statement: Not applicable.

Data Availability Statement: The data presented in this study are available on request from the corresponding author. The data are not publicly available due to the lack of requirements of Gdansk University of Technology and Medical University of Gdansk.

Conflicts of Interest: The authors declare no conflict of interest.

References

1. Marcus, Y. *Deep Eutectic Solvents*; Springer International Publishing: Cham, Switzerland, 2019.
2. Leron, R.B.; Wong, D.S.H.; Li, M.H. Densities of a deep eutectic solvent based on choline chloride and glycerol and its aqueous mixtures at elevated pressures. *Fluid Phase Equilib.* **2012**, *335*, 32–38. [CrossRef]
3. Liu, Y.; Friesen, J.B.; McAlpine, J.B.; Lankin, D.C.; Chen, S.N.; Pauli, G.F. Natural Deep Eutectic Solvents: Properties, Applications, and Perspectives. *J. Nat. Prod.* **2018**, *81*, 679–690. [CrossRef] [PubMed]
4. Abbott, A.P.; Capper, G.; Davies, D.L.; Rasheed, R.K.; Tambyrajah, V. Novel solvent properties of choline chloride/urea mixtures. *Chem. Commun.* **2003**, *9*, 70–71. [CrossRef] [PubMed]
5. Dai, Y.; van Spronsen, J.; Witkamp, G.J.; Verpoorte, R.; Choi, Y.H. Natural deep eutectic solvents as new potential media for green technology. *Anal. Chim. Acta* **2013**, *766*, 61–68. [CrossRef]
6. Sarmad, S.; Mikkola, J.P.; Ji, X. Carbon Dioxide Capture with Ionic Liquids and Deep Eutectic Solvents: A New Generation of Sorbents. *ChemSusChem* **2017**, *10*, 324–352. [CrossRef]
7. Alomar, M.K.; Hayyan, M.; Alsaadi, M.A.; Akib, S.; Hayyan, A.; Hashim, M.A. Glycerol-based deep eutectic solvents: Physical properties. *J. Mol. Liq.* **2016**, *215*, 98–103. [CrossRef]
8. García, G.; Aparicio, S.; Ullah, R.; Atilhan, M. Deep eutectic solvents: Physicochemical properties and gas separation applications. *Energy Fuels* **2015**, *29*, 2616–2644. [CrossRef]
9. Liao, H.G.; Jiang, Y.X.; Zhou, Z.Y.; Chen, S.P.; Sun, S.G. Shape-controlled synthesis of gold nanoparticles in deep eutectic solvents for studies of structure-functionality relationships in electrocatalysis. *Angew. Chem.-Int. Ed.* **2008**, *47*, 9100–9103. [CrossRef]
10. Su, W.C.; Wong, D.S.H.; Li, M.H. Effect of water on solubility of carbon dioxide in (aminomethanamide + 2-hydroxy-N,N,N-trimethylethanaminium chloride). *J. Chem. Eng. Data* **2009**, *54*, 1951–1955. [CrossRef]
11. Li, Z.; Wang, L.; Li, C.; Cui, Y.; Li, S.; Yang, G.; Shen, Y. Absorption of Carbon Dioxide Using Ethanolamine-Based Deep Eutectic Solvents. *ACS Sustain. Chem. Eng.* **2019**, *7*, 10403–10414. [CrossRef]
12. Ahmadi, R.; Hemmateenejad, B.; Safavi, A.; Shojaeifard, Z.; Shahsavar, A.; Mohajeri, A.; Dokoohaki, M.H.; Zolghadr, A.R. Deep eutectic-water binary solvent associations investigated by vibrational spectroscopy and chemometrics. *Phys. Chem. Chem. Phys.* **2018**, *20*, 18463–18473. [CrossRef] [PubMed]
13. Nowosielski, B.; Jamrógiewicz, M.; Łuczak, J.; Śmiechowski, M.; Warmińska, D. Experimental and predicted physicochemical properties of monopropanolamine-based deep eutectic solvents. *J. Mol. Liq.* **2020**, *309*, 113110. [CrossRef]
14. Omar, K.A.; Sadeghi, R. Novel benzilic acid-based deep-eutectic-solvents: Preparation and physicochemical properties determination. *Fluid Phase Equilib.* **2020**, *522*, 112752. [CrossRef]
15. Hayyan, A.; Hadj-Kali, M.K.; Salleh, M.Z.M.; Hashim, M.A.; Rubaidi, S.R.; Hayyan, M.; Zulkifli, M.; Rashid, S.N.; Mirghani, M.E.S.; Ali, E.; et al. Characterization of tetraethylene glycol-based deep eutectic solvents and their potential application for dissolving unsaturated fatty acids. *J. Mol. Liq.* **2020**, *312*, 113284. [CrossRef]
16. van Osch, D.J.G.P.; Dietz, C.H.J.T.; Warrag, S.E.E.; Kroon, M.C. The Curious Case of Hydrophobic Deep Eutectic Solvents: A Story on the Discovery, Design, and Applications. *ACS Sustain. Chem. Eng.* **2020**, *8*, 10591–10612. [CrossRef]
17. Ma, C.; Laaksonen, A.; Liu, C.; Lu, X.; Ji, X. The peculiar effect of water on ionic liquids and deep eutectic solvents. *Chem. Soc. Rev.* **2018**, *47*, 8685–8720. [CrossRef]

18. Kuddushi, M.; Nangala, G.S.; Rajput, S.; Ijardar, S.P.; Malek, N.I. Understanding the peculiar effect of water on the physicochemical properties of choline chloride based deep eutectic solvents theoretically and experimentally. *J. Mol. Liq.* **2019**, *278*, 607–615. [CrossRef]
19. Ghaedi, H.; Ayoub, M.; Sufian, S.; Shariff, A.M.; Murshid, G.; Hailegiorgis, S.M.; Khan, S.N. Density, excess and limiting properties of (water and deep eutectic solvent) systems at temperatures from 293.15 K to 343.15 K. *J. Mol. Liq.* **2017**, *248*, 378–390. [CrossRef]
20. Ghaedi, H.; Ayoub, M.; Sufian, S.; Hailegiorgis, S.M.; Murshid, G.; Farrukh, S.; Khan, S.N. Experimental and prediction of volumetric properties of aqueous solution of (allyltriphenylPhosphonium bromide—Triethylene glycol) deep eutectic solvents. *Thermochim. Acta* **2017**, *657*, 123–133. [CrossRef]
21. Leron, R.B.; Soriano, A.N.; Li, M.H. Densities and refractive indices of the deep eutectic solvents (choline chloride+ethylene glycol or glycerol) and their aqueous mixtures at the temperature ranging from 298.15 to 333.15K. *J. Taiwan Inst. Chem. Eng.* **2012**, *43*, 551–557. [CrossRef]
22. Kumar, A.K.; Shah, E.; Patel, A.; Sharma, S.; Dixit, G. Physico-chemical characterization and evaluation of neat and aqueous mixtures of choline chloride + lactic acid for lignocellulosic biomass fractionation, enzymatic hydrolysis and fermentation. *J. Mol. Liq.* **2018**, *271*, 540–549. [CrossRef]
23. Siongco, K.R.; Leron, R.B.; Li, M.H. Densities, refractive indices, and viscosities of N,N-diethylethanol ammonium chloride-glycerol or -ethylene glycol deep eutectic solvents and their aqueous solutions. *J. Chem. Thermodyn.* **2013**, *65*, 65–72. [CrossRef]
24. Yadav, A.; Pandey, S. Densities and viscosities of (choline chloride + urea) deep eutectic solvent and its aqueous mixtures in the temperature range 293.15 K to 363.15 K. *J. Chem. Eng. Data* **2014**, *59*, 2221–2229. [CrossRef]
25. Li, G.; Deng, D.; Chen, Y.; Shan, H.; Ai, N. Solubilities and thermodynamic properties of CO_2 in choline-chloride based deep eutectic solvents. *J. Chem. Thermodyn.* **2014**, *75*, 58–62. [CrossRef]
26. Haider, M.B.; Jha, D.; Sivagnanam, B.M.; Kumar, R. Thermodynamic and Kinetic Studies of CO_2 Capture by Glycol and Amine-Based Deep Eutectic Solvents. *J. Chem. Eng. Data* **2018**, *63*, 2671–2680. [CrossRef]
27. Trivedi, T.J.; Lee, J.H.; Lee, H.J.; Jeong, Y.K.; Choi, J.W. Deep eutectic solvents as attractive media for CO_2 capture. *Green Chem.* **2016**, *18*, 2834–2842. [CrossRef]
28. Barzegar-Jalali, M.; Jafari, P.; Jouyban, A. Thermodynamic study of the aqueous pseudo-binary mixtures of betaine-based deep eutectic solvents at T = (293.15 to 313.15) K. *Phys. Chem. Liq.* **2022**, 1–16. [CrossRef]
29. Jouyban, A.; Khoubnasabjafari, M.; Vaez-Gharamaleki, Z.; Fekari, Z.; Acree, W.E. Calculation of the viscosity of binary liquids at various temperatures using Jouyban–Acree model. *Chem. Pharm. Bull.* **2005**, *53*, 519–523. [CrossRef]
30. Abu-Bakar, A.S.; Cran, M.J.; Moinuddin, K.A.M. Experimental investigation of effects of variation in heating rate, temperature and heat flux on fire properties of a non-charring polymer. *J. Therm. Anal. Calorim.* **2019**, *137*, 447–459. [CrossRef]
31. Acree, W.E. Mathematical representation of thermodynamic properties. Part 2. Derivation of the combined nearly ideal binary solvent (NIBS)/Redlich–Kister mathematical representation from a two-body and three-body interactional mixing model. *Thermochim. Acta* **1992**, *198*, 71–79. [CrossRef]
32. Jouyban, A.; Fathi-Azarbayjani, A.; Khoubnasabjafari, M.; Acree, W.E. Mathematical representation of the density of liquid mixtures at various temperatures using Jouyban–Acree model. *Indian J. Chem.-Sect. A Inorg. Phys. Theor. Anal. Chem.* **2005**, *44*, 1553–1560.
33. Rodríguez, G.A.; Delgado, D.R.; Martínez, F.; Fakhree, M.A.A.; Jouyban, A. Volumetric properties of glycerol formal + propylene glycol mixtures at several temperatures and correlation with the Jouyban–Acree model. *J. Solut. Chem.* **2012**, *41*, 1477–1494. [CrossRef]
34. Redlich, O.; Kister, A.T. Algebraic Representation of Thermodynamic Properties and the Classification of Solutions. *Ind. Eng. Chem.* **1948**, *40*, 345–348. [CrossRef]
35. Shah, D.; Mjalli, F.S. Effect of water on the thermo-physical properties of Reline: An experimental and molecular simulation based approach. *Phys. Chem. Chem. Phys.* **2014**, *16*, 23900–23907. [CrossRef] [PubMed]
36. Gabriele, F.; Chiarini, M.; Germani, R.; Tiecco, M.; Spreti, N. Effect of water addition on choline chloride/glycol deep eutectic solvents: Characterization of their structural and physicochemical properties. *J. Mol. Liq.* **2019**, *291*, 111301. [CrossRef]
37. Yadav, A.; Kar, J.R.; Verma, M.; Naqvi, S.; Pandey, S. Densities of aqueous mixtures of (choline chloride + ethylene glycol) and (choline chloride + malonic acid) deep eutectic solvents in temperature range 283.15–363.15 K. *Thermochim. Acta* **2015**, *600*, 95–101. [CrossRef]
38. Harifi-Mood, A.R.; Buchner, R. Density, viscosity, and conductivity of choline chloride + ethylene glycol as a deep eutectic solvent and its binary mixtures with dimethyl sulfoxide. *J. Mol. Liq.* **2017**, *225*, 689–695. [CrossRef]
39. Sas, O.G.; Fidalgo, R.; Domínguez, I.; Macedo, E.A.; González, B. Physical properties of the pure deep eutectic solvent, [ChCl]:[Lev] (1:2) DES, and its binary mixtures with alcohols. *J. Chem. Eng. Data* **2016**, *61*, 4191–4202. [CrossRef]
40. Kim, K.S.; Park, B.H. Volumetric properties of solutions of choline chloride + glycerol deep eutectic solvent with water, methanol, ethanol, or iso-propanol. *J. Mol. Liq.* **2018**, *254*, 272–279. [CrossRef]
41. Flory, P.J. Statistical Thermodynamics of Liquid Mixtures. *J. Am. Chem. Soc.* **1965**, *87*, 1833–1838. [CrossRef]
42. Abe, A.; Flory, P.J. The Thermodynamic Properties of Mixtures of Small, Nonpolar Molecules. *J. Am. Chem. Soc.* **1965**, *87*, 1838–1846. [CrossRef]
43. Prigogine, I. *The Molecular Theory of Solutions*; North Holland Publishing Co.: Amsterdam, The Netherlands, 1957.
44. Patterson, D.; Delmas, G. Corresponding states theories and liquid models. *Discuss. Faraday Soc.* **1970**, *49*, 98–105. [CrossRef]

45. Costas, M.; Patterson, D. Volumes of mixing and the P* effect: Part II. Mixtures of alkanes with liquids of different internal pressures. *J. Solut. Chem.* **1982**, *11*, 807–821. [CrossRef]
46. Vaid, Z.S.; More, U.U.; Oswal, S.B.; Malek, N.I. Experimental and theoretical excess molar properties of imidazolium based ionic liquids with isomers of butanol. *Thermochim. Acta* **2016**, *634*, 38–47. [CrossRef]
47. Aangothu, S.R.; Munnangi, S.R.; Raju, K.T.S.S.; Bollikolla, H.B. An experimental investigation of molecular interactions between [Emim][triflate] ionic liquid & 2-alkoxyethanols and theoretical comparison by PFP theory. *J. Chem. Thermodyn.* **2019**, *138*, 43–50. [CrossRef]
48. Chaudhary, N.; Nain, A.K. Physicochemical studies of intermolecular interactions in 1-butyl-3-methylimidazolium tetrafluoroborate + benzonitrile binary mixtures at temperatures from 293.15 to 318.15 K. *Phys. Chem. Liq.* **2020**, *59*, 358–381. [CrossRef]
49. Fatima, U.; Riyazuddeen; Anwar, N. Effect of Solvents and Temperature on Interactions in Binary and Ternary Mixtures of 1-Butyl-3-methylimidazolium Trifluoromethanesulfonate with Acetonitrile or/and N, N-Dimethylformamide. *J. Chem. Eng. Data* **2018**, *63*, 4288–4305. [CrossRef]
50. Fatima, U.; Riyazuddeen; Anwar, N.; Montes-Campos, H.; Varela, L.M. Molecular dynamic simulation, molecular interactions and structural properties of 1-butyl-3-methylimidazolium bis(trifluoromethylsulfonyl)imide + 1-butanol/1-propanol mixtures at (298.15–323.15) K and 0.1 M Pa. *Fluid Phase Equilib.* **2018**, *472*, 9–21. [CrossRef]
51. Haghbakhsh, R.; Raeissi, S. Excess volumes of mixtures consisting of deep eutectic solvents by the Prigogine–Flory–Patterson theory. *J. Mol. Liq.* **2018**, *272*, 731–737. [CrossRef]
52. Shekaari, H.; Zafarani-Moattar, M.T.; Mokhtarpour, M.; Faraji, S. Volumetric and compressibility properties for aqueous solutions of choline chloride based deep eutectic solvents and Prigogine–Flory–Patterson theory to correlate of excess molar volumes at T = (293.15 to 308.15) K. *J. Mol. Liq.* **2019**, *289*, 111077. [CrossRef]
53. Vraneš, M.; Papović, S.; Tot, A.; Zec, N.; Gadžurić, S. Density, excess properties, electrical conductivity and viscosity of 1-butyl-3-methylimidazolium bis(trifluoromethylsulfonyl)imide + γ-butyrolactone binary mixtures. *J. Chem. Thermodyn.* **2014**, *76*, 161–171. [CrossRef]
54. Hawrylak, B.; Gracie, K.; Palepu, R. Thermodynamic properties of binary mixtures of butanediols with water. *J. Solut. Chem.* **1998**, *27*, 17–31. [CrossRef]
55. Benson, G.C.; Kiyohara, O. Evaluation of excess isentropic compressibilities and isochoric heat capacities. *J. Chem. Thermodyn.* **1979**, *11*, 1061–1064. [CrossRef]
56. Brocos, P.; Piñeiro, Á.; Bravo, R.; Amigo, A. Refractive indices, molar volumes and molar refractions of binary liquid mixtures: Concepts and correlations. *Phys. Chem. Chem. Phys.* **2003**, *5*, 550–557. [CrossRef]
57. Nabi, F.; Malik, M.A.; Jesudason, C.G.; Al-Thabaiti, S.A. A review of molecular interactions in organic binary mixtures. *Korean J. Chem. Eng.* **2014**, *31*, 1505–1517. [CrossRef]
58. Garcia, B.; Alcalde, R.; Leal, J.M.; Matos, J.S. Shear viscosities of the N-methylformamide- And N,N-dimethylformamide-(C1-C10) alkan-1-ol solvent systems. *J. Chem. Soc.-Faraday Trans.* **1997**, *93*, 1115–1118. [CrossRef]

Article

Adsorption of *N,N,N′,N′*-Tetraoctyl Diglycolamide on Hypercrosslinked Polysterene from a Supercritical Carbon Dioxide Medium

Mikhail Kostenko [1,*] and Olga Parenago [1,2]

1. Kurnakov Institute of General and Inorganic Chemistry, Russian Academy of Sciences, Leninsky Prospect 31, 119071 Moscow, Russia; oparenago@scf-tp.ru
2. Chemistry Department, Moscow State University, Leninskie Gory 1, Bldg. 3, 119234 Moscow, Russia
* Correspondence: kostenko@supercritical.ru

Abstract: The work considers for the first time the preparation of sorbents based on hypercrosslinked polysterene (HCP) and chelating agent *N,N,N′,N′*-tetraoctyl diglycolamide (TODGA) by impregnation in the supercritical (SC) CO_2 medium. Such sorbents can be applied for further isolation and separation of lanthanides, actinides and other metals. They are usually prepared by impregnation in toxic organic solvents (e.g., methanol, dichloromethane). Our study shows that application of SC CO_2 instead of organic solvents can significantly speed up the impregnation, perfom it in one stage and make the process more eco-friendly. At the same time, the obtained sorbents are close in their parameters to the classical ones. This article presents the results of measuring the TODGA adsorption isotherms on two HCP sorbents (MN202 and MN270) on a wide range of SC fluid parameters. Adsorption measurements were carried out using on-line supercritical fluid chromatography and gravimetry. Based on the sorption capacity parameter, MN202 sorbent was selected as the better carrier for TODGA. An impregnation temperature increase within the range 313–343 K in isochoric conditions (ρ = 0.780 g/mL) reduces the maximum of TODGA adsorption from ~0.68 mmol/g to ~0.49 mmol/g.

Keywords: adsorption; *N,N,N′,N′*-tetraoctyl diglycolamide; hypercrosslinked polystyrene; supercritical fluid; chromatography; gravimetry; isotherm

1. Introduction

Metal extraction from aqueous solutions is a common task both in laboratory practice and in industrial production. This problem is solved quite successfully by using chelating agents. Quite promising among them are diamides, and, in particular, *N,N,N′,N′*-tetraoctyl diglycolamide (TODGA), which can be used for extraction of a variety of rare earth elements and radioactive metals of the actinide family [1–3]. That is why one of the potential applications of TODGA is purification of waste water in the nuclear industry, removing radionuclides [2,4,5]. TODGA is normally used as a solution in aliphatic hydrocarbons [3–5] but for many problems it is more convenient to use adsorbents impregnated by TODGA. Such materials can adsorb metals from acidic aqueous solutions, which allows them to be used in solid phase extraction and extraction of metals from complex mixtures. Previous research [1,6–12] shows that it is possible to use impregnated sorbents based on TODGA and other similar ligands for chromatographic separation of mixtures containing lanthanides, actinides and other metals of valences II, III and IV. The role of carrier matrices in production of sorbents of this type can be played by materials of different kinds, such as silica gels modified by polymers [7], graphene aerogels [13], hypercrosslinked polystyrene (HCP) [1,14], etc. Such carriers are commonly impregnated with TODGA solutions in volatile solvents (methanol, dichloromethane, etc.). The solvents are removed by evaporation (reaching complete deposition of the chelating agent on the carrier) [1,6,8,10–12] or the carrier with a certain amount of the adsorbed chelating agent is filtered out of the solution residue [7,13]. Such approaches to impregnation require volatile toxic organic solvents and, in addition, are rather labor- and time-consuming, which encourages us to use more environmentally friendly and cheaper solvents such as supercritical (SC) CO_2.

In this work, we applied two variants of commercially available HCP as the TODGA carriers. The unique structure of this class of materials provides them with a number of features, such as high porosity, stability within a wide range of conditions and a surface with adsorptive aromatic centers. This allows HCP to act as an adsorbent in a number of industrial, medical and scientific applications [15]. The use of adsorbents of this type in SC fluid media has not been sufficiently studied yet and is of special scientific interest.

The methods of adsorption measurement of SC fluid solution components are largely analogous to those used to measure adsorption from liquid solutions. The methods can be divided into dynamic and static. The dynamic adsorption measurement methods are based on analyzing the substance penetration through an adsorbent layer in a column. There are examples of successful applications of dynamic methods (column breakthrough [16–18] and chromatographic [19–25] methods) to measure adsorption in SC fluid media. However, the accuracy of the dynamic methods is largely dependent on the column packing with an adsorbent due to the hydrodynamic effects [26]. This means that these methods are not reliable in case of HCP studied in our work as we have earlier determined in our laboratory that the use of such materials in SCF leads to changes in the particle size caused by swelling and destruction of the column packing layer [27].

The principle underlying static adsorption measurement methods is reaching the equilibrium of substance distribution between the bulk phase and the adsorbed layer. When measuring adsorption from the liquid phase, an adsorbent sample is introduced into a solution with a pre-known concentration of the adsorbed substance, after which an equilibrium is established in the system at the required temperature. The amount of the adsorbed substance is determined based on the system mass balance [26]. Adsorption measurement from the SCF phase is complicated because the bulk phase is under pressure. For this reason, the experiment is usually conducted by a different technique: after reaching the adsorption equilibrium, the bulk phase is removed from the system and the substance content on the adsorbent surface is determined by a variety of methods (such as gravimetry or desorption into the liquid phase with subsequent spectrophotometric or chromatographic analysis [28–30]). Such approaches are relatively easy to implement but, as is known [28], a pressure relief makes SCF lose their dissolving capacity, which may lead to substance deposition in the sorbent pores. This, in turn, may lead to errors in adsorption measurement. We have earlier [31] proposed a static approach to sorption analysis in SC media based on on-line supercritical fluid chromatography (SFC). A similar approach was applied in several works to determine the substance solubility in SCF [32–35] and distribution coefficients in the liquid-SCF system [36–40], and to control chemical reactions [41,42]. It allows fast quantitative analysis of the considered system components without depressurization or special sample preparation.

In this work, we consider for the first time the preparation of sorbents based on HCP and chelating agent (TODGA) by impregnation in the SC CO_2 medium. Such sorbents can be applied for further isolation and separation of metals. The main purpose of the work was to study the adsorption of TODGA on HCP in the SC CO_2. This is necessary for the development of more eco-friendly methods for producing sorbents based on TODGA and other chelating agents. In the paper, measurements of the TODGA adsorption isotherms on two variants of HCP are presented, the effect of the SC fluid density and temperature on adsorption is estimated, and a brief comparison of the procedure for sorbents preparation in methanol and SC CO_2 media is performed. Besides, special attention is paid by the authors to optimization and comparison of adsorption measurement methods in the considered conditions. The on-line SFC method has been used to measure adsorption only once before so it is relatively new. For this reason, its advantages over the gravimetric method are considered in more detail in the framework of the study.

2. Results and Discussion

2.1. The Experimental Unit Testing and Calibration

Since the experimental unit described in [31] had been significantly modified, before making the main measurements, we conducted preliminary tests to check the stability of temperature maintenance and sealing of all the unit components. The tests were made at 313 K, the autoclave was filled with CO_2 until the pressure reached 20 MPa, after which we monitored the temperature and pressure values for 3 days. The maximum deviation of the temperature value in the autoclave from the pre-set value over the whole period was ±0.4 K. The pressure decrease after 72 h was about 0.15 MPa, which, by the NIST Chemistry WebBook data [43], corresponds to CO_2 leakage in the amount of about 0.20 g (the calculated initial mass was 121.77 g). We considered such results satisfactory enough to proceed with the main experiments.

To calculate the concentrations of TODGA in the autoclave, we had to measure its effective volume. The geometric volume of the autoclave was 150 mL but it also contained the magnetic stir bar, the vial and the support. The autoclave effective volume was measured by filling it step by step with distilled water using 100–1000 µL and 1–10 mL Thermo Scientific Lite (Lenpipet Thermo Scientific, Moscow, Russia) calibrated mechanical pipettes. The measurements were made three times. Taking into account the pipette error, the effective volume of the autoclave was 145 ± 1 mL.

The TODGA retention time in the selected analysis conditions was 1.13 ± 0.02 min. The calibration tests showed high reproducibility of the results obtained in the considered experimental unit. In all the tests, the sample from the autoclave was injected at least twice (Figure 1), to exclude accidental error. Since the TODGA concentration in the autoclave decreased after the sample collection, the area of the next peak was always 0.3–0.5% smaller than that of the previous one. Taking this into account, we were able to estimate the approximate rate of TODGA sample dissolution in the experimental conditions. It was established that increasing the dissolution time from 10 to 60 min did not increase the chromatographic response. This indicates that the TODGA samples completely dissolved in less than 10 min.

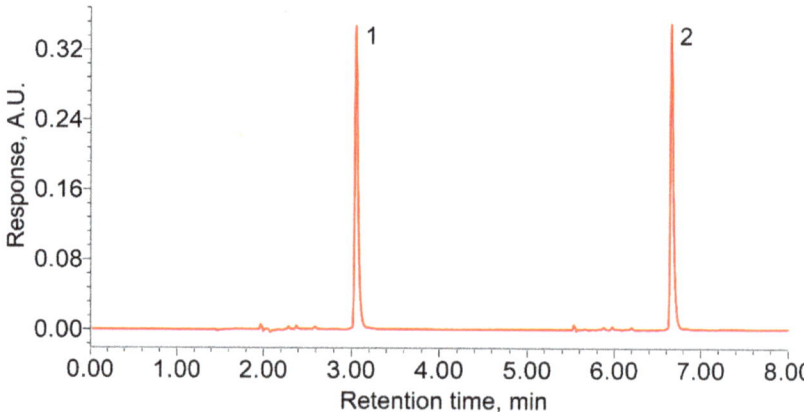

Figure 1. Example of two-time injection of a TODGA sample from the autoclave.

By analyzing a number of TODGA samples, we plotted a calibration dependence of the amount of the substance introduced into the autoclave on the chromatographic peak area (Figure 2). The dependence equation takes the form:

$$n = Resp \cdot 7.8804 \cdot 10^{-8} \qquad (1)$$

where $Resp$ is the chromatographic response (peak area).

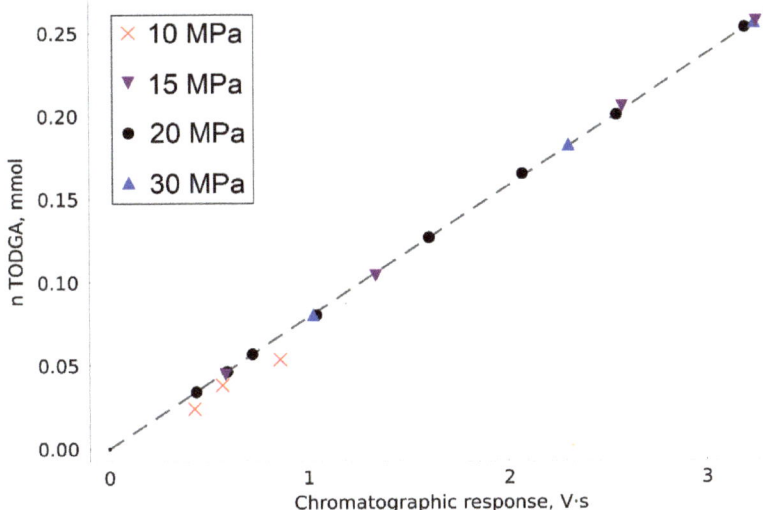

Figure 2. Calibration dependence of TODGA content in the autoclave on the chromatograph response (the dashed line is linear approximation of calibration points).

Most of the calibration points were obtained at a pressure of 20 MPa. However, to confirm its performance in the extended CO_2 density range, we obtained several points in the pressure range from 10 to 30 MPa. When the pressure in the system was reduced to 10 MPa, there were nonreproducible errors in the analyzed points, which are characteristic of conditions of incomplete substance dissolution. This effect was observed even with relatively small concentrations of TODGA. For this reason, the points obtained at 10 MPa were not used to construct calibration dependence.

The calibration experiments confirmed the high accuracy of the measurements made on the modified experimental unit. The maximum absolute deviation of the experimental value (n_i) from the one calculated by calibration ($\hat{n}_i$) was 2.2 µmol, the determination coefficient: $R^2 = 0.9998$. The root mean square error (RMSE) of approximation was calculated as follows:

$$\text{RMSE} = \sqrt{\frac{\sum_i (n_i - \hat{n}_i)^2}{N}} \qquad (2)$$

where N is the number of experimental points. The RMSE value was 0.0012 mmol.

2.2. TODGA Adsorption on MN202 and MN270 from a Solution in SC CO_2

MN202 is known to be applicable as a TODGA carrier. For example, it is used as the basis for producing BAU-1M, a commercial sorbent (Sorbent-Tekhnologii, Moscow, Russia) [1,14]. In addition to MN202, we used the MN270 sorbent with smaller pores and a larger specific surface area. It was interesting for us to check whether it could be applied to solve the problem under consideration. We therefore plotted isotherms of TODGA adsorption on both HCP variants at 313 K and medium pressure of 20 MPa (Figure 3).

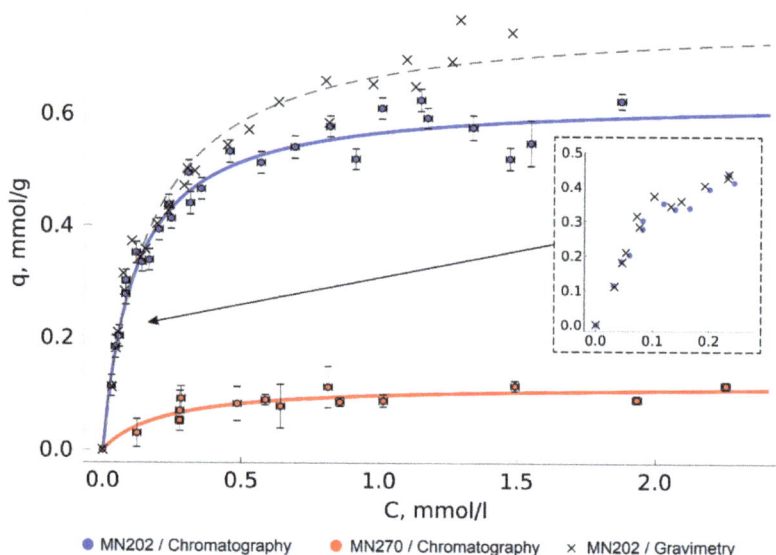

Figure 3. Isotherms of TODGA adsorption on MN202 and MN270 from a solution in SC CO_2 (T = 313 K, P = 20 MPa).

In the first experiments, we estimated the approximate time required for the system to reach the adsorption equilibrium. It was established that as time increased in the sequence of 30, 60 and 120 min, the TODGA content in CO_2 did not change significantly and, consequently, the equilibrium in these conditions was reached in less than 30 min. Nevertheless, to make sure the obtained results were correct, the impregnation time in all the other experiments was 60 min.

The adsorption isotherms are described quite well by the Langmuir model:

$$q = q_s \frac{K \cdot C}{1 + K \cdot C} \qquad (3)$$

where q is the adsorption, q_s is the adsorbed monolayer capacity, K is the adsorption equilibrium constant, C is the TODGA concentration in the bulk phase.

Overall comparison of Langmuir model parameters of TODGA adsorption isotherms is given in Table 1. In the considered conditions, the saturation of the TODGA monolayer on the MN202 surface was reached at approximately 0.64 mmol/g; the mass equivalent of this value, 0.37 g/g, is close to the values of the analogous commercially available sorbents' load [1]. Interestingly, the maximum adsorption on MN270 is much lower, at 0.10 mmol/g. Since the nature of the sorbent surface is identical, this is most probably caused by the difference in the pore size and accessibility. The sorbents' declared characteristics [44,45] show that the only significant difference of MN270 from MN202 is the higher degree of crosslinking, which reduces the mesopores' average size from 220 Å to 80 Å and increases the specific surface area. The specific surface area of the sorbents in a non-swollen state measured by low temperature nitrogen adsorption was 590 ± 40 m^3/g and 980 ± 50 m^3/g for MN202 and MN270, respectively.

Table 1. The Langmuir model parameters of TODGA adsorption isotherms (T = 313 K, P = 20 MPa).

Adsorbent	q_s, mmol/g	K, L/mol	R^2
MN202	0.64	$8.92 \cdot 10^3$	0.9688
MN270	0.10	$4.39 \cdot 10^3$	0.8519

The isotherms show a wide dispersion of experimental adsorption value points, sometimes much higher than the calculated absolute measurement error. This is assumed to be the result of the difference between the sizes of the adsorbent particles (the sorbent particle diameter is from 0.3 to 1.2 mm) and pore accessibility. In the experiments, we used relatively small sorbent samples (about 0.05–0.15 g), which could lead to a considerable spread of the sample specific surface area values in different experiments as the sorbents were somewhat inhomogeneous.

The MN202 sorbent turned out to be a better carrier for TODGA and was used in further experiments.

2.3. Comparison of the Adsorption Measurement Approaches

The values obtained by measuring the adsorption gravimetrically at the initial part of the isotherm are very close to those obtained by chromatographic analysis (Figure 3). As the TODGA concentration in the solution increases, the results of the gravimetric measurements become overestimated. This is the result of the higher intensity of TODGA deposition from a CO_2 solution to the HCP surface after depressurization. This effect leads to increase of the sorbent mass, regardless of the equilibrium-adsorbed TODGA. Some of the adsorption values measured gravimetrically in this work were more than 25% higher than those obtained by chromatography. Presumably, the accuracy of the method can be improved by wrapping the sorbent with a porous material that would reduce the available volume, from which the substance can precipitate on the sorbent under the depressurization. However, the layer of shielding material will reduce the rate of the mass transfer between the sorbent surface and the solution phase, which may considerably increase the time required for the system to reach the equilibrium state and nullify one of the main advantages of the experimental unit—the high measurement speed. Moreover, the material itself can potentially act as an adsorbent and cause error in the adsorption measurement, especially if the adsorption capacity of the main adsorbent is not high. Filter paper is often used to shield sorbents [28–30]. Cellulose, which is the main component of filter paper, is known to be an active adsorbent for many classes of chemical compounds [46–48] and is used as a sorbent (stationary phase) in paper chromatography.

In the present work, gravimetry was used as an additional control method. For this reason, we did not shield the sorbent with any porous materials, in order to minimize the error of the adsorption measurement by the chromatographic method and to minimize the time required for the system to reach the adsorption equilibrium. To limit TODGA deposition on the adsorbent after depressurization, we covered the upper part of the vial with perforated aluminum foil. This version of the gravimetric method had satisfactory results at the initial section of the isotherms, which confirmed that the obtained data were correct.

Thus, on-line SFC analysis proved to be a quick and accurate method for adsorption measurement in this work. The gravimetric method is easier to implement experimentally but it should be used with caution, especially when measuring adsorption from concentrated solutions.

2.4. Fluid Density Effect on TODGA Adsorption

In this work, we found out that change in the fluid medium density by pressure variation in the range of 15 to 30 MPa has no significant effect on TODGA adsorption on MN202 (Figure 4).

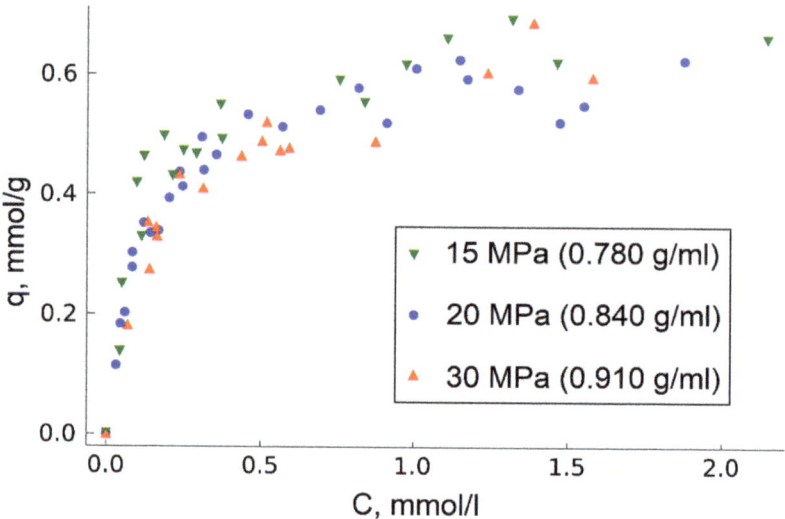

Figure 4. Effect of fluid density on TODGA adsorption (T = 313 K).

This fact is probably associated with the relatively low energy of CO_2-HCP interaction compared to the energy of the sorbent surface interaction with TODGA. For this reason, changes in the density and, hence, amount of CO_2 in the considered system do not produce a significant effect on the TODGA–HCP adsorption equilibrium. This, in turn, makes it possible to produce impregnated sorbents at relatively low pressure, which is especially important when moving from laboratory conditions to industrial production.

2.5. Temperature Effect on TODGA Adsorption

Deviations from the Langmuir model were observed during temperature variation in the range 313–343 K under isochoric conditions (ρ = 0.780 g/mL) (Figure 5).

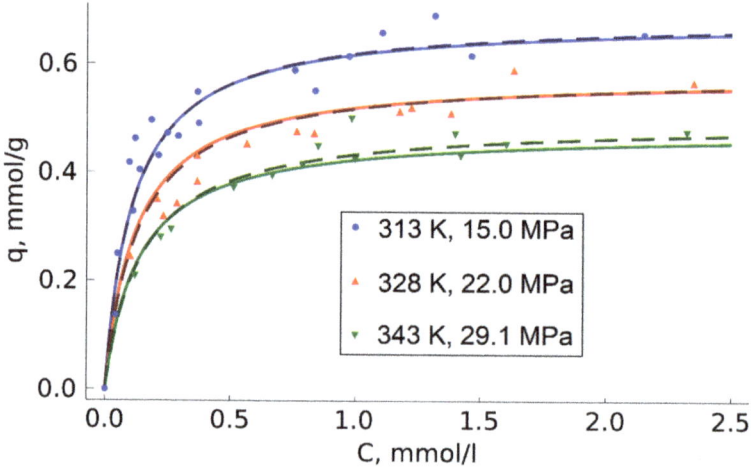

Figure 5. Temperature effect on TODGA adsorption on MN202; approximation of experimental data by Equations (7) (dashed line) and (8) (solid line).

According to the Langmuir model, temperature does not affect the monolayer capacity but can only change the adsorption equilibrium constant influencing the isotherm curvature. However, there was a noticeable reduction in the monolayer capacity when the temperature went up, which was due to the peculiarities of adsorption thermodynamics. Since adsorption is an exothermic process, the temperature increase reduces the maximum equilibrium concentration of an adsorbate on the sorbent surface. For a general description of adsorption, taking into account the influence of temperature, semi-empirical models based on the Langmuir equation are often used [49–51]. Various temperature functions are used in these models instead of the monolayer capacity fixed value, and the equilibrium constant dependence on temperature is described based on its thermodynamic meaning:

$$K = K_0 \cdot exp\left(\frac{E}{R \cdot T}\right) \quad (4)$$

where K_0 is the reference constant (entropy multiplier), E is the heat of adsorption.

In this paper, we used the following model equations to describe the adsorption dependence on concentration and temperature:

$$q = (a - b \cdot T) \frac{K_0 \cdot exp\left(\frac{E}{R \cdot T}\right) \cdot C}{1 + K_0 \cdot exp\left(\frac{E}{R \cdot T}\right) \cdot C} \quad (5)$$

$$q = (a \cdot exp(b \cdot T)) \frac{K_0 \cdot exp\left(\frac{E}{R \cdot T}\right) \cdot C}{1 + K_0 \cdot exp\left(\frac{E}{R \cdot T}\right) \cdot C} \quad (6)$$

where a, b are the empirical coefficients.

Both model equations describe experimental results quite well. The parameters of Equations (7) and (8) obtained by approximation are given in Table 2.

Table 2. Parameters of the model equations for describing the adsorption temperature dependence.

Equation	a, mol/g	b	K_0, L/mol	E, J/mol	R^2
(7)	$2.82 \cdot 10^{-3}$	$6.84 \cdot 10^{-6}$ mol/(g·K)	900.24	6030.05	0.9181
(8)	$20.66 \cdot 10^{-3}$	$-10.86 \cdot 10^{-3}$ K^{-1}	385.20	8170.42	0.9232

There is thus no reason to raise the impregnation temperature when preparing the target adsorbent, which is convenient in terms of process implementation.

2.6. TODGA Adsorption on MN202 from Methanol Solution

In this study, we also built an isotherm of TODGA adsorption on MN202 from a methanol solution at 313 K (Figure 6) by the classical analysis of liquid phase above an adsorbent. In such conditions, the Langmuir model is unsuitable for describing experimental data. This fact is assumed to be associated with significant adsorption ability of the HCP surface towards methanol [52]. Besides, methanol can effectively solvate the TODGA molecule, which follows from complete mutual dissolution of these compounds in the conditions considered in this work. Thus, in the HCP/TODGA/methanol system, more complex interactions can appear in comparison with the HCP/TODGA/CO_2 system, associated with the competition between TODGA and methanol for active adsorption sites on the HCP surface. It should be also taken into account that the degree of swelling of HCP varies in different solvents, which may also lead to some differences in the HCP adsorption properties in methanol or CO_2 media [53,54].

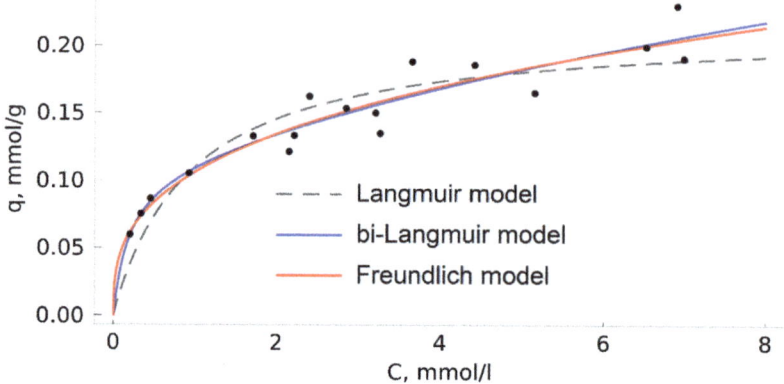

Figure 6. Isotherm of TODGA adsorption on MN202 from a solution in methanol at 313 K.

The Langmuir model is among the simplest ones. It describes in theoretical terms the ideal homogeneous adsorption of gases (adsorption on the surface with a regular arrangement of adsorption centers, identical in energy, with no adsorbate-adsorbate interactions). It is often suitable for describing adsorption from liquid solutions but, in such cases, it should be considered as a semi-empirical model because it contradicts the model postulate about the absence of interaction between the adsorbate particles and between the adsorbate and solvent particles, which leads to misinterpretations of the thermodynamic meaning of the adsorption equilibrium constant [55]. There are numerous empirical and semi-empirical models that take into account adsorption heterogeneity and other factors to describe complex isotherms. In the considered case of TODGA adsorption from a solution in methanol, the bi-Langmuir and Freundlich models demonstrated reasonably good approximations. The parameters of the model equations are given in Table 3.

Table 3. Parameters of the model approximation of the TODGA adsorption isotherm from a solution in methanol.

	Model		
	Langmuir $q=\frac{q_s \cdot K \cdot C}{1+K \cdot C}$	Bi-Langmuir $q = \frac{q_{s1} \cdot K_1 \cdot C}{1+K_1 \cdot C} + \frac{q_{s2} \cdot K_2 \cdot C}{1+K_2 \cdot C}$	Freundlich $q = a \cdot C^{\frac{1}{b}}$
	$q_s = 0.21$ mmol/g, $K = 1.06 \cdot 10^3$ L/mol, $R^2 = 0.8251$.	$q_{s1} = 0.11$ mmol/g, $K_1 = 5.29 \cdot 10^3$ L/mol, $q_{s2} = 0.42$ mmol/g, $K_2 = 45.3$ L/mol, $R^2 = 0.9076$.	$a = 0.107$, $b = 2.97$, $R^2 = 0.9060$.

Unfortunately, the wide points dispersion on the isotherm does not allow us to identify the details of the adsorption mechanism based on the data obtained. It can be cautiously assumed that the applicability of the bi-Langmuir model indicates that there are two or more independent types of adsorption centers on the HCP surface that could be the result of, for example, sorbent swelling or reversible modification of its surface with solvent molecules.

Figure 6 shows that at a TODGA concentration in the methanol solution of about 7 mmol/l, the adsorption value is only 0.20 ± 0.03 mmol/g, with the adsorbent remaining unsaturated. In case of adsorption from CO_2, the isotherm plateau is reached already at 1–1.5 mmol/l. It is thus not advantageous to prepare impregnated TODGA adsorbents from methanol in equilibrium adsorption conditions. Such adsorbents are mainly prepared by complete joint evaporation of the TODGA solution in a volatile solvent in the presence of a carrier matrix. Preparation of impregnated sorbents from a solution in SC CO_2 is characterized by a higher process speed and application of a cheap and nontoxic solvent. The increased diffusion coefficients in the SC medium and the absence of interfacial tension allow the substance to be quickly delivered to the pores of the adsorbent. After the impregnation is completed, CO_2 can be easily removed from the autoclave in the gaseous state, and the obtained product does not require additional treatment, such as drying from a solvent. Of course, it must be taken into account that switching to industrial scale requires solving a number of problems related to the transition to a continuous or half-periodic variant of the process and design of the corresponding experimental unit and technique. Nevertheless, the obtained results confirm the good prospects of the work in this field.

2.7. Comparison of Stability of Impregnated TODGA Adsorbents

Since TODGA is used for metal adsorption from acidic solutions, in this work we compared the resistance of impregnated adsorbents to being washed out by nitric acid. We prepared samples through equilibrium adsorption from SC CO_2 (sample 1) or evaporation from methanol solution (sample 2) with the TODGA content of 0.28 g/g and 0.31 g/g, respectively, according to the method described in the experimental section. After washing out the adsorbents with water and nitric acid solutions, we calculated the decrease in TODGA concentration in the adsorbent samples (Figure 7). The water did not have a leaching effect on impregnated adsorbents; however, as it was expected, in acidic solutions the formation of TODGA complexes with nitric acid [56] led to the extraction of the chelating agent into the solution. The leakage of TODGA and other ligands is a common feature of such sorbents reported in many works [6,12,14]. Despite this fact, they can be successfully applied in various separation processes in practice.

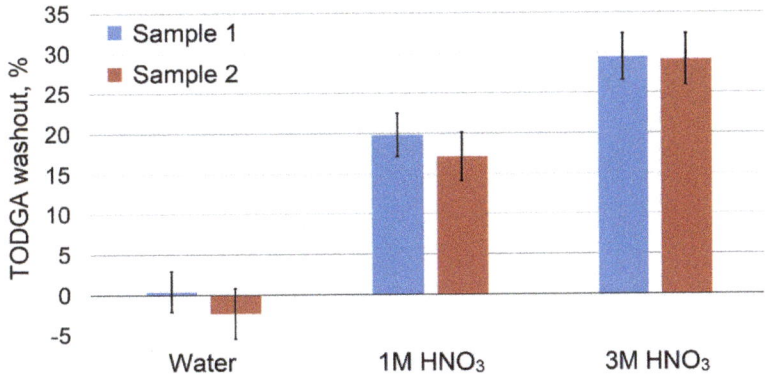

Figure 7. Reduction in TODGA content in sorbents washed with water and nitric acid solutions.

The resistance of adsorbent samples 1 and 2 to being washed out with water and nitric acid solutions was almost identical, taking into account experimental error and an initially higher TODGA content in sample 2. This indicates that the impregnation method does not significantly affect the described property of the obtained product.

3. Materials and Methods

3.1. Materials

The MN202 and MN270 (Purolite, Llantrisant, UK) sorbents were provided by S. Lyubimov (INEOS RAS, Moscow, Russia). Before use, the sorbents were carefully washed with acetone (99.85 wt%, Khimiya XXI Vek, Moscow, Russia) for 2 h at room temperature, with the sorbent/acetone mass ratio approximately equal to 1/10. After the washing, the sorbents were filtered on paper filters and dried in a drying oven at 393 K for 24 h. Then, the sorbent samples were kept in a desiccator in the presence of P_2O_5 (98 wt%, Khimiya

XXI Vek, Moscow, Russia) as the drying agent. Physical and chemical characteristics of sorbents are presented by the manufacturer on the official website [44,45].

TODGA (Figure 8) (≥98%, JSC "Axion—Rare and Precious Metals", Perm, Russia) was not additionally purified before usage.

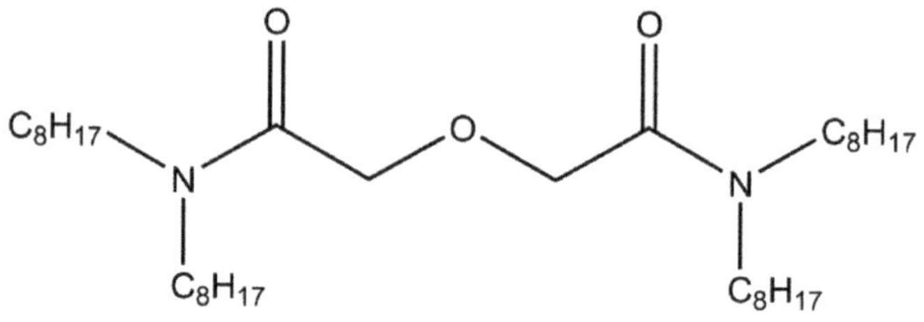

Figure 8. TODGA structural formula.

Methanol (extra pure, Chimmed, Moscow, Russia) was used as the cosolvent in the supercritical fluid chromatography, TODGA impregnation to HCP and re-extraction of the sorbents.

Distilled water obtained using a ListonA1104 distiller (Liston, Zhukov, Russia) and nitric acid (≥64 wt%, Khimiya XXI Vek, Moscow, Russia) were employed to evaluate the stability of impregnated adsorbents in an acidic aqueous medium.

3.2. Equipment

The adsorption measurements in the work were made in a modified version of the experimental unit assembled earlier for direct chromatographic analysis of solutions in SC fluids [31] (Figure 9). Namely, we used a 150 mL autoclave designed for void volume minimization. A C-MAG HS 7 magnetic stirrer (IKA, Staufen, Germany) was employed to mix the medium in an autoclave. Thermostatic control utilised an electric heating jacket and based on the data from a thermocouple placed inside the autoclave, with TRM202 (Owen, Moscow, Russia) as the controller. The thermocouple and the controller were calibrated by the readings of a tested liquid thermostat, with the deviation of the readings within the range 298–348 K not exceeding 0.5 K. All the sampling lines were minimized in length and volume, and temperature was controlled using a liquid thermostat consisting of a submersible M02 unit (Termex, Tomsk, Russia). The pressure in the system was measured by an APZ-3420 electronic transducer (Piezus, Moscow, Russia) with the maximum absolute error of ±0.1 MPa.

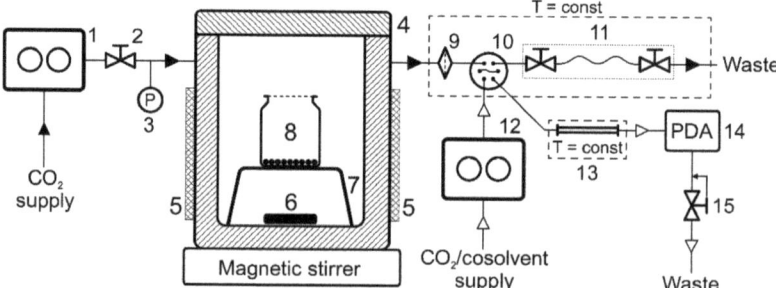

Figure 9. Scheme of the experimental unit: 1—CO_2 pump, 2—valve, 3—pressure transducer, 4—autoclave, 5—heating jacket, 6—magnetic stir bar, 7—support, 8—glass vial with an adsorbent, 9—in-line filter, 10—6-port valve with a sample loop, 11—sampling device, 12—chromatograph pump, 13—chromatographic column, 14—detector, 15—automatic back pressure regulator.

The principle of the unit operation consists in sampling a certain amount of the medium from the autoclave, where the process under study is realized, into the chromatograph sample loop for subsequent analysis. The volume of the sample loop in all the experiments was 10 µL. The sample collection into the loop was made by expanding the volume of the medium under pressure and allowing its flow into a special capillary with a needle valves on each end (sampling device). The capillary volume was 250 µL. Supercritical fluid chromatography enables a sample to be analyzed directly, without depressurization or additional sample preparation, which reduces the possibility of introducing additional error.

A Waters Acquity UPC^2 chromatograph (Waters, Milford, MA, USA) was used as the analytical instrument in the experimental unit. The chromatograph consisted of:

- CO_2 and cosolvent pump (Acquity ccBSM);
- Acquity UPLC autosampler;
- column thermostat (Acquity Column Manager);
- diode-array detector (Acquity UPC^2 PDA);
- flow control unit and a back pressure regulator (Acquity Convergence Manager).

The sorbent and TODGA samples were weighed on an Ohaus Pioneer PX225D semi-microbalance (Ohaus, Parsippany, NJ, USA).

The adsorbent specific surface area values were determined by the low temperature nitrogen adsorption method using an ATX-06 sorption unit (Katakon, Novosibirsk, Russia) by the BET model. Five points were measured within the range of nitrogen partial pressure values 0.05–0.25. The removal of the adsorbed moisture from the samples before the surface area measurement was carried out in a dry helium flow at 393 K and took 60 min.

3.3. Quantitative Analysis of TODGA

The TODGA samples were analyzed by the SFC method in the following conditions: column—Luna C18-2 (150 × 4.6 mm, 5 µm, Phenomenex, Torrance, CA, USA), column temperature—308 K, mobile phase flow rate—3 mL/min., mobile phase composition—CO_2/methanol (95/5 vol%), back pressure in the system—10.5 MPa. The detection was conducted at the wavelength of 215 nm.

The quantitative determination of the TODGA content in the autoclave was carried out using a calibration dependence. A TODGA sample in a glass vial was placed into the autoclave, the system was thermostatically controlledand CO_2 was fed by a Supercritical 24 pump (Teledyne SSI, State College, PA, USA) until the target pressure value was reached. The carbon dioxide feeding rate was about 10 mL/min. in liquid state under the pump head cooling regime. The system was intensively stirred until the sample dissolved in SC CO_2. Sampling from the autoclave was then conducted. To do that, we opened the first valve of the sampling device and let part of the autoclave medium flow under pressure through the sample loop and fill it. By turning the 6-port valve, we injected the sample into the flow of the chromatograph mobile phase, where we performed analysis under the conditions described earlier. We then closed the first valve of the sampling device and opened the second one to remove the sample residue. Before the next sample was injected, we returned the valves to their initial positions. Based on the data obtained, we plotted the calibration curve of the TODGA amount in the autoclave on the chromatographic peak area.

3.4. Measurement of TODGA Adsorption from a SC CO_2 Solution

Two approaches to TODGA adsorption measurement in an SC CO_2 medium were applied in the work. The first approach had been earlier described by us [31] and consists of direct chromatographic analysis of the solution bulk phase being in a thermodynamic equilibrium with the adsorbent. Based on the system's material balance and the residual TODGA concentration in CO_2 determined by the analysis, we calculated the adsorption value by the formula:

$$q = \frac{n_0^{TODGA} - n^{TODGA}}{m_0^{ads}} \qquad (7)$$

where n_0^{TODGA} is the initial TODGA amount in the solution, n^{TODGA} is the TODGA amount in the solution after adsorption equilibrium is reached, m_0^{ads} is the adsorbent mass.

The alternative approach to adsorption measurement used in the work was the gravimetric method [28,30], calculating the amount of the adsorbed substance by measuring the changes in the adsorbent mass during adsorption in static conditions:

$$q = \frac{\left(m^{ads} - m_0^{ads}\right)}{m^{ads} \cdot M^{TODGA}} \quad (8)$$

where m^{ads} is the adsorbent mass after the experiment, M^{TODGA} is the TODGA molar mass.

Both approaches were realized simultaneously in the same apparatus. Adsorbent and TODGA samples in a glass vial covered with perforated aluminum foil were placed into the autoclave. The autoclave was sealed and the system was thermostatically controlled. CO_2 was pumped into the experimental unit at the rate of about 10 mL/min under the pump head cooling regime. After reaching the target temperature and pressure values, the system was stirred intensively until thermodynamic equilibrium was reached. We then sampled the fluid phase for TODGA SFC analysis. Next, the autoclave was depressurized, the sorbent was taken out of the autoclave and weighed. This allowed us to make a direct comparison of the results obtained in identical conditions applying the two approaches.

3.5. Measurement of TODGA Adsorption from a Methanol Solution

Adsorbent, methanol and TODGA samples weighed on a semi-microbalance were placed into a 40 mL vial, which was sealed, carefully mixed and placed into a liquid thermostat at 313 K for 24 h. Then 2-3 mL of liquid phase samples were collected and filtered through a syringe filter (PTFE with the pore diameter of 0.2 μm). The TODGA concentration in the sample was determined by the SFC method in accordance with the calibration curve prepared in advance. The adsorption was calculated based on the TODGA mass balance between the bulk and adsorbed phases:

$$q = \frac{\left(C_0^{TODGA} - C^{TODGA}\right) \cdot V_{MeOH}}{m_0^{ads}} \quad (9)$$

where C_0^{TODGA} and C^{TODGA} are the TODGA concentrations in methanol at the beginning of the experiment and after the adsorption equilibrium is reached, respectively, V_{MeOH} is the methanol volume.

3.6. Preparation and Comparison of Impregnated Adsorbents' Stability

Sorbents based on MN202 impregnated with TODGA were obtained by the following methods:

(1) Impregnation from a liquid solvent. Sorbent and TODGA samples (4.00 and 1.32 g, respectively) were placed into a 100 mL round-bottom flask in order to prepare a species with the chelating agent content of about 30 g/g. After that, 20 mL of methanol were added to the mixture under constant stirring. The flask was covered with a lid and stored for 2 h. The methanol was then distilled off by a vacuum rotary evaporator at a temperature of 353 K for 3 h. The target product was taken out of the flask and placed in a weighing bottle, which was then sealed.

(2) Impregnation from SC CO_2. A glass vial with sorbent and TODGA samples was placed into the autoclave. The sample mass values were calculated based on adsorption isotherms plotted in advance to prepare impregnated forms with the TODGA content of about 30 g/g. CO_2 was then fed into the system under constant stirring to reach the target pressure value of 20 MPa. The medium temperature was 313 K. The system was kept for 1 h, after which chromatographic analysis of the medium above the sorbent was performed. The autoclave was then depressurized, the product was taken out of the autoclave and was placed in a weighing bottle, which was then sealed.

The actual final content of TODGA in both samples was determined gravimetrically as the difference between the sorbents' masses after and before impregnation. The stability of the obtained sorbents in HNO_3 aqueous solutions was determined by submerging a sample (0.500 ± 0.001 g) in 10 mL of water or a nitric acid solution with the concentration of 1 or 3 M. The samples were kept at room temperature (295 ± 2 K) for two days under regular stirring. The liquid phase was then poured out, the samples were placed on filter paper and the residual moisture was removed by air-drying. The TODGA content in the samples after the treatment was checked by re-extracting TODGA from HCP by methanol with subsequent chromatographic analysis. Samples of 0.060 ± 0.005 g were used for the re-extraction. Each sample was washed with methanol three times in portions of 15 mL each for 8–12 h, after which all the extracts were placed in a 50 mL volumetric flask and diluted with methanol to scale. Using SFC, we found the TODGA concentration in the obtained solutions and its mass fraction in the impregnated sorbent species after the treatment.

4. Conclusions

The present work studied the adsorption of TODGA, a common chelating agent, on HCP in an SC CO_2 medium. The measurements were made by two methods: on-line SFC and gravimetrical analysis. Both methods showed almost identical results, but the first proved to be more reliable when measuring adsorption from concentrated solutions.

Isotherms of TODGA adsorption on the MN202 and MN270 sorbents were measured at 313 K and 20 MPa. Based on the capacity of the sorbents determined by isotherm approximation, the MN202 sorbent was chosen as a suitable carrier for TODGA.

Varying the fluid density during impregnation within the range 0.78–0.91 g/mL in isothermal conditions did not have a significant effect on TODGA adsorption. In contrast, the temperature parameter has a significant influence on the MN202 TODGA adsorption capacity. Semi-empirical models based on the Langmuir equation were used to make a general description of the adsorption dependence on concentration and temperature. The temperature increase from 313 K to 343 K at the medium density of 0.78 g/mL changed the maximum amount of adsorbed TODGA from ~0.68 mmol/g to ~0.49 mmol/g.

In conditions of equilibrium adsorption from methanol it is inappropriate to prepare impregnated TODGA/HCP adsorbents, due to the small slope value of the isotherm. The SC CO_2 application in the impregnation procedure allowed us to reduce the process time and eliminate toxic organic solvents. The samples of the sorbent impregnated in methanol and SC CO_2 have identical chemical resistance to nitric acid solutions, so there is no difference between two impregnation media with respect to this property.

Author Contributions: Conceptualization, M.K. and O.P.; methodology, M.K.; investigation, M.K. and O.P.; writing—original draft preparation, M.K.; writing—review and editing, O.P.; supervision, O.P. All authors have read and agreed to the published version of the manuscript.

Funding: This work was supported by the Ministry of Science and Higher Education of the Russian Federation as part of the State Assignment of the Kurnakov Institute of General and Inorganic Chemistry of the Russian Academy of Sciences.

Institutional Review Board Statement: Not applicable.

Informed Consent Statement: Not applicable.

Data Availability Statement: Not applicable.

Acknowledgments: The authors express their gratitude to Sergey Lyubimov from INEOS RAS (Moscow, Russia) for providing the HCP samples, Andrey Shadrin from JSC Bochvar VNIINM (Moscow, Russia) for providing TODGA.

Conflicts of Interest: The authors declare no conflict of interest.

Sample Availability: Not available.

References

1. Baulin, V.E.; Baulin, D.V.; Usolkin, A.N.; Ivenskaya, N.M.; Vlasova, N.V.; Kozlov, P.V.; Remizov, M.B.; Chuchlanseva, E.V.; Tsivadze, A.Y. сраоо-хроаораφесое еее еоа аоосх расоро с рееесорео, репоах N,N,N',N'-eпаоoao. *Sorbtsionnye i Khromatograficheskie Protsessy* **2018**, *18*, 717–725.
2. Dicholkar, D.D.; Kumar, P.; Heer, P.K.; Gaikar, V.G.; Kumar, S.; Natarajan, R. Synthesis of N,N,N',N'-Tetraoctyl-3-Oxapentane-1,5-Diamide (TODGA) and Its Steam Thermolysis-Nitrolysis as a Nuclear Waste Solvent Minimization Method. *Ind. Eng. Chem. Res.* **2013**, *52*, 2457–2469. [CrossRef]
3. Metwally, E.; Saleh, A.S.; Abdel-Wahaab, S.M.; El-Naggar, H.A. Extraction Behavior of Cerium by Tetraoctyldiglycolamide from Nitric Acid Solutions. *J. Radioanal. Nucl. Chem.* **2010**, *286*, 217–221. [CrossRef]
4. Wang, Z.; Huang, H.; Ding, S.; Hu, X.; Zhang, L.; Liu, Y.; Song, L.; Chen, Z.; Li, S. Extraction of Trivalent Americium and Europium with TODGA Homologs from HNO_3 Solution. *J. Radioanal. Nucl. Chem.* **2017**, *313*, 309–318. [CrossRef]
5. Ansari, S.A.; Pathak, P.N.; Manchanda, V.K.; Husain, M.; Prasad, A.K.; Parmar, V.S. N,N,N',N'-Tetraoctyl Diglycolamide (TODGA): A Promising Extractant for Actinide-Partitioning from High-Level Waste (HLW). *Solvent Extr. Ion. Exch.* **2005**, *23*, 463–479. [CrossRef]
6. Xu, Y.; Wei, Y.; Liu, R.; Usuda, S.; Ishii, K.; Yamazaki, H. Adsorption Characteristics of Trivalent Rare Earths and Chemical Stability of a Silica-Based Macroporous TODGA Adsorbent in HNO_3 Solution. *J. Nucl. Sci. Technol.* **2011**, *48*, 1223–1229. [CrossRef]
7. Watanabe, S.; Arai, T.; Ogawa, T.; Takizawa, M.; Sano, K.; Nomura, K.; Koma, Y. Optimizing Composition of $TODGA/SiO_2$-P Adsorbent for Extraction Chromatography Process. *Procedia Chem.* **2012**, *7*, 411–417. [CrossRef]
8. Kim, S.-Y.; Xu, Y.; Ito, T.; Wu, Y.; Tada, T.; Hitomi, K.; Kuraoka, E.; Ishii, K. A Novel Partitioning Process for Treatment of High Level Liquid Waste Using Macroporous Silica-Based Adsorbents. *J. Radioanal. Nucl. Chem.* **2013**, *295*, 1043–1050. [CrossRef]
9. Usuda, S.; Yamanishi, K.; Mimura, H.; Sasaki, Y.; Kirishima, A.; Sato, N.; Niibori, Y. Chromatographic Separation Behaviors of Am, Cm, and Eu onto TODGA and DOODA(C8) Adsorbents with Hydrophilic Ligand–Nitric Acid Eluents. *Chem. Lett.* **2013**, *42*, 1220–1222. [CrossRef]
10. Ge, G.; Yuanlai, X.; Xinxin, Y.; Fen, W.; Fang, Z.; Junxia, Y.; Ruan, C. Effect of HNO_3 Concentration on a Novel Silica-Based Adsorbent for Separating Pd(II) from Simulated High Level Liquid Waste. *Sci. Rep.* **2017**, *7*, 11290. [CrossRef]
11. Wu, H.; Kawamura, T.; Kim, S.-Y. Adsorption and Separation Behaviors of Y(III) and Sr(II) in Acid Solution by a Porous Silica Based Adsorbent. *Nucl. Eng. Technol.* **2021**, *53*, 3352–3358. [CrossRef]
12. Horwitz, E.P.; McAlister, D.R.; Bond, A.H.; Barrans, R.E. Novel Extraction Chromatographic Resins Based on Tetraalkyldiglycolamides: Characterization and Potential Applications. *Solvent Extr. Ion Exch.* **2005**, *23*, 319–344. [CrossRef]
13. Chen, M.; Li, Z.; Geng, Y.; Zhao, H.; He, S.; Li, Q.; Zhang, L. Adsorption Behavior of Thorium on N,N,N',N'-Tetraoctyldiglycolamide (TODGA) Impregnated Graphene Aerogel. *Talanta* **2018**, *181*, 311–317. [CrossRef] [PubMed]
14. Milyutin, V.V.; Khesina, Z.B.; Laktyushina, A.A.; Buryak, A.K.; Nekrasova, N.A.; Kononenko, O.A.; Pavlov, Y.S. Chemical Durability and Radiation Resistance of Sorbents Based on N,N,N',N'-Tetra-n-Octyldiglycolamide. *Radiochemistry* **2016**, *58*, 59–62. [CrossRef]
15. Davankov, V.; Tsyurupa, M.; Ilyin, M.; Pavlova, L. Hypercross-Linked Polystyrene and Its Potentials for Liquid Chromatography: A Mini-Review. *J. Chromatogr. A* **2002**, *965*, 65–73. [CrossRef]
16. Macnaughton, S.J.; Foster, N.R. Supercritical Adsorption and Desorption Behavior of DDT on Activated Carbon Using Carbon Dioxide. *Ind. Eng. Chem. Res.* **1995**, *34*, 275–282. [CrossRef]
17. Benkhedda, J.; Jaubert, J.-N.; Barth, D.; Zetzl, C.; Brunner, G. Adsorption And Desorption Of M-Xylene From Supercritical Carbon Dioxide On Activated Carbon. *Sep. Sci. Technol.* **2001**, *36*, 2197–2211. [CrossRef]
18. Ushiki, I.; Ota, M.; Sato, Y.; Inomata, H. Measurements and Dubinin–Astakhov Correlation of Adsorption Equilibria of Toluene, Acetone, n-Hexane, n-Decane and Methanol Solutes in Supercritical Carbon Dioxide on Activated Carbon at Temperature from 313 to 353 K and at Pressure from 4.2 to 15.0 MPa. *Fluid Phase Equilibria* **2013**, *344*, 101–107. [CrossRef]
19. Glenne, E.; Öhlén, K.; Leek, H.; Klarqvist, M.; Samuelsson, J.; Fornstedt, T. A Closer Study of Methanol Adsorption and Its Impact on Solute Retentions in Supercritical Fluid Chromatography. *J. Chromatogr. A* **2016**, *1442*, 129–139. [CrossRef]
20. Enmark, M.; Forssén, P.; Samuelsson, J.; Fornstedt, T. Determination of Adsorption Isotherms in Supercritical Fluid Chromatography. *J. Chromatogr. A* **2013**, *1312*, 124–133. [CrossRef]
21. Kamarei, F.; Tarafder, A.; Gritti, F.; Vajda, P.; Guiochon, G. Determination of the Adsorption Isotherm of the Naproxen Enantiomers on (S,S)-Whelk-O1 in Supercritical Fluid Chromatography. *J. Chromatogr. A* **2013**, *1314*, 276–287. [CrossRef] [PubMed]
22. Vajda, P.; Guiochon, G. Effects of the Back Pressure and the Temperature on the Finite Layer Thickness of the Adsorbed Phase Layer in Supercritical Fluid Chromatography. *J. Chromatogr. A* **2013**, *1309*, 41–47. [CrossRef] [PubMed]
23. Enmark, M.; Samuelsson, J.; Forss, E.; Forssén, P.; Fornstedt, T. Investigation of Plateau Methods for Adsorption Isotherm Determination in Supercritical Fluid Chromatography. *J. Chromatogr. A* **2014**, *1354*, 129–138. [CrossRef]
24. Vajda, P.; Guiochon, G. Modifier Adsorption in Supercritical Fluid Chromatography onto Silica Surface. *J. Chromatogr. A* **2013**, *1305*, 293–299. [CrossRef]
25. Vajda, P.; Guiochon, G. Surface Excess Isotherms of Organic Solvent Mixtures in a System Made of Liquid Carbon Dioxide and a Silicagel Surface. *J. Chromatogr. A* **2013**, *1308*, 139–143. [CrossRef] [PubMed]
26. Singh, J.K.; Verma, N. (Eds.) *Aqueous Phase Adsorption: Theory, Simulations and Experiments*, 1st ed.; CRC Press: Boca Raton, FL, USA; Taylor & Francis: New York, NY, USA, 2019; ISBN 978-1-351-27252-0.

27. Tsyurupa, M.P.; Blinnikova, Z.K.; Il'in, M.M.; Davankov, V.A.; Parenago, O.O.; Pokrovskii, O.I.; Usovich, O.I. Monodisperse Microbeads of Hypercrosslinked Polystyrene for Liquid and Supercritical Fluid Chromatography. *Russ. J. Phys. Chem.* **2015**, *89*, 2064–2071. [CrossRef]
28. Smirnova, I.; Mamic, J.; Arlt, W. Adsorption of Drugs on Silica Aerogels. *Langmuir* **2003**, *19*, 8521–8525. [CrossRef]
29. Lovskaya, D.; Menshutina, N. Alginate-Based Aerogel Particles as Drug Delivery Systems: Investigation of the Supercritical Adsorption and In Vitro Evaluations. *Materials* **2020**, *13*, 329. [CrossRef]
30. Caputo, G. Supercritical Fluid Adsorption of Domperidone on Silica Aerogel. *ACES* **2013**, *3*, 189–194. [CrossRef]
31. Kostenko, M.O.; Ustinovich, K.B.; Pokrovskii, O.I. Online Monitoring of Adsorption onto Silica Xerogels and Aerogels from Supercritical Solutions Using Supercritical Fluid Chromatography. *Russ. J. Inorg. Chem.* **2020**, *65*, 1577–1584. [CrossRef]
32. Johannsen, M.; Brunner, G. Solubilities of the Xanthines Caffeine, Theophylline and Theobromine in Supercritical Carbon Dioxide. *Fluid Phase Equilibria* **1994**, *95*, 215–226. [CrossRef]
33. Li, B.; Guo, W.; Ramsey, E.D. Determining the Solubility of Nifedipine and Quinine in Supercritical Fluid Carbon Dioxide Using Continuously Stirred Static Solubility Apparatus Interfaced with Online Supercritical Fluid Chromatography. *J. Chem. Eng. Data* **2017**, *62*, 1530–1537. [CrossRef]
34. Li, B.; Guo, W.; Song, W.; Ramsey, E.D. Determining the Solubility of Organic Compounds in Supercritical Carbon Dioxide Using Supercritical Fluid Chromatography Directly Interfaced to Supercritical Fluid Solubility Apparatus. *J. Chem. Eng. Data* **2016**, *61*, 2128–2134. [CrossRef]
35. Li, B.; Guo, W.; Ramsey, E.D. Measuring the Solubility of Anthracene and Chrysene in Supercritical Fluid Carbon Dioxide Using Static Solubility Apparatus Directly Interfaced Online to Supercritical Fluid Chromatography. *J. Chem. Eng. Data* **2018**, *63*, 651–660. [CrossRef]
36. Ramsey, E.D.; Minty, B.; McCullagh, M.A.; Games, D.E.; Rees, A.T. Analysis of Phenols in Water at the Ppb Level Using Direct Supercritical Fluid Extraction of Aqueous Samples Combined On-Line With Supercritical Fluid Chromatography–Mass Spectrometry. *Anal. Commun.* **1997**, *34*, 3–6. [CrossRef]
37. Li, B.; Guo, W.; Ramsey, E.D. Determining Phenol Partition Coefficient Values in Water- and Industrial Process Water-Supercritical CO_2 Systems Using Direct Aqueous SFE Apparatus Simultaneously Interfaced with on-Line SFC and on-Line HPLC. *J. Supercrit. Fluids* **2019**, *152*, 104558. [CrossRef]
38. Li, B.; Guo, W.; Ramsey, E.D. Simultaneous Determination of Partition Coefficients of Three Organic Compounds Distributed as Mixtures between $SF-CO_2$ and Water Using Interfaced SFE-SFC Apparatus. *J. Supercrit. Fluids* **2020**, *159*, 104759. [CrossRef]
39. Voshkin, A.; Solov'ev, V.; Kostenko, M.; Zakhodyaeva, Y.; Pokrovskiy, O. A Doubly Green Separation Process: Merging Aqueous Two-Phase Extraction and Supercritical Fluid Extraction. *Processes* **2021**, *9*, 727. [CrossRef]
40. Gromov, O.I.; Kostenko, M.O.; Petrunin, A.V.; Popova, A.A.; Parenago, O.O.; Minaev, N.V.; Golubeva, E.N.; Melnikov, M.Y. Solute Diffusion into Polymer Swollen by Supercritical CO_2 by High-Pressure Electron Paramagnetic Resonance Spectroscopy and Chromatography. *Polymers* **2021**, *13*, 3059. [CrossRef]
41. Li, B.; Guo, W.; Chi, H.; Kimura, M.; Ramsey, E.D. Monitoring the Progress of a Photochemical Reaction Performed in Supercritical Fluid Carbon Dioxide Using a Continuously Stirred Reaction Cell Interfaced to On-Line SFC. *Chromatographia* **2017**, *80*, 1179–1188. [CrossRef]
42. Li, B.; Guo, W.; Ramsey, E.D. Monitoring the Progress of the Acetylation Reactions of 4-Aminophenol and 2-Aminophenol in Acetonitrile Modified Supercritical Fluid Carbon Dioxide and Pure Acetonitrile Using on-Line Supercritical Fluid Chromatography and on-Line Liquid Chromatography. *J. Supercrit. Fluids* **2018**, *133*, 372–382. [CrossRef]
43. Linstrom, P. *NIST Chemistry WebBook*; NIST Standard Reference Database 69; NIST: Gaithersburg, MD, USA, 1997.
44. Purolite Purolite Product: Macronet™ MN202. Available online: http://www.purolite.com/product/mn202 (accessed on 5 November 2021).
45. Purolite Purolite Product: Macronet™ MN270. Available online: http://www.purolite.com/product/mn270 (accessed on 5 November 2021).
46. Suhas; Gupta, V.K.; Carrott, P.J.M.; Singh, R.; Chaudhary, M.; Kushwaha, S. Cellulose: A Review as Natural, Modified and Activated Carbon Adsorbent. *Bioresour. Technol.* **2016**, *216*, 1066–1076. [CrossRef] [PubMed]
47. Engin, M.S.; Uyanik, A.; Cay, S.; Icbudak, H. Effect of the Adsorptive Character of Filter Papers on the Concentrations Determined in Studies Involving Heavy Metal Ions. *Adsorpt. Sci. Technol.* **2010**, *28*, 837–846. [CrossRef]
48. Harris, R.C. The Adsorption of Protein by Filter Paper in the Estimation of Albumin in Blood Serum. *J. Biol. Chem.* **1939**, *127*, 751–756. [CrossRef]
49. Hindarso, H.; Ismadji, S.; Wicaksana, F.; Mudjijati; Indraswati, N. Adsorption of Benzene and Toluene from Aqueous Solution onto Granular Activated Carbon. *J. Chem. Eng. Data* **2001**, *46*, 788–791. [CrossRef]
50. Fianu, J.; Gholinezhad, J.; Hassan, M. Comparison of Temperature-Dependent Gas Adsorption Models and Their Application to Shale Gas Reservoirs. *Energy Fuels* **2018**, *32*, 4763–4771. [CrossRef]
51. Giraudet, S.; Pré, P.; Le Cloirec, P. Modeling the Temperature Dependence of Adsorption Equilibriums of VOC(s) onto Activated Carbons. *J. Environ. Eng.* **2010**, *136*, 103–111. [CrossRef]
52. Rosenberg, G.I.; Shabaeva, A.S.; Moryakov, V.S.; Musin, T.G.; Tsyurupa, M.P.; Davankov, V.A. Sorption Properties of Hyper-crosslinked Polystyrene Sorbents. *React. Polym. Ion Exch. Sorbents* **1983**, *1*, 175–182. [CrossRef]

53. Davankov, V.A.; Pastukhov, A.V.; Tsyurupa, M.P. Unusual Mobility of Hypercrosslinked Polystyrene Networks: Swelling and Dilatometric Studies. *J. Polym. Sci. Part B Polym. Phys.* **2000**, *38*, 1553–1563. [CrossRef]
54. Tsyurupa, M.P.; Blinnikova, Z.K.; Proskurina, N.A.; Pastukhov, A.V.; Pavlova, L.A.; Davankov, V.A. Hypercrosslinked Polystyrene: The First Nanoporous Polymeric Material. *Nanotechnol. Russ.* **2009**, *4*, 665–675. [CrossRef]
55. Guiochon, G.; Shirazi, D.G.; Felinger, A.; Katti, A.M. *Fundamentals of Preparative and Nonlinear Chromatography*, 2nd ed.; Academic Press: Boston, MA, USA, 2006; ISBN 978-0-12-370537-2.
56. Bell, K.; Geist, A.; McLachlan, F.; Modolo, G.; Taylor, R.; Wilden, A. Nitric Acid Extraction into TODGA. *Procedia Chem.* **2012**, *7*, 152–159. [CrossRef]

Article

Ionic Liquid Extraction Behavior of Cr(VI) Absorbed on Humic Acid–Vermiculite

Hsin-Liang Huang *, P.C. Lin, H.T. Wang, Hsin-Hung Huang and Chao-Ho Wu

Department of Safety, Health and Environmental Engineering, National United University, Miaoli 36063, Taiwan; sandy23063051@hotmail.com (P.C.L.); flower611003@yahoo.com.tw (H.T.W.); hsinhunghuang@gmail.com (H.-H.H.); a445dd@gmail.com (C.-H.W.)
* Correspondence: hlhuang@nuu.edu.tw; Tel.: +886-37-382277

Abstract: Cr(VI) can be released into soil as a result of mining, electroplating, and smelting operations. Due to the high toxicity of Cr(VI), its removal is necessary in order to protect ecosystems. Vermiculite is applied in situations where there is a high degree of metal pollution, as it is helpful during the remediation process due to its high cation exchange capacity. The Cr(VI) contained in the vermiculite should be extracted in order to recover it and to reduce the impact on the environment. In this work, adsorption equilibrium data for Cr(VI) in a simulated sorbent for soil remediation (a mixture that included both humic acid (HA) and vermiculite) were a good fit with the Langmuir isotherm model. The simulated sorbent for soil remediation was a favorable sorbent for Cr(VI) when it was in the test soil. An ionic liquid, [C_4mim]Cl (1-butyl-3-methylimidazolium chloride), was studied to determine its efficiency in extracting Cr(VI) from the Cr- contaminated simulated sorbent in soil remediation. At 298 K and within 30 min, approximately 33.48 ± 0.79% of Cr(VI) in the simulated sorbent in soil remediation was extracted into [C_4mim]Cl. Using FTIR spectroscopy, the absorbance intensities of the bands at 1032 and 1010 cm^{-1}, which were attributed to C-O bond stretching in the polysaccharides of HA, were used to detect the changes in HA in the Cr-contaminated simulated sorbent for soil remediation before and after extraction. The results showed that Cr(VI) that has been absorbed on HA can be extracted into [C_4mim]Cl. Using ^{1}H NMR, it was observed that the 1-methylimizadole of [C_4mim]Cl played an important role in the extraction of Cr(VI), which bonded with HA on vermiculite and was able to be transformed into the [C_4mim]Cl phase.

Keywords: hexavalent chromium; humic acid; vermiculite; ionic liquid; NMR

1. Introduction

Arsenic, cadmium, chromium, mercury, nickel, lead, zinc, and copper are the major metals that are often found in contaminated soil. According to studies conducted by the Taiwan Environmental Protection Agency (EPA), thirteen out of sixty-two remediation sites are presently contaminated with metals: mainly nickel, chromium, lead, and zinc, followed by copper and cadmium [1].

Cr is naturally present in its oxidized state of Cr(III), whereas divalent, tetravalent, and pentavalent Cr are unstable in the environment. Cr(III) is one of the essential elements involved in protein metabolism in animals and humans [2]. Cr(VI) is genetically toxic to cells; it has been shown to be a carcinogen as it affects the functions of deoxyribonucleic acid [3]. It is possible for Cr(VI) to be absorbed into edible plants or vegetables, resulting in them having reduced root and coleoptile growth [4,5]. Moreover, the Cr(VI) concentration in plants depends on the soluble fraction of it that is present in the soil [6,7]. Once it has entered the food chain, Cr(VI) may cause harm to both animal and human organisms when they ingest these affected edible plants or vegetables. Animal studies by the Institute of Labor, Occupational Safety and Health, Taiwan Ministry of Labor have shown that Cr(VI) can cause malignant tumors, while Cr(III) does not [8]. The United States EPA has classified Cr(VI) as a Group A carcinogen in humans, while Cr(III) is classified as a Group D unclassified substance. Therefore, the removal of Cr(VI) from contaminated soils is essential to avoid its toxic impact on ecosystems. High concentrations of Cr(VI) are often found to bind to humic acid (HA), vermiculite, and kaolin clay [9]. Vermiculite, which has a high cation exchange capacity, is a group of hydrated laminar minerals. It is often used as an additive to improve soil structure. Vermiculite is also used as a sorbent in order to absorb the metals that are present in contaminated soil. This results in the accumulation of high concentrations of chromium in vermiculite [10,11]. It has also been reported that Cr(VI) can be absorbed by HA. As the pH increases from 1 to 7, the Cr(VI) adsorption to HA decreases from 100% to 34% [12]. Moreover, the efficiency of Cr(VI) removal increases at higher temperatures [13]. In addition, in one study using peanut shells as a sorbent, the Cr(VI) removal ratios increased from 50% to 90% in an aqueous solution when the reaction time was increased from one to five hours [14]. Remediation technologies for soil that has been contaminated with metals include biological methods, soil washing, solidification/stabilization, and extraction [15–17]. Soil washing is one of the most commonly used techniques that can be applied for the remediation of metal-contaminated soils. Cleaning agents, such as surfactants, are used to remove metals from contaminated soil by extracting these toxins into a liquid phase [18]. Soil washing can also be applied for the recovery of the metals that were used in the sorbents for soil remediation.

Ionic liquids (ILs) have special chemical and physical properties, including high thermal stability, negligible vapor pressure, a broad liquid phase range, and excellent electric conductivity [19]. ILs can be used to replace conventional organic solvents and their impact on the environment is considered to be minimal; therefore, they have also been referred to as green solvents [20]. ILs are made of various ions, thereby possessing the properties of salts. Varying the compositions of anions and cations results in different ILs that can be used for different applications. ILs that remain in a liquid phase at room temperature are called room temperature ILs (RTILs) [21]. Kozonoi et al. utilized 1-butyl-3-methylimidazolium nonafluorobutanesulfonate ([bmi][NfO]) to extract Cs^+, Na^+, Li^+, Sr^{2+}, Ca^{2+}, and La^{3+} ions from aqueous solutions, with extraction ratios of 5, 24, 39, 79, 81, and 98%, respectively [22]. Metal ions with a higher valence can be more easily extracted with ILs. In addition, the extraction efficiencies of metal ions, such as copper, lead, and sodium, are greater when ILs are used than those are achieved with regular organic solvents, e.g., chloroform [23].

The structure of the Cr(VI) complexes that have been formed in the sorbent for remediation in contaminated soil are too complex to reveal the mechanism of absorbance phenomena. In order to understand the effects of Cr(VI) that has been absorbed on vermiculite and that has been extracted with the ionic liquid, a mixture comprising both humic acid (HA) and vermiculite was prepared to simulate the sorbent for remediation in a contaminated soil sample. The adsorption equilibrium data for Cr(VI) in the simulated sorbent were established. During the extraction process, 1-butyl-3-methylimidazolium chloride ([C_4min]Cl) was used to extract Cr(VI) from the Cr-contaminated simulated sorbent for soil remediation, and the extraction mechanism was explored.

2. Results and Discussion

2.1. Adsorption of Chromium Species on the Simulated Sorbent in Soil

Table 1 lists the absorption efficiencies of Cr(VI) to the HA, vermiculite, and the simulated sorbent for soil remediation. As shown in Table 1, HA had a high Cr(VI) absorption capacity. The absorption efficiency of Cr(VI) onto HA was greater than that of Cr(VI) onto vermiculite (81.32 ± 1.05% vs. 64.47 ± 1.62%). The absorption efficiency of Cr(VI) onto the simulated sorbent for soil remediation that included both HA and vermiculite was 91.78 ± 1.82%, showing that HA and vermiculite had a synergistic effect on the absorption efficiency of Cr(VI) into the sorbent for soil remediation.

Table 1. Absorption efficiencies of Cr(VI) onto sorbents.

Absorbed Cr(VI) on:	Absorption Efficiency (%)
humic acid	81.32 ± 1.05
vermiculite	64.47 ± 1.62
simulated sorbent in soil remediation	92.11 ± 2.26

The speciation of chromium on the simulated sorbent in soil remediation was studied using XANES spectroscopy (see Figure 1). The pre-edge intensity of the $3d$ elements with T_d symmetry was greater than those with O_h symmetry. The intense peaks for the tetrahedral species of the $3d$ transition metals in the pre-edge range were attributable to the p component in the d-p hybridized orbital. The number of d-electrons that the tetrahedral species has affects the intensity of the pre-edge peak [24,25]. The existence of Cr(VI) and Cr(III) in the simulated sorbent in soil remediation was observed by the pre-edge feature that was centered at 5993–5994 eV in the XANES spectra and are distinctive of the deconvolution that takes place during component fitting. In the simulated sorbent in soil remediation, Cr(VI)-HA and Cr(VI)$_{ads}$ were the main species that were present, as seen in Figure 1. It was also determined that about 11% of the Cr(VI) compound was reduced to Cr(III). Cr(VI) can interact with the carboxyl groups of HA, resulting in the reduction of Cr(VI) [26,27].

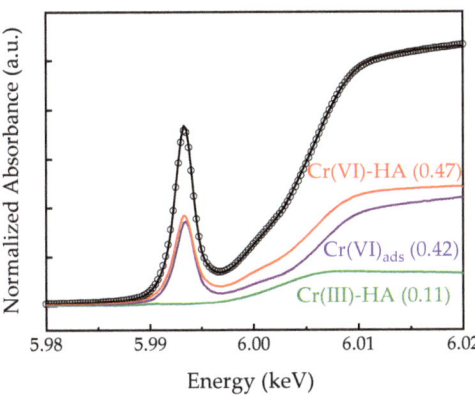

Figure 1. The least-square fitted XANES spectra of chromium in HA–vermiculite.

2.2. Adsorption Equilibrium of Chromium Species onto Simulated Sorbent for Soil Remediation

In the adsorption experiments, the concentration of the adsorbed Cr(VI) that was present in the simulated sorbent in soil increased as C_e increased (see Figure 2). In Table 2, the model parameters for Cr(VI) absorption onto the simulated sorbent for soil remediation were estimated by fitting the experimental data. According to the values of the correlation coefficient (R^2), the Cr(VI) absorption onto the simulated sorbent for soil remediation tended to follow the Langmuir and Freundlich isotherm equations in the C_0 range of 1000–8000 mg/L. The Langmuir isotherm fit the experimental data better than the Freundlich isotherm. In the Langmuir isotherm model, the monolayer saturation capacity of the Cr(VI) that was present in the simulated sorbent in soil remediation was 5.57 mg/g. The calculated R_L value from the Langmuir isotherm model was 0.139, which indicated favorable Cr(VI) absorption onto the simulated sorbent for soil remediation. Moreover, the value of $1/n$ in the Freundlich isotherm model was 0.325, which also indicated that Cr(VI) was favorably absorbed on the simulated sorbent for soil remediation.

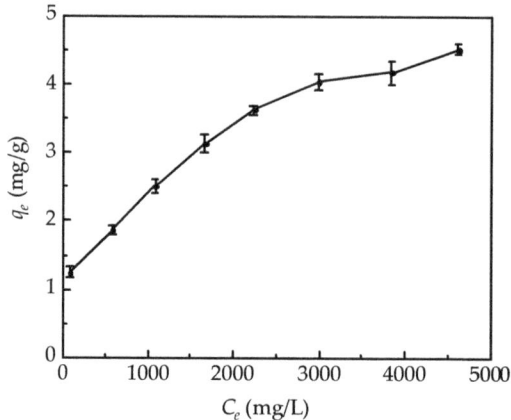

Figure 2. Experimental adsorption equilibrium results for the absorption of Cr(VI) onto the simulated sorbent for soil remediation. Error bars show the standard deviation of five replicates.

Table 2. Langmuir and Freundlich isotherm parameters for Cr(VI) absorption on the simulated sorbent in soil remediation.

Absorption Isotherm	Parameters		
Langmuir model	q_m (mg/g) 5.57	K_L (1/g) 0.00619	R^2 0.996
Freundlich model	K_f (Lmg^{n-1}/g^n) data 0.553	$1/n$ 0.325	R^2 0.989

2.3. Extraction of Chromium Species from Cr-Contaminated Simulated Sorbent for Soil Remediation with [C$_4$mim]Cl

The compound [C$_4$mim]Cl, a hydrophilic IL, can be used to extract HA and is able to intermix with Cr(VI) [28,29]. Thus, in this study, [C$_4$mim]Cl was used to extract Cr(VI) from the Cr-simulated sorbent in soil remediation. The results show that approximately 33.48 ± 0.79% of the Cr(VI) was extracted into [C$_4$mim]Cl (see Table 3). To understand whether different matrices affected the extraction efficiency of Cr(VI), HA and vermiculite were also tested. The results show that the extraction efficiencies of Cr(VI) compared to those of HA and vermiculite were about 82.85 ± 0.96 and 21.97 ± 1.11%, respectively, showing that HA and vermiculite affected the extraction efficiency. In a similar study, approximately 70% of Cr(VI) that had been chelated with HA in a mesoporous sorbent was able to be extracted into [C$_4$mim]Cl [30]. Therefore, the extraction efficiency of Cr(VI) in HA was higher than that in the vermiculite and in the simulated sorbent for soil remediation.

Table 3. Extraction efficiencies of Cr(VI) from sorbents into [C$_4$mim]Cl.

Extracted Cr(VI) in [C$_4$mim]Cl from:	Extraction Efficiency (%)
humic acid	82.85 ± 0.96
vermiculite	21.97 ± 1.11
simulated sorbent in soil remediation	33.48 ± 0.79

2.4. FTIR Analysis

To further understand the extraction mechanism of [C$_4$mim]Cl, the structures of different matrices were tested both before and after extraction using FTIR spectroscopy. As shown in Figure 3, the band at 1088 cm^{-1} was attributed to the C-O group stretching in the ester. The two bands found at 1032 and 1010 cm^{-1} corresponded to the C-O group stretching in polysaccharides [31]. In Figure 3a,b, the broadened peak at 1088 cm^{-1} and the red shifts (ν, cm^{-1}: 1032→1011 and 1010→960) were found because of the interaction between HA and vermiculite in the simulated sorbent for soil remediation. Moreover, the bands at 1011 and 960 cm^{-1} were also shifted to 999 and 958 cm^{-1}, respectively, in the Cr-simulated sorbent for soil remediation (see Figure 3b,c). It was clear that the Cr(VI) had been absorbed onto the C-O bonds of HA interacted with vermiculite. In Figure 3d, a band shift (999→1001 cm^{-1}) was identified when [C$_4$mim]Cl was used to extract Cr(VI) from the Cr-contaminated simulated sorbent for soil remediation, showing that the Cr(VI) that had been absorbed on HA was able to be extracted into [C$_4$mim]Cl. Furthermore, the changes in the peak at 1088 cm^{-1} were barely observable during extraction with [C$_4$mim]Cl (see Figure 3b–d). A slight perturbation was observed in the vermiculite during extraction with [C$_4$mim]Cl.

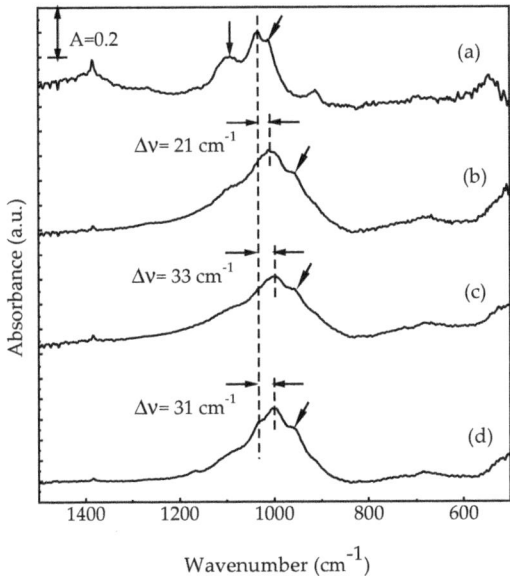

Figure 3. Fourier-transform infrared spectra of (**a**) humic acid, (**b**) simulated sorbent for soil remediation, and Cr–contamined simulated sorbent for soil remediation (**c**) before and (**d**) after extraction with [C$_4$mim]Cl.

2.5. ^{1}H NMR

The ^{1}H NMR spectra of [C$_4$mim]Cl were also measured (see Figure 4). The yield of [C$_4$mim]Cl was 98.2%. The structure and impurities of [C$_4$mim]Cl are shown in Figure 4a. The ^{1}H NMR analysis of [C$_4$mim]Cl revealed values of δ 9.66 (s, 1 H), 8.00 (t, J = 1.6 Hz, 1 H), 7.87 (t, J = 1.6 Hz, 1 H), 4.23 (t, J = 7.2 Hz, 2 H), 3.09 (s, 3 H), 1.75 (m, 2 H), 1.22 (m, 2 H) and 0.84 (t, J = 7.6 Hz). In Figure 4b, interactions occurred between [C$_4$mim]$^+$ and the different extracts. Therefore, a downshift of the protons in the imidazole ring (δ 9.66→9.69) was observed. The chemical structure of the imidazole ring in [C$_4$mim]$^+$ was slightly disturbed in the presence of vermiculite (see Figure 4c). However, HA and Cr(VI) were the main species that were observed to interact with the imidazole ring in [C$_4$mim]$^+$ because the same field shifts (δ 9.66→9.68 and 9.66→9.72) were also obtained in Figure 4d,e. Note that less 1% of the Cr(III) compounds was able to be extracted into [C$_4$mim]Cl [32]. Moreover, the intensities at δ 3.69 were diminished due to interactions between the methyl protons in [C$_4$mim]$^+$ and the extracts (vermiculite, humic acid, and Cr(VI)), as seen in Figure 4. The 1-methylimidazole in [C$_4$mim]Cl played an important role in extracting the Cr(VI), which bonded with HA on the vermiculite, transforming it into the [C$_4$mim]Cl phase. Weaker interaction between [C₄mim]Cl and the vermiculite was shown to affect the extraction efficiency of Cr(VI).

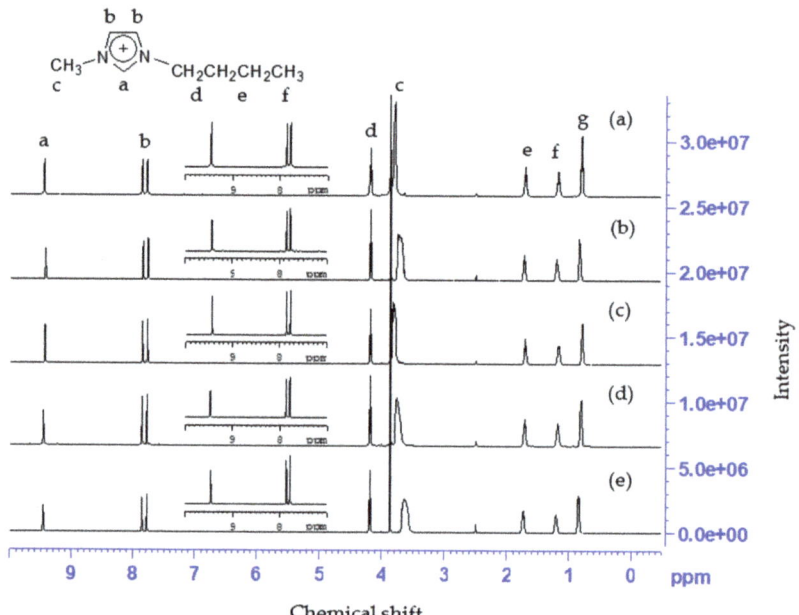

Figure 4. The ^{1}H NMR spectra of (**a**) [C₄mim]Cl, (**b**) Cr-contaminated simulated sorbent in soil remediation–extracted [C₄mim]Cl, (**c**) vermiculite-extracted [C₄mim]Cl, (**d**) humic acid–extracted [C₄mim]Cl, and (**e**) Cr(VI)–extracted [C₄mim]Cl.

3. Materials and Methods

3.1. Preparation of Simulated Sorbent for Soil Remediation and Cr-Contaminated Simulated Sorbent for Soil Remediation

In order to synthesize the simulated sorbent for soil remediation, 1.5 g of HA (humic acid sodium salt, Sigma-Aldrich, St. Louis, MO, USA), 6 g of vermiculite (Aldrich, St. Louis, MO, USA), and 50 mL of H$_2$O were stirred in a 150 mL beaker for 1 d, filtered, dried at 343 K, and ground. To prepare the Cr(VI)-contaminated HA, vermiculite, and simulated sorbent for soil remediation, 7.5 g of either HA, vermiculite, or simulated soil were incubated with 10 mL of 1000 mg/L of the Cr(VI) solution for 1 h at 298 K. The 1000 mg/L Cr(VI) solution was prepared from 0.25 g of K$_2$Cr$_2$O$_7$ (99%, Sigma-Aldrich, St. Louis, MO, USA) in 250 mL of H$_2$O at 298 K. To calculate the adsorption efficiencies, the adsorbed Cr(VI) on humic acid, vermiculite, and simulated sorbent for soil remediation was digested and the chromium concentrations were measured by AA (Hitachi Z-5000, Hitachi Instruments Co., Tokyo, Japan).

3.2. Adsorption Isotherm

For the adsorption isotherm experiments, 2000, 3000, 4000, 5000, 6000, 7000, and 8000 mg/L Cr(VI) solutions were prepared using methods that were similar to those used for the preparation of the 1000 mg/L Cr(VI) solution was. Samples of 10 mL of each of these different concentration solutions was mixed with 7.5 g of each simulated sorbent for soil remediation in test tubes; the test tubes were then shaken at 298 K for 4 h. All of the experiments were run in five replicates.

The Langmuir and Freundlich adsorption equations were used to explain the adsorption isotherms:

Langmuir model:
$$q_e = (q_m K_L C_e)/(1 + K_L C_e)$$

where q_e (mg/g) is the equilibrium concentration of the absorbed chromium in the simulated sorbent for soil remediation, C_e (mg/L) is the equilibrium concentration of the chromium in the solution, and K_L (1/mg) and q_m (mg/g) are the constants.

$$R_L = 1/(1 + K_L C_o)$$

where R_L is the equilibrium parameter, and C_o is the initial chromium concentration in the solution (mg/L). The value of R_L suggests the tendency of the isotherm to be irreversible ($R_L = 0$), favorable ($0 < R_L < 1$), linear ($R_L = 1$), or unfavorable ($R_L > 1$).

Freundlich model:
$$q_e = K_F C_e^{1/n}$$

where q_e (mg/g) is the equilibrium concentration of the absorbed chromium in the simulated sorbent for soil remediation, C_e (mg/L) is the equilibrium concentration of the chromium in the solution, and K_F and n are the constants that are associated with the adsorption capacity ((mg/g)(L/mg)$^{1/n}$) and the adsorption intensity, respectively.

3.3. Synthesis of [C$_4$mim]Cl

To prepare [C$_4$mim]Cl, equal numbers of moles of 1-methylimidazole (99%, Sigma-Aldrich, St. Louis, MO, USA) and 1-chlorobutane (99%, Alfa-Aesar, Kendal., Germany) were mixed, stirred, and refluxed in a 250 mL flask at 343—353 K for 4 d. After cooling, 30 mL of ethyl acetate (99.9%, J.T. Baker, Phillipsburg, NJ, USA) was added to remove the unreacted 1-methylimidazole. Moreover, any unreacted matter was removed from the [C$_4$mim]Cl by means of a rotary evaporator (N-1300VF, EYELA, Tokyo, Japan).

3.4. Extraction of Cr(VI) from Sorbents with [C$_4$mim]Cl

In order to test the extraction efficiency of the Cr(VI) by [C$_4$mim]Cl, 0.4 g of Cr(VI)-contaminated HA, vermiculite, and the simulated sorbent for soil remediation were incubated with 1.5 g of [C$_4$mim]Cl in 0.5 mL of H$_2$O in glass tubes that were each shaken for 30 min. Afterwards, the mixture was filtered to separate the [C$_4$mim]Cl from the solids. The solids were washed with deionized water several times to remove any residual [C$_4$mim]Cl that was present in the HA, vermiculite, and the simulated sorbent for soil remediation. The resulting [C$_4$mim]Cl was removed using H$_2$O and a rotary evaporator. In order to clarify the influence of [C$_4$mim]Cl on the extraction process, H$_2$O was used as the extraction solution without [C$_4$mim]Cl using similar extraction steps. All of the experiments were run with five replicates. Moreover, a blank experiment was also carried out. The extraction efficiencies of Cr(VI) from the Cr(VI)-contaminated HA, vermiculite, and the simulated sorbent for soil remediation into [C$_4$mim]Cl were estimated by the following equation:

$$\text{Extraction efficiency (\%)} = (C_1 - C_2)/C_1 \times 100\%$$

where C_1 is the initial chromium concentration in the humic acid, vermiculite, and the simulated sorbent for soil remediation (mg/g), and C_2 is the chromium concentration in the humic acid, vermiculite, and the simulated sorbent for soil remediation after extraction with [C$_4$mim]Cl (mg/g).

3.5. Concentrations of Cr(VI) Absorbed on Sorbents and Extracted into [C$_4$mim]Cl

Samples of 0.5 g of Cr(VI)-contaminated HA, vermiculite, or simulated sorbent for soil remediation and the resulting [C$_4$mim]Cl were each digested in 6 mL of HCl (37%, Riedel-de Haën, Seelze, Germany) and 2 mL of HNO$_3$ (65%, Merck, Darmstad, Germany) using a microwave digestion system (CEM MARS 6, Mattews, NC, USA). The digestion temperature was increased from room temperature to 448 K by means of a 10 K/min heating rate that took place over 10 min [33]. The final total volume was made up to 50 mL. An AA spectrometer was applied to determine the chromium concentrations in the Cr(VI)-contaminated HA, vermiculite, or the simulated sorbent for soil remediation and the resulting [C$_4$mim]Cl. The Cr concentration of the calibration curves ranged from 0.1–5.0 mg/L, with a correlation coefficient > 0.9995. Each sample was measured three times, and the mean was calculated automatically using AA.

3.6. Spectroscopic Analysis

The Cr K-edge XAS (X-ray absorption spectroscopy, 16A1, National Synchrotron Radiation Research Center, Hsinchu, Taiwan) spectra of the chromium on the simulated sorbent for soil remediation was recorded on the Wiggler beam line (16A1) at the Taiwan National Synchrotron Radiation Research Center. The electron storage ring was operated at an energy of 1.5 GeV and at a current of 300 mA. A chromium foil absorption edge at 5989 eV was used to calibrate the photon energy. To measure the Cr K-edge absorption spectra, the fluorescence mode on a Lytle detector was used. The XANES (X-ray absorption near edge structure) spectra of chromium model compounds, such as $CrCl_3 \cdot 6H_2O$, K_2CrO_7, $Cr(NO_3)_3$, $Cr(OH)_3$, Na_2CrO_4, Cr_2O_3, CrO_3, Cr(VI)-HA, Cr(III)-HA, Cr(VI)$_{ads}$ (by impregnation of K_2CrO_7 (3 wt%) on vermiculite), Cr(VI)$_{ads}$ (by impregnation of $CrCl_3 \cdot 6H_2O$ (3 wt%) on vermiculite), Cr(VI) ion (prepared by dissolution of 0.5 g K_2CrO_7 in 50 mL H_2O), Cr(III) ion (prepared by dissolution of 0.5 g $CrCl_3 \cdot 6H_2O$ in 50 mL H_2O), and Cr foil were also measured. Cr(VI)-HA was prepared by mixing 0.5 g of K_2CrO_7 and HA in 50 mL of deionized water, and then filtering and drying the mixture at 343 K. To prepare Cr(VI)-HA and Cr(III)-HA, 7.5 g of HA was incubated with 10 mL of 1000 mg/L Cr(VI) and Cr(III) solution, respectively, for 1 h at 298 K, and the mixture was then filtered and dried at 343 K.

The solid samples and KBr (99.5%, Panreac, Barcelona, Spain) were uniformly mixed and ground at a ratio of 1 to 100. The mixture was pulverized in an agate mortar, and the mixed dyes were prepared in quantities of 5–8 tons. An FTIR spectrometer (Thermo Nicolet 6700, Thermo Fisher Scientific, Waltham, MA, USA) was used to investigate the structures of the HA, vermiculite, and Cr-simulated sorbent from soil remediation before and after extraction in the wavenumber range of 4000–400 cm^{-1}, with 32 scans, and a 4 cm^{-1} resolution. The ^{1}H NMR spectra of the [C$_4$mim]Cl were also determined on a Bruker Avance 300 spectrometer (Bruker, Billerica, MA, USA) with tetramethyl silane (TSM) as an internal standard (acquisition time = 1.373 s, actual pulse repetition time = 2 s, number of scans = 32, and excitation pulse angle = 30°).

4. Conclusions

The absorption efficiencies of Cr(VI) onto HA, vermiculite, and the simulated sorbent for soil remediation were 81.32 ± 1.05%, 64.47 ± 1.62%, and 91.78 ± 1.82%, respectively. The simulated sorbent for soil remediation that contained HA and vermiculite had a higher Cr(VI) adsorption efficiency. The experimental absorption data were fitted better by the Langmuir isotherm model than by the Freundlich isotherm model. It was also found that Cr(VI) was favorably absorbed onto the simulated sorbent for soil remediation. During the extraction of Cr(VI) from the Cr-contaminated simulated sorbent for soil remediation with [C$_4$mim]Cl, about 33.48 ± 0.79% of the Cr(VI) was extracted into [C$_4$mim]Cl. The results of the FTIR spectra showed that Cr(VI) could adsorb onto HA, which interacted with vermiculite in the simulated sorbent for soil remediation. After extraction, the absorbed Cr(VI) could be extracted from the simulated sorbent for soil remediation into [C$_4$mim]Cl. An interaction between the 1-methylimidazole of the [C$_4$mim]Cl and Cr(VI) was also observed during extraction by ^{1}H NMR. The [C$_4$mim]Cl and vermiculite had weaker interaction, which effected the Cr(VI) extraction efficiency. This work exemplifies that the ^{1}H NMR technique can reveal the changes that take place in [C$_4$mim]Cl during extraction of chromium species from Cr-contaminated simulated sorbent into [C$_4$mim]Cl.

Author Contributions: Conceptualization, H.-L.H.; methodology, H.-L.H., P.C.L. and H.-H.H.; software, H.-L.H.; validation, H.-L.H. and H.-H.H.; formal analysis, H.-L.H., P.C.L., H.T.W. and H.-H.H.; investigation, H.-L.H. and H.-H.H.; resources, H.-L.H.; data curation, H.-L.H., P.C.L., H.T.W. and C.-H.W.; writing—original draft preparation, H.-L.H.; writing—review and editing, H.-L.H. and H.-H.H.; visualization, H.-L.H.; supervision, H.-L.H.; project administration, H.-L.H.; funding acquisition, H.-L.H. All authors have read and agreed to the published version of the manuscript.

Funding: This research was funded by Ministry of Science and Technology ROC, grant number MOST 109-2622-E-239-004-CC3, and National United University, grant number 109-NUUPRJ-13.

Institutional Review Board Statement: Not applicable.

Informed Consent Statement: Not applicable.

Data Availability Statement: Not applicable.

Acknowledgments: The authors thank Ling-Yun Jang of the Taiwan National Synchrotron Radiation Research Center for their XAS experimental assistance. The authors gratefully acknowledge the use of the NMR 000700 equipment belonging to the Core Facility Center of National Cheng Kung University.

Conflicts of Interest: The authors declare no conflict of interest.

Sample Availability: Samples of the compounds are available from the authors.

References

1. Taiwan Environmental Protection Agency. Soil and Groundwater Pollution Remediation Fund Management Board. Available online: http://sgw.epa.gov.tw/public/Default.aspx (accessed on 21 November 2019).
2. Petersen, C.M.; Edwards, K.C.; Gilbert, N.C.; Vincent, J.B.; Thompson, M.K. X-ray Structure of Chromium(III)-containing Transferrin: First Structure of a Physiological Cr(III)-binding Protein. *J. Inorg. Biochem.* **2020**, *210*, 111101. [CrossRef]
3. Mishra, M.; Sharma, A.; Shukla, A.K.; Pragya, P.; Murthy, R.C.; de Pomerai, D.; Dwivedi, U.N.; Chowdhuri, D.K. Transcriptomic Analysis Provides Insights on Hexavalent Chromium Induced DNA Double Strand Breaks and their Possible Repair in Midgut Cells of Drosophila Melanogaster Larvae. *Mutat. Res./Fundam. Mol. Mech. Mutagen.* **2013**, *747*, 28–39. [CrossRef]
4. Kleiman, I.D.; Cogliatti, D.H. Chromium Removal from Aqueous Solutions by Different Plant Species. *Environ. Technol.* **1998**, *11*, 1127–1132. [CrossRef]
5. Mandiwana, K.L.; Panichev, N.; Kataeva, M.; Siebert, S. The Solubility of Cr(III) and Cr(VI) Compounds in Soil and their Availability to Plants. *J. Hazard. Mater.* **2007**, *147*, 540–545. [CrossRef] [PubMed]
6. Mukhopadhyay, N.; Aery, N.C. Effect of Cr (III) and Cr (VI) on the Growth and Physiology of Triticum Aestivum Plants during Early Seedling Growth. *Biologia* **2000**, *55*, 403–408.
7. Mahajan, P.; Batish, D.R.; Singh, H.P.; Kohli, R.K. beta-Pinene Partially Ameliorates Cr(VI)-inhibited Growth and Biochemical Changes in Emerging Seedlings. *Plant Growth Regul.* **2016**, *79*, 243–249. [CrossRef]
8. Hsiech, C.M.; Wu, Y.H.; Lin, Y. The Analysis of Chromate Acid in the Workplace Environment by Solid Phase Extraction. *Int. J. Occup. Saf. Health* **2006**, *14*, 264–274.
9. Srinivasan, R. Advances in Application of Natural Clay and Its Composites in Removal of Biological, Organic, and Inorganic Contaminants from Drinking Water. *Adv. Mater. Sci. Eng.* **2011**, *2011*, 872531. [CrossRef]
10. Abollino, O.; Giacomino, A.; Malandrino, M.; Mentasti, E. The Efficiency of Vermiculite as Natural Sorbent for Heavy Metals. Application to a Contaminated Soil. *Water Air Soil Pollut.* **2017**, *181*, 149–160. [CrossRef]
11. Malandrino, M.; Abollino, O.; Buoso, S.; Giacomino, A.; La Gioia, C.; Mentasti, E. Accumulation of Heavy Metals from Contaminated Soil to Plants and Evaluation of Soil Remediation by Vermiculite. *Chemosphere* **2011**, *82*, 169–178. [CrossRef] [PubMed]
12. Li, Y.; Yue, Q.Y.; Gao, B.Y.; Li, Q.; Li, C.L. Adsorption Thermodynamic and Kinetic Studies of Dissolved Chromium onto Humic Acids. *Colloids Surf. B* **2008**, *65*, 25–29. [CrossRef] [PubMed]
13. Albadarin, A.B.; Mangwandi, C.; Al-Muhtaseb, A.H.; Walkera, G.M.; Allena, S.J.; Ahmad, M.N.N. Kinetic and Thermodynamics of Chromium Ions Adsorption onto Low-cost Dolomite Adsorbent. *Chem. Eng. J.* **2012**, *179*, 193–202. [CrossRef]
14. Dubey, S.P.; Gopal, K. Adsorption of Chromium(VI) on Low Cost Adsorbents Derived from Agricultural Waste Material: A Comparative Study. *J. Hazard. Mater.* **2007**, *145*, 465–470. [CrossRef] [PubMed]
15. Parus, A.; Framski, G. Impact of O-alkyl-pyridineamidoximes on the Soil Environment. *Sci. Total Environ.* **2018**, *643*, 1278–1284. [CrossRef] [PubMed]
16. Mulligan, C.N.; Yong, R.N.; Gibbs, B.F. An Evaluation of Technologies for the Heavy Metal Remediation of Dredged Sediments. *J. Hazard. Mater.* **2001**, *85*, 145–163. [CrossRef]
17. Paria, S.; Yuet, P.K. Solidification-stabilization of Organic and Inorganic Contaminants using Portland Cement: A Literature Review. *Environ. Rev.* **2006**, *14*, 217–255. [CrossRef]

18. Mulligan, C.N.; Yong, R.N.; Gibbs, B.F. Surfactant-enhanced Remediation of Contaminated Soil: A Review. *Eng. Geol.* **2001**, *60*, 371–380. [CrossRef]
19. Patel, D.D.; Lee, J.M. Applications of ionic liquids. *Chem. Rec.* **2012**, *12*, 329–355. [CrossRef]
20. Earle, M.J.; Seddon, K.R. Ionic Liquids. Green Solvents for the Future. *Pure Appl. Chem.* **2000**, *72*, 1391–1398. [CrossRef]
21. Yue, C.; Fang, D.; Liu, L.; Yi, T.F. Synthesis and Application of Task-Specific Ionic Liquids used as Catalysts and/or Solvents in Organic Unit Reactions. *J. Mol. Liq.* **2011**, *163*, 99–121. [CrossRef]
22. Kozonoi, N.N.; Ikeda, Y. Extraction Mechanism of Metal Ion from Aqueous Solution to the Hydrophobic Ionic Liquid, 1-butyl-3-methylimidazolium nonafluorobutanesulfonate. *Monatsh. Chem.* **2007**, *138*, 1145–1151. [CrossRef]
23. Lertlapwasin, R.; Bhawawet, N.; Imyim, A.; Fuangswasdi, S. Ionic Liquid Extraction of Heavy Metal Ions by 2-aminothiophenol in 1-butyl-3-methylimidazolium hexafluorophosphate and their Association Constants. *Sep. Purif. Technol.* **2010**, *72*, 70–76. [CrossRef]
24. Yamamoto, T. Assignment of pre-edge peaks in K-edge X-ray absorption spectra of 3d transition metal compounds: Electric dipole or quadrupole? *X-Ray Spectrom.* **2008**, *37*, 572–584. [CrossRef]
25. Elias, V.R.; Sabre, E.V.; Winkler, E.L.; Andrini, L.; Requejo, F.G.; Casuscelli, S.G.; Eimer, G.A. Influence of the hydration by the environmental humidity on the metallic speciation and the photocatalytic activity of Cr/MCM-41. *J. Solid State Chem.* **2014**, *213*, 229–234. [CrossRef]
26. Wittbrodt, P.R.; Palmer, C.D. Reduction of Cr(VI) by soil humic acids. *Eur. J. Mech. A/Solids* **1997**, *48*, 151–162. [CrossRef]
27. Huang, S.W.; Chiang, P.N.; Liu, J.C.; Hung, J.T.; Kuan, W.H.; Tzou, Y.M.; Wang, S.L.; Huang, J.H.; Chen, C.C.; Wang, M.K.; et al. Chromate reduction on humic acid derived from a peat soil—Exploration of the activated sites on HAs for chromate removal. *Chemosphere* **2012**, *87*, 587–594. [CrossRef] [PubMed]
28. Zhang, Z.; Liu, J.F.; Cai, X.Q.; Jiang, W.W.; Luo, W.R.; Jiang, G.B. Sorption to Dissolved Humic Acid and its Impacts on the Toxicity of Imidazolium Based Ionic Liquids. *Environ. Sci. Technol.* **2011**, *45*, 1688–1694. [CrossRef]
29. Sun, P.; Liu, Z.T.; Liu, Z.W. Particles from Bird Feather: A Novel Application of an Ionic Liquid and Waste Resource. *J. Hazard. Mater.* **2009**, *170*, 786–790. [CrossRef]
30. Huang, H.L.; Huang, H.H.; Wei, Y.J. Reduction of Toxic Cr(VI)-humic acid in an Ionic Liquid. *Spectrochim. Acta B At. Spectrosc.* **2017**, *133*, 9–13. [CrossRef]
31. Zhang, J.; Chen, L.; Yin, H.; Jin, S.; Liu, F.; Chen, H. Mechanism Study of Humic Acid Functional Groups for Cr(VI) Retention: Two-dimensional FTIR and C-13 CP/MAS NMR Correlation Spectroscopic Analysis. *Environ. Pollut.* **2017**, *225*, 86–92. [CrossRef] [PubMed]
32. Huang, H.L.; Wei, Y.J. Speciation of chromium compounds extracted from ZSM-5 into and ionic liquid. *J. Appl. Spectrosc.* **2021**, *88*, 332–336. [CrossRef]
33. ISO. Soil Quality-Determination of Cadmium, Chromium, Cobalt, Copper, Lead, Manganese, Nickel and Zinc in Aqua Regia Extracts of Soil-Flame and Electrothermal Atomic Absorption Spectrometric Methods. ISO 11047. 1998. Available online: https://www.iso.org/standard/24010.html (accessed on 21 November 2019).

MDPI
St. Alban-Anlage 66
4052 Basel
Switzerland
www.mdpi.com

Molecules Editorial Office
E-mail: molecules@mdpi.com
www.mdpi.com/journal/molecules

Disclaimer/Publisher's Note: The statements, opinions and data contained in all publications are solely those of the individual author(s) and contributor(s) and not of MDPI and/or the editor(s). MDPI and/or the editor(s) disclaim responsibility for any injury to people or property resulting from any ideas, methods, instructions or products referred to in the content.

www.ingramcontent.com/pod-product-compliance
Lightning Source LLC
LaVergne TN
LVHW070718100526
838202LV00013B/1121